# INTRODUCTION

The term "true science" must mean mankind's unbiased all-encompassing search for all the aspects that account for not just what we each physically sense, but our entire experience; an impossible task for a fallen mankind. And "true religion" must mean GOD'S one perfect belief and created way of life. But this world is filled with factions of false partial science and worldly religion, arrayed against GOD'S perfect created order. There is one objective truth to all reality, and this truth seeks out its own eternal family. We appeal to each soul's desire to know that at the end of all possible learning, there can be only one truth and one understanding. And in that one same spirit, we share our eternal scientific spiritual perspective and insight into that most profound of all questions; being is there one perfect researchable scientific model of reality and truth, that does *necessarily* require our Eternal Creative FATHER to exist there above us all. We ask and hope for everyone to please play, not the age-old devil's advocate, but an honest loving advocate of the good and the true. Can we all please allow ourselves to know that there is so much more to reality, than just the created physical aspects we sense and feel.

Author, Lawrence Michael Stack: Theorist, Philosopher, Metaphysician, and Theist.



ISBN 978-0-578-74408-7 (UNIVERSAL/CUSTOM)
LCCN 2020915264

Other books, currently available by this author

2. The Eternal Perspective
3. All Conscious Formulations
4. Live to know the GOD of Prophecy

*Lower case italics, implies mild emphasis added.*

# 1 THE THEORY OF CREATED CONSCIOUS PERCEPTION

We begin by asking what empirical proof, defining the true foundation of all perceived reality, has mankind truly found. Don't be impressed by technological advancement, as such advancements do not require a full understanding of reality; all they require is a minimal physical model. Mankind has learned to use the detectable quantum and physical aspects of reality in impressive ways, yet they do not sense reality's true foundations. To eliminate its bias in our discussion, we must set technology aside, and focus only on the *theoretical* search to know reality. The world still asks, what was the switch that turned on the universe, and all physical life; and our enigmatic awareness of belief? And who or what flipped that switch? Many past and present minds have tried to unravel the foundations of reality. Yet *all* those souls remain surrounded by those same original unanswered questions about the enigmatic forceful orders of reality and our mysterious limited self-awareness of it. Many revered intellects have dedicated their lives in search of these greatest of understandings; yet have all left, or leave us without answers. They leave behind speculations and haunting frustration, and lament how their physical laws fail to explain so much. Like what breathed the fires of life into reality; and why their search, can never declare from what and where comes that vexing ghost in the machine? And how with every stare into the deepest appearances of the quantum/classical laws, gazing back is the felt glare of some unknown controller. How in their life's search, confronting the deepest questions, that like a compass always points north, the over-all evidence always allows that one *un*acceptable answer. To this day, there remain the questions about our miraculous conscious realities. Being what *is* our conscious understanding, and our awareness of qualia, knowledge, and belief; and how can these qualities ever be physical aspects? What are our created perceptions made of, where are they, how do we create and store them, and how do we re-find and recreate them? Why do we see everywhere in everything, that the essence of every conscious desire has its determined and determinate influence upon all quantum and classical realities? These mysteries have been studied in materialist ways for so long, without advancing a single answer. The answers to all these conscious questions are the key to understand, not just the physical surface of reality, but all its deep foundations. Right up front, I say by reading and holding the full firm understanding of this book, as it is meant; the existence of GOD is shown to be required. That is, up to and beyond the extent, that *anything* else has been proven before the world. We make the strongest empirical, logical, and spiritual arguments for the true conscious *and* physical existence of GOD that can be made. We reveal a consonance between all force, energy, matter, awareness, reason, understanding, logic, purpose, and belief. But to do this, we must make our scientific arguments, while also aligning their correlating spiritual points. We do not argue for change to the current physical model of reality. But for a new larger faster conscious/physical model of reality. In metaphysical nomenclature, the terms Dualist or Dualism, refer to souls who see all the interwoven existential aspects and reasons for why and how the Conscious and Physical aspects, must be two distinct and separate, yet interacting qualities. In this world, we see great efforts to blur these distinctions. *Pure* Dualism is central to this book. Please notice how when the worldly sciences lecture us on the foundations of reality, they speak in subtle vagaries in between, above, below, and beside the matter. And they never define, or need *any* near complete understanding of anything foundational. They are content to live in their never-ending questions of never truly knowing; and never realizing why they need the obscurity and failure of their ever-changing physical theories. Ardent materialists say everything that occupies space, is matter or energy. But what law or principle says non-material, non-energetic, *forceful* conscious souls cannot occupy, relate, and interact with the known aspects of reality, since we find endless evidence of it being so. And another strange aside to the materialist's beliefs, is they live in a profound glaring contradiction. As on one hand, they say all things *sensed* are force, energy, or matter. Yet for our own created perceptions of imagery, and created representations of sound, taste, smell, and touch we all physically and consciously *sense*; for these they use the word illusory. They do not *want* quantum particle wave aspects to be the basis of our created perceptions. We physically sense our created visual perception, and our auditory creations, and others. But what is it that *must* form their physical ethereal structures and quality that allow us to sense and experience their existence within our conscious estate? These created perceptions we see, hear, taste, touch, and smell, do they not actually exist? Are they physically non-existent and formed from nothing? What is nothing? And how do we create, sense, and become aware of something made of nothing? We hear, see, and sense our created perception in many physical ways; so, they must be real and physical. These Materialist contradictions, cannot be seen as having lawful standing. Our perceptions are surely most ethereal in every sense; yet the quantum arena is the realm of such ghostly aspects, and a realm of strange and barely defined manifestations and occurrences. Though all things within the quantum realm are most ethereal, they are known to be barely real and of true physical consequence. Consider the visual you now currently experience; you see its structure, depth, colors, and organization. But how might you create this, minus some form of projected patterned forceful information? We will explain in detail, how perceptions are created by every soul's projected patterned field of conscious forceful desire. Which then align these three-dimensional structures of quantum particles and waves in the same way as were all things created to be seen, or otherwise sensed to exist. We *do* consciously sense our created substantive perceptions, so they are real. *We* create particle/wave-based images, using projected patterned 3D fields of lesser and greater force quanta. Yet materialist say perceptions don't physically exist. The *only* way we can create our ethereal particle-wave-based perceptions, is by projected patterns of quantized force that actively align our ethereal particle/wave-perceptions. Materialist *must* reject created particle wave-based perceptions. Because to admit we create real quantum particle perceptions, takes us all down a single path to a forceful ruling Conscious Observer Effect. Again, we will show in great detail how our soul creates *internal* representations of the *external* physical reality, using projected patterned information that mirrors those realities, sensed through the body's neural interface relay systems. Now physical and or conscious information, are described as *any* pattern rising above pure randomness. Any sensed pattern, is a form of information that all souls can recreate within their limits. Willful *forceful* created particle-wave based perceptions, reveal that *all* physical order and complexity are consciously forcefully imposed, and do not come by physical reality itself. So, an important question to ask is at the beginning of anything's existence, which came first: external force, or internal energy? It is always external projected patterns of force. Energy first appears after force creates and moves a physical aspect. Only projected force, initially attracts, repels, and aligns; and only force can create and destroy physical complexity. We all know creating or destroying anything, even our three-dimensional particle-wave-based perceptions, is defined as work, and every work requires initiating force. Science measures the moving and stored effects of force, as energy. But the first moving of any *non-energized* physical aspect, as with conscious awareness, can't be physically explained or defined. A force to work Principle says, that consciously creating our representative particle-wave-based streaming perceptions of reality, *requires* our work of projecting 3D patterned fields of consciously generated force quanta. These force patterns are blueprints for our perception's created particle wave-based structure. But our created perception itself, is *not* in any way our awareness. Created perceptions are a product of Dualism. And this work defines our soul as a logical inherent projector of newly generated patterned fields of force. But what is our generating conscious *source* for our projected three-dimensional force patterns. Study and the process of elimination, reveals that every mirrored field of force we project to create perceptions, *must* be founded in every conscious desire to exist. Our *every* free-will desire, when studied, contain projected *intentional intended* aspects, and has its *forceful* feel of purposeful desired outcome. Also, our willful *desire* to exist is the only quality we see projected and woven through our every single thought. As with all force, our desire is seen in wanting something to come to us (force attraction) or for it to go away from us (force repulsion). Our every forceful desire has these aspects. *New* generated projected forceful desire to exist, is the only possible source. Materialist talk about incoming sensed physical information, and all observed neural correlations. But never about where, how, or from what are our perceptions made. Perceptions are a created-phenomena, occurring *after* all those neural correlations. But creating them, is still a step that can never be equated with the next step of our then, being aware of our created perception. Consider the projected patterns of information that form a camera's captured images. They are physically captured and stored, with no self-awareness. A transmitter then sends those patterns to our TV, which re-projects the information of images and sound patterns, having no self-awareness, to our sensing eyes and ears. And again, these patterns of information, having no self-awareness, are sent from our physical interfacing relay, or brain/body, to our soul. Which again, forcefully re-projects those sensed mirrored patterned fields of force quanta, with no self-awareness, into the quantum medium of the aether (to be defined in detail) which creates our internal multi-sensory particle-wave-based perceptions. And only then, our soul indwells our created perception to sense and find *our* awareness of it. By our soul's forceful desire to exist and use sensed information to create multi-sensory particle-wave based perceptions; this shows our souls are intrinsic, and inherently desiring generating forceful projecting creative entities of awareness. This conscious force projection Principle, says such intrinsic inherent conscious attributes must exist for us to create our subjective perceived physical existence. Science knows force, only by its affects; they do not know what force is, or *where* it originates. So, they can *never* say it *does not originate* within each souls generated projected conscious desires. Where does force originate? Has it always just been there? How can that be? We sense certain external information, and we consciously exist within our created perception of that information. If we perceive nothing, we temporarily fail to consciously exist (to be explained). The human body's entire neural system is a sensing processing 3D interface, relaying information about the external reality to our soul. A soul senses patterns and frequencies radiating within all the body's lively aspects, down to each microtubule structure in each cell. We create, indwell, and experience our perceptions. Yet our awareness is such a far higher distinct order of phenomena above our created perceptions. By a few holographic principles, we see how in a logical mirrored manner, we desire, generate, and project these 3D patterned fields of quantized force *within* our soul's forceful dimensions and reach. And through a quantum process, our projected fields create and align a mirrored web of particle entanglements within the wave function to form our streaming perceptions of experiential awareness. We indwell and are aware *of* the visual image through a certain format within this structure; and our other senses through other formats layered in together. So, we see a far greater Uncertainty Principle and Observer Effect; being far more than a simple case-by-case quantum measurement. Our projected forceful desires, known as the Observer Effect, ruling over the Uncertainty Principle, is ubiquitous and relative. This seems a huge stretch at this point, but please be patient, as at minimum it takes our 4 books of around 2000 pages to properly convey. Projected conscious forceful desire to exist, is the soul's generated mechanism that breathes life's fire, and is the ghost in the machine, the unknow controller, and that feared conscious answer. And as a soul's created 3D particle-wave-based perceptions epitomize the Conscious Observer Effect; each soul's followed awareness of them, epitomizes the creative awareness of reality within GOD, *and* us all (to be broadly explained). Whenever *we* speak of information, we are talking about the patterned fields of conscious language being sensed by, or projected into all creations by all souls within the aether. A quantum physical symmetry is created to mirror each projected forceful pattern of desire, bringing our existence by every sensed perception. And all that fallen mankind *thinks* they know, is at all levels, approximal, uncertain, and asymmetric. Pure symmetry exists within every original creation. But in this world, symmetries are attacked as soon as they are created. This world's godless ways and beliefs, turn all created symmetry into asymmetry. Materialist must work within their limited classical and quantum horizons; and are hindered further by their godless perspective of those horizons. Materialist try to define reality from a temporary partial small slow physical perspective, being a worthless pursuit. An eternal dynamic life, can endure only beyond the ways of this world, and in the pure ways of GOD. Pure dynamic symmetry can never come by chance. Materialist dwell in their slow-small-interpretations of just 6% of the physical half of reality, so they failed before they started. Materialist believe in singularities, but the singularity they define, exist outside of purity. Their singularities require internal flaws, fluctuations, and anomalies to enable their theories; so, they define a singularity as being imperfect. But all created moments of experience in our perceived existence, come by way of pure Dualism, so, we see only *two* qualities underly reality: being the top-down, endless fields of *newly* generated lesser to greater force quanta, projected in every soul's desire to exist. And a bottom-up, nearly infinite number of non-conductive points of pure physical singularities. These points are eternal and perfect, and become aligned in complexity, only by projected field patterns of force quanta capturing and entangling them. These singular qualities are reality's pure eternal building blocks, being far too minute for mankind to comprehend. These singular qualities cannot internally hold or emit charge or force. Imagine shrinking down so far, to hold a single eternal physical spherical quality and aspect of reality. There is the Dualism of that which truly knows, and that which knows nothing. The creative conscious artist, and the non-conscious physical medium or canvas. We each are consciously aware of, and exist only through this Dualism that creates our own perceptions. We may say force and matter may still exist if separated, *but* no one can know they separately exist in an experienced certainty, as only our forceful entangled complexity of the two, bring our experiential awareness. This Conscious Force over Physical Reality Principle, states our awareness of reality resides in creating and indwelling our created subjective perceptions of said reality. And temporary non-existence, resides in the uncertain state of our soul, in lacking all created perception. There is in every separation of our soul from this required physical information, a state of our soul's uncertainty; yet our probabilities and possibilities remain. The separated conscious and physical parts may be said to almost exist, or to be in a state of waiting to exist (to be explained). Now for the next step to truly understand reality, we add a huge new observer's frame of reference to Einstein's relativity. This greater observer's frame of reference is required to account for the created perceptions of all vastly different size conscious souls. We call this affect, the "Observers Frame of Refence Variable". For all different conscious size perspectives of observation, we must see how pure Dualism that creates our perceptions, applies to the entire universe. The universe is the precept upon precept, line by line, layered, forcefully desired and projected particle-wave based firm vast 3D perception, perceived by FATHER. And within his created particle-wave-based universe, he has partitioned *out of* and from *within* his soul, the eternal gift of different sized souls. Our own distinct minute three-dimensional portion, or space of a forceful conscious desirous soul (to be explained) is who we each consciously are. So, we see immense, to *far* lesser size and forceful souls. Let's apply the greater *conscious size* relativity to the three created physical dimensions of height, width, and depth. Now the conscious dimensions of each soul are: one, being each conscious entities 3D spherical size or space; two, a *gifted* minute soul's position and rate of momentum through GOD'S created external universe, defined in Einstein's relativity; and three, is not any linear arrow of time, but each soul's continuum of their own created expanding spherical moment. These 3-dimensions of each soul, define their *physical* perspective. So, these are the *six* combined true conscious and physical dimensions of each soul's reality. To demonstrate these perspectives of vast differing conscious and body sizes, existing within *one* shared physical reality. We first see the soul and body of FATHER, being at *minimum*, seven times that of the universe, making his observational frame of reference for all physical to conscious aspects, vastly different from mankind's. From fallen mankind's, small slow physical perspective, looking out, the patterns of slow photons that show us our moving hand, have an *appearance* of being extremely fast, yet require billions of years to cross the universe. Yet FATHER, being larger, consciously *and* physically than the universe, from his far larger, looking in, observational frame of reference, using far faster light particles, perceives slowly moving *his* hand across the entire universe, in what to him is a tiny moment. One of these two conscious frames of physical reference, created reality, and it is not mankind's. Reality has at least two smaller far faster light particles than photons. And the larger a soul consciously and physically is, the less relative the slow large photons are to one's perception. So, not just theoretical Tachyons, but *at least* three luminary particles, form the true spectrum of light, far beyond mankind's sensing ability. For FATHER to see his external universe all at once as it currently is, requires an immense speed of light. GOD created variations of light particles and their speeds, to match the sizes of conscious/physical entities who need them to see. A vast immense angel perceiving life in Heaven relies on a certain super-luminary particle; but if he is translated down to the size of us on earth, he may use photons. Please remember that chronologically, FATHER created all light *before* he filled the heavens with stars. So, at first, he created all light to radiate and fill his aether. Now by a relative to soul size perspective, we see a godless fallen mankind's view of the past, using photons, can appear to *them*, *as if,* the past was a longer continuum, compared to FATHER'S true perspective of the same. If mankind, were physically not more massive, but our body mass and soul were inflated to a fourth the size of our solar system. Our new perspective says that for accurate relative perceptions, we would require a faster light. Otherwise, as we move our vast long arm, we must wait hours to see our arm slowly move in an unrelated way. These faster light particles and other things we will expand upon, show how measuring the universe's age by expansion and photons is impossible. Now, we see a faster greater observer frame of reference can be held by us, as it is not purely physical. It can be a desired accepted held understanding, and it can be applied to *our* awareness. See the universe from the outside like FATHER, seeing it all at once by the fastest particles of light. We see the universe is smaller than FATHER, and photons are the slowest light particles that form a relative subjective, and not an objective appearance for mankind. So, if FATHER zooms into the universe until he sees the earth turning before his sight, has he reduced his *conscious* perspective? No. FATHER'S is eternal in the reality he is physically and consciously larger than the universe, his perspective cannot shrink, but it can focus. So, we see that though the two conscious and physical halves of reality interact and connect in many ways, we find that differing conscious sizes and speeds, within one physical reality, says they must see things differently. Yet a slow small fallen perspective of mankind, can by pure desire, adopt FATHER'S greater conscious perspective. 2nd Peter 3:8, (KJV), "*BUT, BELOVED, BE NOT IGNORANT OF THIS ONE THING, THAT ONE DAY IS WITH THE LORD AS A THOUSAND YEARS, AND A THOUSAND YEARS AS ONE DAY.*" Isaiah 48:13 (KJV) "*MINE HAND ALSO HATH LAID THE FOUNDATION OF THE EARTH, AND MY RIGHT HAND HATH SPANNED THE HEAVENS: WHEN I CALL UNTO THEM, THEY STAND UP TOGETHER*". Isaiah 45:12

(KJV) "*I HAVE MADE THE EARTH, AND CREATED MAN UPON IT: I, EVEN MY HANDS, HAVE STRETCHED OUT THE HEAVENS, AND ALL THEIR HOST HAVE I COMMANDED.*" These were FATHER'S, in that moment, physical observations. Now we see this Conscious Size and Speed, to Physical Frame of Reference Principle, allows for *violations* of the Lorentz Invariance and all mankind's laws and standards; but *only* beyond mankind's limited frame of reference (to be explained). Based on conscious and physical size and speed of each entity, there is a strict, filtered, restrained, spectrum of comprehension as to the smallest and largest perceptions that can be created and indwelled by each size entity. Each soul is limited by its creative size, speed, and forceful *reach*, in weaving the two halves of reality. Every soul, at our being gifted, starts *at* and grows *out* from the center of our size defined spectrum. So, if fallen mankind physically senses things above or below their smallest to largest boundary of created perceptive ability, the laws of their own *physical* relativity must be stretched or dilated below their experiential frame of reference. Even seeing things too far away through telescopes using photons, only shows them things that do not exist as they see them. Any reality mankind senses beyond their slow small limits, is seen by them as violating their laws formed from their slow small frame of reference. Reality beyond the limited perspective of mankind's conscious and physical size, can never be understood by them on their own. Einstein showed how position and momentum are relative to a soul's subjective perception, and that if we reach the speed of photons, this alters our perception in predictable ways. Relativity, and Quantum Mechanics, show fallen mankind's slow small limit of comprehension is far below the sum of all reality. In knowing GOD'S plan, we see *why* such boundaries are there. We see the quantum realm and the aether, extend past mankind's conscious and physical comprehension. We also see how an expanded conscious and physical size, *or,* an increased conscious and physical momentum, bring the same relative, yet subjective perceptual affects. This is predicted in GOD'S true model. We all create and indwell our own conscious-physical existence, within our own 3 conscious, plus the 3 external physical dimensions. This subjective Conscious Entity size and speed to Physical Relativity Principle, shows within these six dimensions, that the observed laws and constants mankind says govern reality, are subjective and relative only to their limited perspective of photons and space, through their slow small awareness. Fallen mankind's laws, constants, theories, and material scientific proffers, do not govern the sum of reality; they only define their perceived perspective of reality. Further, their laws do not even govern their perspective of reality, their laws are formed and governed *by* a small slow fallen perspective. There are no physical laws over reality, there are only FATHER'S true objective conscious laws, and lesser false subjective conscious laws. Now the measured variable speed of all processed fallen human experience, achieved by Dualism, through a fallen soul and body, according to a Caltech study, says a human brain only processes single thoughts, and at a slow 10 bit per second rate. This is largely because a fallen physical body's nerve signals travel at only 175 mph. We also know through all five senses, a human body only senses up to a billion bits of information per second, out of hundreds of billions of bits of information that act within and upon a human body in each second of existence. Consider a bug crawling across the ground, we know in a minute, slow, and insufficient way, it experiences something; but we also know it cannot in any way, ever truly know anything about reality. *Fallen* mankind is so superior in experiential awareness to a bug, yet they too are still now far too small and slow in their experiential perspective to sense or know the greatest or smallest layers of conscious and physical reality. *Fallen* mankind must accept that they do not in any experiential relatable or relative way, interact with all the conscious and physical layers that exist, in any deep or broad enough way to truly know anything. *Fallen* mankind cannot grasp why the one truth and understanding of all reality, can only be seen and known in its fullness, *and* at the actual speed in which all these greater and lesser realities exist. To truly understand all complex layers of conscious and physical reality from mankind's slow small fallen perspective, is impossible. Fallen mankind is now stuck down low in the minutia and confusions of hell in their insufficiencies. Now in what we call each flash, or moment of human experiential reality, while the fallen conscious awareness of mankind, slowly crawls along at 10 bits per second, the far faster higher reality of complex orchestrations of physical life, occur in the background. And if we scale the speed and size of conscious/physical knowing awareness, and experience, from that bug up to mankind, and then all the way up to FATHER'S forceful conscious control over his creations of life, and all reality.

We then see in a separate way, that what fallen mankind's science *errantly* calls the sub-conscious, are the forceful desires of FATHER constantly sent down, as up to over 400 billion bits of patterned forceful instruction per second to sustain all body functions of physical life, which requires much of the brain to process. Yes, FATHER says mankind rebelled and fell from the heights and speed of GOD'S eternal adjoined oneness long ago. And we are also told, if we, as fallen loving souls, turn back to learn grow and run to the open arms and shared mindful belief and perspective of GOD from which mankind fell; we can gradually re-join, grow, and once again share in GOD'S far greater experiential spiritual speed and capacity of understanding. Let me add, that in growing to know any adequate level of GOD'S conscious experiential speed of comprehension, it sets you apart in a great lonely way from this world in communicating all understanding with most fallen souls, including the lost souls you love, just as they are separated from truly communicating with GOD. Now that 10-bit to 400-billion-bit ratio, does not define the full difference between the experiential speed of the mind of GOD and fallen souls. It is a glimpse at a bit of the ongoing work FATHER invests in the Dualism that sustains reality, and each fallen soul to make their final choice as to returning to, or rejecting GOD. In all of these things so far, we see why mankind can't consciously kill Schrodinger's external real physical cat, by opening a physical box that is external to us; it being just our choice, in FATHER'S ongoing created perception. Yet in *our* own consciously created particle-wave-based perceptions of doing so, we do instantly kill or let live that cat at our perceived opening of that box. Using a 4D model, over the true 6D model, drops mankind from a top-down eternal perspective, to a bottom-up destructive, degrading, timely perspective. For a true understanding, we must *always* see reality from our being outside of, larger, and faster than all things while looking in and down, using the fastest light, just as FATHER does. We need to see the aether and universe as when it was all first created, and compare it all to how it is now. For our understanding of reality, we must pass the bounds of what fallen science can know. Once you truly see FATHER'S created reality, you see $E= MC^2$, the Lorentz Invariance, Maxwell's laws, Heisenberg's Uncertainty, Schrodinger's equation, and all mankind's models, equations, laws, and principles, can still work in their same imperfect ways; as long as you work inside mankind's small, slow, timely physical perspective. But once your work and perspective take you to or beyond those fallen perceptual boundaries, well, some physical functions may continue to work a little further with added variations. But at a soon certain point, they, along with all materialism, fail completely. Fallen science speculates beyond mankind's limited perceptual boundary, but they can never form experiments to prove false dreams beyond their conscious/physical boundaries, such as multi-verse or quantum particle gravity. In GOD'S creation, gravity and all forces are his quantized patterns, projected to form a desired reality. If you *understand* just these first few things we have covered, in the *way* they are meant, you start to see the wondrous conscious reality that reaches far beyond mankind's fallen understanding. You know that if you exclude a single relative aspect from any equation, you falsify the answer and every accepted understanding. If you remove particle-wave based created perceptions, or three or more light speeds, or our conscious size relative perspective of awareness, from reality's full equation, you falsify all science. Mankind can never on their own, find that great part of the equation they lost in their fall from GOD. Our created perceptions also show we experience expanding overlapping shared conscious boundaries, within this aether. We sense our conscious estate grow and shrink within our own size spectrum, to create the size and substance of each perception. The breathing dimension of our soul, rests in between our smallest to largest created perceptions we each can create, being set by the limited sensing abilities of our body/interface/relay system, *and* perspective of our adjoined soul. If we physically look up at the stars, we don't project our created perception all the way out to those stars. We only sense waves of photons at the outer edge of our souls reach. Using information sensed by our body/interface, we create perceptions and are aware of nearby things. Yet we encompass our awareness of them, where we and they are and occur; not within our tiny fleshly neural interface. If I stub my toe, I project and create my layered perceptions of feelings and awareness within my soul, and my environment. But my perceived awareness of pain is where it occurred in my toe. If in a room, my perception stops at the walls surface. If I see someone in a field, I project within my limitations an expanded created perception that overlays them and my surroundings where they are, and not within my brain. Within the spectrum of our creative range, we create particle-wave representations of quite minute things, up to capturing areas of GOD'S perceived universe. Another aspect to ponder, is our dream fantasy state of awareness, has exactly the same conscious size limitations, which we will explain later. To detect our projected force patterns is impossible, because the nearer an object is to our reaching soul, the more the strength of our created perceptions overlay the actual external reality in being aligned to what is there already. But for our created dream perceptions, such limitations may not exist. We all have experienced expanded feelings in beholding something rare or great in scale, like the Grand Canyon or looking from a planes window. New wonder, color, and beauty, influence our perception of existing at our sudden awareness of any new strong experience. Great moments of fear or thrill alter us consciously to expand, narrow, or alter, our experience beyond our common perceptions of momentum, place, and feelings. Now remember science finds all their concept of a linear continuum called time, can*not* be defined in a fashion that works universally from all size or momentum perspectives. And their many malleable concepts of time present other physical and conscious considerations and problems. Now materialist study only the sensed external existence, assuming it is key to defining reality; but they assume this without rhyme or reason. Materialist refuse to admit that our conscious perspectives and desires, are key factors in creating and defining our own reality. They believe their *subjective* sense of physical reality, must be the objective reality. Now the undeniable evidence for me, that FATHER consciously projects his layered forceful patterns from out there to create and sustain *his* vast particle-wave-based perception we call the universe; is that we, from way down in here, project our conscious force patterns to create *our* particle-wave-based, yet far more ethereal perceptions. And that all souls combined, wanting different things, we could never in unison project and create what more firmly exists. *All* reality, external to our internal created perceptions of it, under mankind's slow small physical understanding, lacks any true form or real beginning. A Materialist understanding relies entirely upon the current temporary state of physical evidence and physical assumptions, and physical miracles. Now, we see quanta of understanding: up and down are two quanta, left and right are two, and near and far are two. So, holding these six quanta of knowledge, grows our soul's size of understanding to then comprehend reality's dimensional aspects. The limited physical evidence alone, is *extremely* subjective under a slow small biased interpretation. At our *gifted* inception of being minute three-dimensional spaces of awareness that coincided with GOD aligning our soul to our physical body's initial first cells conception, which caused our soul's first minute forceful spark of awareness. We began our life minutely perceiving just sparks, blurs, and blips; yet we grew to learn and build our desired belief of understanding. But if we grow our conscious size and complexity, minus GOD'S omniscient guidance, using desired suppositions of false understanding; reality then becomes individual and malleable, appearing as many possibilities and probabilities. By *desire*, many pleasing suppositions can congeal to appear as a probability, even if false. Most souls, have in this way, desired, supposed, and built up their *individual* beliefs. But knowing there can only be one truth to reality, this should appear to us all as mass desired biased confusion and delusion, yet it is called normal individualism. In fallen individuals, we see only misaligned frames of reference, whose understanding is so minute and wrong, compared to omniscience. FATHER has a pure understanding of all he has perceived, and all that perpetuates life eternal. So, imagine if we all resisted false speculation, and learned from GOD alone. Every learned aspect of omniscient understanding, moves a soul further into life eternal, but every accepted false slow small physical supposition, drags a soul down into the hellscape of time. Most every soul continues to vastly over-estimate who and what they physically are; and vastly under-estimate who and what they are meant to consciously and spiritually be. Below GOD, all eternal souls, regardless of our gifted size, create out from one common center. Being a common starting point for every gifted soul's perceptive ability. This means the higher faster larger souls, see what the fallen slower smaller lower souls see, and more; but the small cannot see all that the larger see. Visualize conscious Russian nesting dolls, looking outward, all having that one common center, but differing outer boundaries to their 3D perspective. Now, for all who discard an aether in their model, we ask you to continue with an open mind. We ask you to see that by there being a *need* for a substantive aspect to space itself, this fact raises questions materialist don't care to consider; like how the aether itself could come to be. To hold a godless aether-less belief of reality, you must form your model starting from nothing, or a physical miracle that can't be explained. Throughout this book, from many perspectives, we will share strong relative reasons, that hopefully cause you to see this aether. We see that FATHER provided within all soul's forceful reach, an unseen quantum aether medium of the finest structure that allows us to forcefully project, and create the ethereal perceptions of our existence. An aether that permeates and mediates all created realities, that from mankind's fallen perspective, in any un-utilized state, is beyond fallen mankind's detection and lawful comprehension. Existing from below all quantum realm boundaries, up through all the firmer existence in between, and on through the vast quantum Heavens and beyond. So, it is everywhere anything exists external to FATHER'S personage. These next pages describe GOD'S consciously projected aether's structural design. But the aethers aspects are so many, and must be explained throughout the whole of this book, with each point made in its place of broader context for proper understanding. Within the aether's forceful projected creation, we see a near infinite sum of identical spherical purely physical point singularities that form FATHER'S body. These points are in size, far below mankind's perspective, or comprehension. The aether's creation was not a bang, but a very different big moment that started all we physically sense external to our soul. If you are an unbelieving mind, please try to see the physical relevance of these things you try to see as theological. Before the beginning of what now is, there was only One vast Soul of forceful projecting awareness, and his vast body of purely physical points. These two qualities have always been the knowing conscious force and malleable body that formed all perceptions within the one dreaming creative self-aware soul of FATHER. At a past point of unimaginable eternal loneliness, being the only soul to exist, a plan and choice were made to eternally partition out varying three-dimensional free-willed measures of FATHER'S soul, in order to not be alone. Before gifting new souls, FATHER perceived for himself an external semblance, having the means to convey and receive love; being a face, eyes, mouth, ears, etc. Then after *speaking forth* the soul and points of his SON, and then the soul of HOLYSPIRIT. FATHER *spoke* again in a great projected desired intended force and points to form his vast spherical aether, having about 137 times the number of points, as used in creating all the particles in the universe. FATHER spoke these singularities into his hands to be his spherical medium for creating all further complexities that are a universe and physical life. Before this, all FATHER'S perceptions were created within his self. These singular qualities, or *points*, were spoken forth as the substantive ethereal breath and words of GOD. This forceful spoken act, caused all these points to forcefully consist in a desired vast spherical fluid crystalline state, ready to accept all new creative complex patterns of force. All rising physical complexity was added by each pattern FATHER spoke into this spherical wonder. And within this great sphere, there are set forceful limits to the complexity, size, consistency, or purpose, of the creations perceived by each variable projected force pattern. We equate this sphere's singularities as being a quantum, or Planck, base-line, aligned by a most basic field, in a fluid uniform semblance of spatial consistency, like an ethereal spherical fluid crystal. This vast grid structure, in which all greater creations now exist and move, is an unknown variable in secular theories. In all, it is a gift of an excitable medium forming space itself; which the universe and we indwell. This vast structure enables all desired forced conscious creations. And within it, FATHER perceived all greater particles, all spin, rotation, orbits, all complex quantum and classical aspects, and all physical life. FATHER'S desired patterned force/physical point based aether, enables all radiation, all point attractions, repulsions, excitations, and all forcefully created three-dimensional systems. The aether's fluid vast minimal spherical field, allows all firmer fields of projected force convergence, as its points are realigned to all superior stronger fields. The aether enables FATHER'S external search for love, through this entangled conscious force process, and all souls created perceptions to exist. FATHER'S aether, or firmament, was forcefully *and* physically spoken forth. His creations are entangled by and to his desire, as his forceful patterned words of desire were spoken, like "let there be this and that…". And as all his patterned forceful desires were projected into his aether, a mirrored complex creation appeared. When a desire of FATHER, *forcefully* says, "let there be…" it must then exist, just as *we* forcefully project to create our every real perception within the same aether. Now there is no physical starting point where our forceful conscious patterns of information radiate outward, as all patterns of creative force, are projected and entangled in place. Now to define mankind's fallen perspective of reality, you must *abstractly* see growing levels of reality, as created realities build more and more, to grow large enough to exist within mankind's perceivable photon spectrum. "3-D" is a truly subjective and size relative term. Every three-dimensional perception is created and must be seen and known as being relative to the size and force of its creator. FATHER'S three-dimensional perspective, vastly exceeds ours and the great angels also, in all ways. All lesser perspectives can never know the complete perspective of FATHER. Now within the language of physics, the term *Locality* is a term used to describe things sensed in the classical cause and affect way, that materialist hope will describe all reality. They hope all things are affected only by the physical energized things, near enough to exert a forceful influence upon them. But Non-Locality, and Quantum Mechanics, destroyed that hope. Now, all beliefs as to the conscious and physical affects between souls and reality, are also divided. Quantum non-locality, also implies the past and future are entangled to the present moment. But in GOD'S model, the past and future can go no further than being seen as the last lagging and leading aspects of our *current* conscious moment of reality. A real distant past or future moment can never exist. All conscious/physical reality, exists only within the defined created *experiential* moment of now; all concepts of any arrow or direction of time are false. Most souls see good reasons for us all to grow in one true understanding, yet most still grow individually, and don't see why believing in time is ungodly. And though science has no truly defined terms for time, it can be difficult to accept that time is an illusion. But if time were a fourth dimension of a purely physical reality, time *and* that physical reality could not act and be seen in such different and subjective physical and conscious ways. All that ever exists to us, exists *only* in our current moment. Continuity exists within each soul, but a belief in linear time blurs one's every understanding. Materialist have failed to define a form of physical consciousness; and completely failed to disprove our non-physical forceful creative soul. They fail to explain this physical reality using their malleable illusions of time. They fail to explain our consciousness and the quantum classical realities, in any *consistent* way that incorporates time. Just these few indices alone, show mankind's desired accepted definition of a 4D reality must be false. Mankind's *fallen* experiential awareness, is an ongoing process of a slowing shrinking, current moment, in being separate and contrary to GOD'S expanded accelerated current moment. We move in three dimensions: up and down, side to side, and back and forth. We sense and reach out into a vast shared external moment perceived by GOD. Each newly partitioned soul starts to create as a new ever-expanding spherical moment of existence within FATHER'S spherical ever-expanding moment. This many tiny free-will experiential moments, existing within GOD'S vast experiential free-will moment Principle, not only explains, but predicts all symmetries and asymmetries. To know *eternity*, we see how GOD'S forceful desires created or perceived all the original external realities, during the first six rotations of the earth. In this, FATHER perceived all that mankind senses, and far beyond; and it is all known within GOD'S experiential moment. FATHER'S aether, was the beginning of the all-encompassing moments of a physical creation he built-up in layered perfection. But then *contrary* projections of individual forceful rebellion were introduced, and the vast perfect *physical* creation became altered. FATHER'S creation's unfolding moment, is now being broken down by all contrary forceful desires. If we desire and choose to live outside of GOD'S omniscient created perfection, we create and add forceful disorder, destroying its perfection. Now in that time does not exist for GOD'S Children, physical locality and quantum non-locality hold different-meanings for us. Reality is to most souls, an unknowable puzzle, because reality can only be known when *all* pieces are *properly* assembled, viewed, and understood as a whole; being beyond fallen mankind's small slow timely perspective. For most souls, reality is a strangeness, that has parts they accept, and parts they reject. Evcn as souls reflect on their body's physical death, their acceptance of the full truth of GOD is rarely desired. Science is always so busy trying to figure out how to physically take reality apart, they can't see the truth is actually about how reality was all consciously put together. To visualize an eternal conscious soul is difficult for most. Each eternal soul is a defined yet growing or shrinking size and space of an eternal other-worldly aware and desiring generator and projector of new patterns of force quanta. A soul in an inactive state of complete rest of receiving no information, so, projecting no force pattern, creating no perceptions, and experiencing no awareness, temporarily ceases in its created experiential existence. Every soul and the aether are eternal, being beyond all effects of physical entropy. But by rebellion bringing physical entropy, the created second half of our dual existence, being a sensing body and this world and universe, all became a temporary story. Our soul can be dormant or active, it can grow or diminish, we can know truth or confusion, but we can never cease in the possibilities, probabilities, and certainties, that we will forever re-find our awareness. As we sense and then create perceptions to exercise our awareness, our desires fill and define our soul's current malleable space and size, as our patterns of generated force are entanglement in place. Classical views of moving information, raise paradoxes at the appearance of entangled creativity and non-locality. Consciously projected defined fields of force, are like Newtonian Gravity, consciously filling a space to align and entangle all aspects within its reach. But most souls weaken FATHER'S creation, by projecting interference patterns of desire *into* his prior desired patterns. Now if a physicist measures a particle, they then know the complete and omniscient information defining the state of that quantum system. So, at that moment, that systems entropy becomes zero, which a soul can consciously maintain, as found in the Quantum Zeno Affect. So, by it being a fully known consciously entangled information, it became a micro-*omniscient* system, having zero contrary or missing information. When contrary forces of disorder, appear within a perfect system information, there is corruption to that informational system. Original created omniscience, equals perfection. Under dualism of consciously forced determined complexity, any corruption of information, equals system entropy. Un-defiled and enforced projected omniscience, is the only way of eternal physical stability. The universe as a whole, was perceived into its omniscient informed existence by FATHER, in designed layered and projected entangled ways, as the earth rotated before his face, six times. But his projected omniscient information that formed it, is now rejected by most, and now absorbs countless individual contrary force patterns of belief. So, FATHER'S original projected perfect physical creation, is falling and no longer perfect. Now in *our* soul's forceful creations, our own projected information is miraculous, compared to all classical forms of communications. But our projected force patterns of desire, not only create our perceptions, they affect other quantum entangled aspects within our overlaying reach. Conscious quantum non-local creative entanglement, occurs at the fastest speed of light, which can cross the universe in a millisecond. And we see *zero* evidence of any *physical* contribution to this process, at any scale. Materialist are proud of controlling some particle entanglements in their experiments. This, while the number of particles we entangle in one moment of our streaming multi-sensory perception, is immense. The universe itself is a substantive, now devolving, once perfect perception of FATHER, a now corrupt creation he desires less and less, and more and more wants to forget. Now these force patterns we all project, are specifically structured, in finely defined complex ways, forms, varied values, and shapes, all having specific purpose and reason as to affecting and experiencing our awareness of reality. To best see how we create complex perceptions, visualize *our* projected 3-D patterned fields of generated desired force quanta, forcefully aligning a mirrored 3-D structural physical design within our larger wave creation of our feeling and emotion. Visualize how our detailed variable force wave patterns, first align larger particles. These rising waves of point singularities we align start a process mankind calls a collapsing wave function. We see a soup of created complexity is taken out from all the aether's possibility. Our new created perception, must mirror our projected 3D pattern in its various strengths of form and purpose of crests, and deeply defined complexities within every wave structure. And as all more forceful particles of firmer substantive structure rise, they each form a wave crest as they bend stretch and raise the possibilities of the aether towards new larger entangled spherical systems. These forced condensing in place whirlpools of points, within our created wave, congeal and rise to a level of building for us a particle we can actually physically and consciously sense. To visualize how our created particles appear within a wave, imagine a certain number of points forming distinct in place drops of water, as their surface breaks free from their surrounding 41,250° spherical surface tension, deep within a glass of water. We see all *our* new created larger particles, forming within *our* newly forced field, called a wave function. The swirling ends of each larger particles force curving around field of points, to be forcefully broken free from the wave to join together and form a distinct spherical field of points we call a particle. These spherical fields of points, can then build layers one upon another to make larger particles. These layered spherical fields, align out from the original central concentration of projected conscious force that anchors it all, creating the newly forced symmetry of a new designed entangled omniscient system of information and purpose. Now if any created particle is hit with over-riding force, it can be altered. A particle can grow in layers, or be absorbed, or broken into smaller spherical systems, that for a certain moment remain purposely or consciously entangled. This entire complex pattern of maybe millions of created greater particles and all the associated lesser particle wave information, are entangled together to form the over-all information, we in our soul, sense and indwell as our created representative perception. Each particle can also be forcefully reduced down to the aethers original singularities to be gracefully re-absorbed. Now by layer upon layer of sense upon sense force creations, we see a great spectrum of possible particles. But the particles forming *our* perceptions, are far smaller than the greater particles and elemental structures created by FATHER in his external perceptions, called the Periodic Table, and all its lesser subatomic particles. These ascending orders of particle size and speed, are formed in a prime number ascension of order. And we see the smallest, fastest, most abundant of all light particles created by FATHER, which mankind cannot detect, crosses the entire universe in a millisecond. We find the indivisible prime of *one*, being the single points that form the aether itself. And to sustain FATHER'S and our perception of his creation, FATHER created *at least* three much larger and slower, too much smaller and faster light particles for GOD, and the Angels and Mankind's far smaller/slower bodies to sense. We see FATHER created layered spherical field particles like bosons and quarks, being big enough that mankind can sense in their limited ways. All particles in their omniscient *original* fusing creation, existed in *zero* entropy or fission. Now for each soul, regardless of size or speed, our largest created structural patterns of particles, being just our physical side of our perceptions, remain as saved in the aether; but the associated lesser forced wave properties of our senses and emotions fade. Only our largest created particles and strongest aspects, form our basic memorialized perceptions. So, if and when recalled, only our *current* emotions, desires, and beliefs, will be sensed as its new surrounding waves. For memory, we need only the patterned larger particles and aspects of each perception to enable our recreative memory process. This Dualism of desired entanglement is everywhere, even far beyond mankind's perceivable comprehension. So, we see no gravity, or bent space, pulling or pushing our feet to the ground, or holding planets in their orbits. We see GOD'S desired, projected forceful creation of all bodies being forcefully entangled in their place and purpose of creation. This is a universal, yet subjectively aspect of the weaker smaller, to stronger larger, Conscious Force Entanglement Principle. If properly understood, this model eliminates *all* relativity and gravity problems. Now nothing is ever annihilated as particles are broken down below mankind's comprehension, or all the way back into the aether's grid. The points forming the aether are eternal, indestructible, and are internally unaffected by any enveloping force that aligns them to build complexity. Forcefully moving and aligning, is all any soul below FATHER, can do with these true singularities. How many of these points did FATHER use to make each photon. We estimate it took around 13 or more successive layers of spherical fields of points to form all bosons and quarks; being among FATHER'S most massive created particles, that form the elements in mankind's Standard Model. If you can see it, the existence of not just flat or curved 3D fields, but many layered spherical 3D fields over forcefully enveloped and aligned singularities, asks and answers all old and new questions. The true created physical beginning, required one immense desired generated force, spoken out in a vast soup of countless singularity points to form the spherical aether. The over-all force required, projected, and stored in creating and sustaining the entire aether is staggering. And imagine the next forceful waves of certainty projected throughout the aether, as GOD said; "let there be light". Now we call GOD'S created model "Universal Entangled Dualism, or UED"; and we call his Quantum Mechanical Interpretation "Wave and Particle Detachment Theory, or WPDT". These now being a true foundation and model to build upon and understand all the true far greater realities. FATHER spoke forth his aether of all possibility to fill and surround every soul's forceful conscious reach; yet it is set below their greatly limited physical detection. Now with ever-more souls existing, and their ever-increasing use of quantum wave technologies across the spectrum. The aether is experiencing ever-increasing states of hot-spot excitations around this earth. But we know it survives nuclear explosions and the existence of Super Novas and Black Holes, so we see no worries as to the singularities or the aether's structure. But might we be over-using certain specific frequencies of force that bind it, causing micro-tears in reality; again, very doubtful. But mankind can only observe certain wave excitations and FATHER'S immense particle creations. Consider how the near speed of light physical length contraction, and time dilation, are just the illusory effects of our approaching the limit of the information mankind's bodies provide, that enable our perceptive abilities. And please see that Special Relativity, nor Michelson-Morley, nor any other experiment failing to detect a Materialists proffer of an aether wind, can never detect or disprove FATHER'S forcefully determined point singularity-based aether. Mankind vaguely sees something they call dark matter and energy. These being just the undetectable particle and point residue of dissolving photons and other particles in an aether once greatly filled with light. By Materialist excluding all known conscious factors, they falsify every equation and result. If we have three conscious affects, and three physical affects, totaling 6 affects, on each side of one line; but they only accept and add the three physical affects from one side to the three on the other, to get six physical affects. This math appears correct in being taken out of context, but in context it is incomplete and incorrect. Materialist see only physical reductionism, while we see rising creationism, being so much more than the physical appearance alone. Now the concept of a spherical singularity has long been proffered as the beginning of the Universe. It is also proffered as the center of a Black Hole. What could be the center of a spherical singularity, or its surface? What fully defines a Singularity? A singularity must be one singular uniform quality, flawless and absolutely perfect; thereby being eternal. So, what could arise within a perfect spherical quality to trigger a Big Bang. A Big Bang would require an inner flaw or anomaly, which would declare it was not a true singularity. A spherical singularity must be eternal. For their Big Bang, did a fake singularity just rest there for a micro-aspect of time, set against absolute nothingness all around it? How is absolute nothingness defined so we might consider the effects of nothing touching everything? How can nothingness exist, and be defined? Is nothingness non-exist*ence*? Is it non-exist*ent*? What affect might nothing have upon something, and visa-versa? Now if a singularity exists at the center of a black hole, yet emits Hawking Radiation, it is no perfect spherical singularity of quality and form. This can be no more than an extreme process of accelerated, then stalled, and reaccelerated entropy and energy. This is not a true singularity of quality and form. Now mankind has never been able to reconcile gravity into any spherical singularity, it being GOD'S quantized external entangling over-all force of desire upon all dynamic physicality. Gravity is a fatal flaw in all mankind's models and theories, it is an irreconcilable defect and a death blow to a Big Bang Theory. *Why* do Materialist think and hope a singularity of form and quality existed at the beginning of the universe, or exists now at the center of a Black Hole? Yet we see countless minute true singularities of one pure physical quality and form, do exist as FATHER'S body, to form the structure of his own perceived complex created experiential reality. But created complexity can never rise up from within any *one* single true singularity. All complex reality is let loose, and is created by the dual creative aspects of FATHER'S forceful designing desires over his body of countless singularities. FATHER'S body of physical singularities are not numbered as infinite, yet they are to us innumerable, being the sum of his current body, his aether, and all it holds. Do we equate the aether to the physical body of GOD? In some lesser way, yes. Now the extent to which mankind over-estimates the firm physical reality is astounding; and the extent to which they under-estimate the quantum realm is far more. Their denial of the creative primacy of Dualism, is every Materialist's greatest crime against science. Can you please see that you can never begin to understand the ever-*decreasing* purpose and mechanisms of reality seen in our lawful conscious *un*-awareness of sleep or death; until we fully understand and accept the ever-*increasing* purpose and mechanisms of reality found only in our lawful vivid awareness of eternal life. Mankind will never understand reality in their falling linear way; reality can only be known in GOD'S rising spherical way. Visualize trying to segment a spherical and a linear, past, present, and future. And try to visualize your current perceived moment of experience in both, to see what they might look like? The linear current moment is an undefinable center segment of a linear line. But the spherical is the entire surface of a soul's expanding sphere? In a linear view, *our* past and future, are seen as part of that linear progression that contains all reality's past, yet they can never define how small or large the segment of each soul's current moment might be, between the past and future. But in the spherical view, we see each soul's current moment covers their entire surface, and all they may currently experience. Each soul's current sphere, holds all the information of their own past, and is not a part of all external reality in general. A soul exists in eternity, or it gets lost in an illusion of time. Can we, minus proven rules of understanding, just define, shrink, or expand, the now moment in a linear view of what we call our present? If so, can we expand it until it holds our entire past? And how far can we push it forward? This conscious flailing, makes no sense, there are no true founding or boundary principles, laws, or reasons, for time. A true view of a soul's now moment, says an eternal soul's spherical surface, reflects the *sum* of their current believed and sensed physical and emotional awareness, set by each soul's size and belief. Now as a conscious exercise, stare at any physical object and ask yourself, do you know that you only see and are aware of your created perception of that object? Or do you believe you see the object itself? And imagine you had a vast number of pure singularities to create things with, and you knew the quantum formula, to create a pencil using your creative perception, you could forcefully do so. It would just be a firm external creation of your forceful will. You just visualize and project your forceful pattern of desire into your medium, and a pencil appears before your eyes. This is what each soul of GOD does to create their perceptions, according to their force potential and reach. Now what defines reality? An atom, being made of subatomic particles, is 99.9999999% empty space. If we expand an atoms nucleus billions of times up to the size of a peanut, the outer electron shell of that atom, would be the size of a football stadium. An atoms physical mass to empty space ratio, it is 99.9999999% empty space, and so is a human body and every physical reality. Yet all are filled by GOD'S desired projected force that defines all consistency. Science also finds the mass to emptiness ratio for FATHER'S entire universe is far more extreme; being 99.9999999999999% empty space, compared to the created mass within it. Do you see how FATHER creates so much from so little, as he forcefully wills and defines all mass within his aether's emptiness of space. But always remember that this is a constant ongoing sustaining labor of love on FATHER'S part; so please don't work against him. Now we set a graph on the next page to visualize the sum of the aether and reality to help distinguish what mankind physically detects, from what they can't detect. A spherical model would be better, where the lowest section of the graph is a center ball, representing all pure *un*utilized aether. The center of the graph, is the thin 6% of physical reality mankind senses, laid over the core. The top of the graph, is the quantum Heaven, mankind can't sense, forming a third thick layer of a three-layered sphere.

This upper portion, represents everything in creation that is faster or larger or more ethereal than mankind can physically sense, test or understand. To include GOD'S interactive appearance and all quantum ethereal life and the Heavens. Yet GOD still consciously and physically works and exists throughout all 3 portions.

This upper bracket line represents the upper boundaries of all mankind can physical sense and comprehend

This middle portion, represents all that mankind can physically sense and test and understand. Mankind's conscious and physical size, originates at, and physically looks out from the center point between these two brackets.

This lower bracket line, represents the Quantum Planck Boundaries; where mankind's physical perceptive capabilities breakdown and fail.

This portion represents, all manifest things, smaller and slower than mankind can physically sense or test or understand. We divide this portion out, only to illustrate mankind's intermediate situation and limited spectrum. It need also be understood, that this portion represents the bare created and un-utilized Aether, which permeates and allows everything that is not a soul, in all three portions. It is all of GOD and his creation.

Do you see how and why the ever-increasing despair and destruction permeating this fallen creation, is equal to all the individual contrary projected forces, that work against GOD'S created purpose? This conscious and physical devolving of mankind is irreversible at this point, and is not even seen by most. And each soul falls into their own created abyss, as their dark understanding increases within them. Yet in all of this, they temporarily sense pleasure as their evil ways separate them from GOD into unseen insanity. Now do not tell me the sciences have built up true knowledge or undeniable proof, for all godless scientist, have only their own desired subjective interpretations of knowledge, hence all their ever-failing theories, and their detractors. This world does not know the truth, so, this world is not true or real in any eternal knowing sense. It subjectively appears real to the many, as it is being remade in their desired image. GOD has, and does all they can, but it is now a runaway free-will affect that can't be stopped; but it can be escaped. We must leave all insane individual ways of life behind, and only grasp GOD'S created ways, seeing every prior aspect of false belief within us, must be rejected and replaced. Mankind's theories always change as they consider the physical qualities, they barely sense. I see their interpretations of limited physical data; I see their assumptions and their required founding physical miracles. I see their desires for it all to be made in a purely physical way that oddly comforts them. I see unreasoned beliefs they form, to sever conscious force from physical creation. Long ago the labels of philosophy and ideology were coined and placed upon studies of the conscious and its affects upon reality, to hide the truth that the conscious and the physical are woven together to create reality. Materialist proffer hollow multiverse theories, not found in any measure of reality, but conjured up to temper their desperate souls. They need to believe that the final truth is beyond mankind's reach, as their dreams are both chasing, and being chased by their fallen godless desires. Physical desires lead the world nowhere, as they blindly run fast and far astray. Multiverse theory is a black hole of belief, where every believed possibility, births only more unanswerable questions. Such souls are strangely content and assured by the minutia of data that can in any way be interpreted to help substantiate their desire for GOD to not be there. They appear confident in their public denials of GOD'S existence; but other thoughts taunt and haunt them when alone in the dark. A great question is why should all the proof of all reality being purely physical, be so elusive, and fleeting. This proof, seems alive like a hunted animal running, circling, and hiding from a multitude of trackers. So careful of being found, so much mystery, birthing so many un-provable theories. Written thousands of years ago. Eccl. 3:10-11, "*I HAVE SEEN THE TRAVAIL, WHICH GOD HATH GIVEN TO THE SONS OF MEN TO BE EXERCISED IN IT. HE HATH MADE EVERY THING BEAUTIFUL IN HIS TIME: ALSO HE HATH SET THE WORLD IN THEIR HEART, SO THAT NO MAN CAN FIND OUT THE WORK THAT GOD MAKETH FROM THE BEGINNING TO THE END*." But GOD'S Children are shown all things. Now in conscious determinism, we too find that electrons and *then* photons, by design, were created sequentially; both being perceived on that first, of six days in which FATHER'S universe was perceived into its layered complex existence. By science declaring a required order of creation for particles and elements, in *no* way discredits a consciously deterministic model. Now materialist can't know what lies beyond their perceivable horizon, so, it is impossible for them to know the sum of reality. Scripture says in the beginning, on the *first* day, FATHER forcefully perceived a watery sphere, rotating in his created light before his face. Earth being unfinished, then rotated six full turns during his layered creation process, all from his *eternal* perspective, not mankind's fallen small slow perspective. These contrary perspectives of reality, share nothing as to why, how, how long ago, or how long the creation took. There are inflation models that estimate the size of reality, after their theorized inflation epoch, to be the size of a small galaxy before its current ongoing expansion. We do not endorse these, but it shows that mankind's model allows many possible outcomes. What Scripture says, is that from the first rotation of a far less complex elemental sphere, being the start of the earth, that FATHER'S complex universe was perceived and completed in layers by the conclusion of earth's sixth rotation. So, the seventh rotation he blessed as FATHER rested in all its perfection. In the 1950s and 60s, before super computers, cosmologists reverse engineered their Big Bang in every possible way, using chalk, on chalk-boards. Trying to mathematically crush all mass, energy, and force, into a singularity; but every effort failed before their eyes. They were stunned that so many intermediate calculations implied they were on track to a singularity; yet no true mathematical singularity was found. Are you old enough to remember that in the mid-twentieth century, a tiny error made by one cosmologist, made their big bang slow for a fraction of a millisecond, then greatly increase for a tiny fraction, and then slow again. And this incredible error produced a false infinite answer, that led to Inflation Theory by predicting a runaway endless creation of reality. Science abhors all infinite mathematical answers as being meaningless mistakes; *except* for this one. Such a tempting error, like a piece of candy before the wide eyes of a child. Yes, trying to build a theory for reality upon a math error, is insane. Now add the inexplicable gap between life and non-life, add the mind body problem, add the Fine-Tuning Problem, the Conscious Observer Effect, the Observers Frame of Refence Paradox, the Measurement Problem, add their *un*unified Physical Model, and the Double Slit Experiment, the Quantum Zeno Effect, Non-Locality, a Homogeneous Universe, all their Paradoxes; and again, how does force originate; and lastly add the fact that they have founded every theory upon unexplainable physical miracles. By all of this, GOD should be so evident to all souls having so many problems filling their desired belief. Yet in their core desire against GOD, they grasp at the darkness for *any* godless notion to worship, and forget all those problems. Scripture said millennia ago in 2nd Tim. 3:7, "*EVER LEARNING, AND NEVER ABLE TO COME TO THE KNOWLEDGE OF THE TRUTH.*" Their endless problems are declared null, against the desired belief they seek, and inflation theory is accepted at their leading edge of inquiry. Secular science is no longer about finding a physical beginning, it's now about escaping and hiding in self-deceit from the true conscious beginning. In all their error and blindness, *why* is there this insane rule and ardent divide? Why is it their supreme rule, that any pre-existing conscious forceful determinism, can never be allowed in physics? Does their aversion to newly generated conscious force, come purely by their aversions to the existence of GOD? If they could truly separate GOD from conscious primacy, could they accept it in great joy? Now science knows that our perceiving objects approaching us, or retreating from us, render different perspectives. But they don't see how consciously finding GOD increases our perspective, and running from GOD slows our perspective and limits our ability to see reality. So, two godless scientists stand up, one says we may never find the answers to reality, the other says we are on track to find all such answers. This endless *subjective* perspective of belief Principle, permeates fallen mankind. *If* the very beginning and very end of anything are unknown, the understanding of its middle is beyond reach and must forever-change. And so, all epistemological methods are wrong, and must also forever change. These facts declare a true and *certain* unreliability to mankind's every current understanding of reality. This Eternal Principle states that as mankind's incomplete knowledge expands, it *must* always fail and be adjusted to allow for their next and next false theory. Epistemology's definition has morphed in recent decades; do any ask why? No. They believe they look for the truth; but in all reality, they only look for their next lie. In their desire to believe they are going in the right direction; everything is thrown at the wall to find anything that temporarily appears to stick. This, while pure Dualist see this evident fluid dance between the conscious and the physical, with neither, ever found out of step. And as it must, the primacy of conscious force always leads in this dance, as the physical always follows. In this world, I have never seen an act of physical heroism, or an act of physical violence. I've never seen physical acts of selfishness, or charity. As a Law, I see only conscious acts of physical violence and conscious acts of physical heroism. I see everything everyone has ever done, has been a consciously motivated intended forceful act as we create our streaming perceptions of it all. This Conscious Primacy Principle exists throughout all reality. In its incompleteness, secular science is in all ways, just another imperfect practicing belief of religion and politics, controlled by a false consensus of hope, faith, and desire, founded in a false and incomplete understanding. All lost souls require abstracts to believe in for their awareness of life, as they refuse most parts of the one pure conscious truth. So, they go in other directions looking for beliefs to replace what they find undesirable. By GOD'S ways being unacceptable and undesired to the many, they become unbelieved. Now the material sciences need to admit *why* there are two sets of laws of consistencies governing the bicameral realms of reality. That reason being, there are two realms of lively existence: one is the upper vast quantum ethereal heavenly realm, existing everywhere in its ethereal consistencies, being a vast ethereal Kingdom full of ethereal bodies, and *all* conscious life. And two, is the lower smaller firmer condensed physical realm with its deteriorating laws of consistency and physical life. Both existing in all their certainties, and being created within the possibilities of the aether. Now a certainty can exist eternal, surrounded by uncertainty; but no certainty can endure eternal, if an uncertainty arises within. We see a conscious and ethereal substantive realm; and one firmer purely physical realm existing together, overlayed in one aether. These realms are two, yet one, they are separate, yet joined within the aether. Their two sets of conscious and physical laws can never be unified in any purely physical model. Though left and right, and up and down, are conceptually adjoined; you can never truly unify or separate them. You can only forever search for the infinite finer and finer defining of the line between their difference and their oneness. Just as no soul can ever place any true divide of complete separation, between any two aspects of reality. These two interacting realms are like two dynamic waterfalls having differing forms of water, passing and existing together at all points, yet separate in purpose and form of perception. But in reality's last eternal divide, we see one conscious and physical waterfall rising up in its Dualism, as the other falls down forever, separated in purpose and form, never to hurt the other. The only true ever-growing existential divide to ever exist, as each side forever leaves the other in all understanding of reality. For there to be a real true complete eternal separation between any two conscious or physical things that exist, requires a true division of every relative aspect of reality that exists between the two, which requires two separate realities. So, if a soul eternally falls into the diminishing ways of division, they must forever grow in separation. Mankind's slow small understanding of reality and existence are bent and twisted by their many differing theories, causing great loss and sadness. We cannot in *any* act of certainty say physicality exists, until we *consciously* perceive its information. The only proper perspective and understanding of the concept of reality and our existence, is defined by both sides of the coin; and we do see both sides exist. One side being our conscious projected force patterns of desires to exist in our awareness. The other side is all the physical possibility within the aether, from which we create our perceptions. There is no other logical and complete way to grasp the concept of our created *perceived* existence without both sides coming together. Our soul must sense and be aware of something; only this brings certainty to our soul of *our* existence. And the physical must be perceived by the conscious to have any realized certainty that *it* exists. This required Dualism underlies reality. Do our created sensed perceptions passively form themselves from nothing? No. They are aggressively and forcefully projected into their greater 3D created certainty by our generated logical forceful patterns of projected desire to exist. Every other quantum mechanical interpretation, has no answers. Any true quantum mechanical interpretation, must answer 3 *physically* unanswerable questions about our created reality. One, how can we get physical certainty from a probability? Two, how can we get a physical probability from an undefined possibility? And three, how can we get a physical possibility from nothing? We sense external mathematically patterned quantized information through our interfacing relaying body, which we then understand and interpret through our biased desire as we indwell our forcefully created ethereal representations of that same information; all having our *desired* biased touch. Ergo, to degrees we can all desire to see things differently. Perceptions are ethereal manifestations that can only be described as our own created subjective, yet substantive certainties. And mathematics is just a form of conceptual conscious summation, to help us understand GOD'S vast projected patterns that we ponder through the sum of our acquired logic. So, when we look at the periodic table of known physical elements, we find zero awareness, belief, or understanding; or in the elemental particles that form the elements. By weaving compounds together, we take no first step towards creating self-awareness. No soul can add two physical components together, and then three in every possible way, then four in every possible way, and so on to infinity. And point to any level of their own *consciously* organized physical complexity, and say here I see a true intelligible reason for a new ineffable miracle called self-awareness to suddenly strongly, just emerge. And *again*, even this useless conceptual endeavor first *required* your conscious awareness. We see FATHER projected his force to form all chemical electromagnetic reactions to aid in physical life; yet at most all points throughout all the complex ways of life, we see a far more understanding guidance than electro chemical reactions alone could ever orchestrate. We see a certainty of awareness and knowing woven throughout the complex process of life. We see a great reasoned purpose is *required* to orchestrate physical life. And from the first appearance of such a thing as this, and every appearance everywhere of everything such as this, comes this question; how can the physical *know* and take on this continued *guided* purpose to perpetuate life? Where is born its understanding to guide all such complex orchestrations of hope? How does it start, and why should it ever, ever stop? Is this forceful understanding guidance, projected consciously from without, or physically from within? Can and do we each just assume and believe this to be one way or the other? And for each souls required believing in something, do we all *assume* this to be one way or the other? Surely our own beliefs declare our own desires in this matter. It will take so many words and avenues of reason to illuminate all the vantage points required to fully appreciate this truth; and we so try. But a conscious paradox says it is never just what one says, but what others hear. The Free-Will Uncertainty Principle states, we can never know what happens to *our* meaning, once our words leave our soul and enter another. We point out here, that even though we've just begun, with so much more to come. If you properly see and understand these first few articles that show perceptions are forced particle-wave based creations. And see how all we now perceive second-hand, was first perceived to exist external to us. Then empirically, you must see that "GOD *Must* Exist Necessarily". Consider something that always interested me, being that if you continually, *one* by *one*, deny all the many evidences of conscious guidance from the full equation of reality; GOD slowly disappears before your eyes more and more at each step. But if you allow, *one by one*, the introduction of the guidance of conscious force projections, up to include our created perceptions and all firmer physical aspects; then the supreme conscious force of GOD does gradually and necessarily appear *more and more* with each step, and it cannot be avoided. So, this is then a real qualitative quantitative equation; and all things ordered or measurable, must be considered scientifically. Now in physical science, there are these two terms of cause and effect, being a forced energetic action of quality and quantity as a *cause*, and the seen influence upon the next event, being its *effect.* These two are applied to all physical theories as the two required pillars on which they build their materialist beliefs. But projected conscious forceful desires of purpose, being the cause; is disallowed in materialist theory. Road blocks of bias are erected everywhere, as unreasoned dishonest restrictions control the scientific narrative. They force every attempt to show how conscious forceful desires align all physical creations, to present our arguments using conscious effects alone, without a conscious cause. Yet we all see conscious cause and effect constantly, everywhere. In Dualism, we must be allowed to show conscious force cause. But that cause is our projected forceful *motive* of desire. So, the forceful motivational reasons for the way the creation once was, and the way it is now, must be heard and considered to find all true applicable final conclusions. Only then can honest jurors, judge its true scientific merit. In everyday life we apply conscious desire to understand the many occurrences in our existence. A parent, spouse, friend, sibling, prosecutor, defender, or judge, will ask many questions to determine the reason and purpose for why one did, or didn't do something. Only then upon establishing motive, physical opportunity, and evidentiary continuity, is there seen a required *legal* preponderance of admissible evidence. The law of true understanding among conscious creative and destructive creatures, requires we allow and show motive or intent, to support all physical evidence. Why deny this required aspect of intent or motivation in science. Desired forceful motivations are seen everywhere in everything. If one holds to a materialist perspective, it has been built and woven together by the many threads of their past perceived and interpreted experiences, interlaced through past beliefs to gradually weave the ever-changing fabric of their current belief. So, honest integrity *requires,* that any jurist holding a materialist perspective, for them to consider *this* case in all its strength. They must allow us to weave our many threads together *with* motive, all within their sight, and wait to see our final fabric. Our many threads in this book, must not be individually judged, or excluded as being contrary to a current material belief. Each thread must be honestly heard and held as a piece of evidence, and weighed and judged as a whole fabric at the conclusion of our presentation. Then, all physical evidence and shown motives, must completely describe and align with *all* observations and occurrences in true logic, reason, and purpose. Only then can an honest verdict be rendered, as to its probabilities. Now the single points of the aether, beyond GOD'S original spoken forceful alignment, have no force themselves to further align, or include themselves in any rising complexity. Yet our soul's projected forceful patterned desire, can capture and align them into larger perceivable three-dimensional certainties. The true full understanding of the Conscious Observer Effect, sees each soul's ability to forcefully create particle-wave based substantive perception. Not just the larger particles, but the lesser fields or wave properties are also forcefully raised and entangled together. Each soul's projected force, create waves rising up out of the aether's possibilities, to then create GOD'S greater, and our lesser size particles. Wave properties are not just forced energy, they are forced true ethereal condensing point certainties. This fact must be woven through every part of our belief, including our own created perceptions. If you slightly believe in GOD, please ask yourself, are the ubiquitous fluid wave excitations, and collapsing of larger quantum realities, just strange leftover oddities in GOD'S creation. Or is this the seen creative manner of *GOD'S BEING*. Are these created waves and particles the *evidence* of GOD'S vast, and our lesser projected forceful creative works. In understanding conscious forced wave functions, and the forced certainty of all perceptions, we then see the edges and foundations of GOD'S, and our own created physical manifestations. Romans 1:20-22, "*BECAUSE THAT WHICH MAY BE KNOWN OF GOD IS MANIFEST IN THEM; FOR GOD HATH SHEWED IT UNTO THEM. FOR THE INVISIBLE THINGS OF HIM FROM THE CREATION OF THE WORLD ARE CLEARLY SEEN, BEING UNDERSTOOD BY THE THINGS THAT ARE MADE, EVEN HIS ETERNAL POWER AND GODHEAD; SO THAT THEY ARE WITHOUT EXCUSE: BECAUSE THAT, WHEN THEY KNEW GOD, THEY GLORIFIED HIM NOT AS GOD, NEITHER WERE THANKFUL; BUT, BECAME VAIN IN THEIR IMAGINATIONS, AND THEIR*

*FOOLISH HEART WAS DARKENED.*" Mankind has come full circle by technology, curiosity, and imagination, to peer into the ethereal quantum realm in wonder. But they do not see that for them, looking into that realm is like looking through a dark distorted glass, and there is a vivid mirror on the other side. Can the rebellious of mankind overcome their distorted belief, and come to see their souls in that quantum mirror. 1st Cor. 13:12, "*FOR NOW WE SEE THROUGH A GLASS, DARKLY; BUT THEN FACE TO FACE: NOW I KNOW IN PART; BUT THEN SHALL I KNOW EVEN AS ALSO I AM KNOWN.*" Now when we consider GOD'S far firmer, vast built-up complex layers of conscious force creation, the countless layers of forced wave/particle collapse required to build the universe as a once perfect complex created perception, takes on a strong meaning. But if you deny GOD, all force, all particles, and all complexity, must have come by an unknown, and by chance, which can never explain anything. But we see all GOD'S desired spin, rotations, and orbits, are projected to force each particles definition and purpose, and all larger firmer realities to serve their intended purpose, right where they exist. So, their fruitless search for a physical force particle *disappears*, as the truth of all conscious forceful entangled creation *appears*. How can pure randomness become complex on its own? It can't. We create our own 3D particle wave-based dreams inside the immense firmer forcefully created 3D perception of GOD. We are all, divided out, tiny, highly restricted creative gods. FATHER gifted and divided our souls, out from his Soul. And we exist deep within the immense firm physical perception of the one true GOD. Psalms 82:6-7, "*THEY KNOW NOT, NEITHER WILL THEY UNDERSTAND; THEY WALK ON IN DARKNESS: ALL THE FOUNDATIONS OF THE EARTH ARE OUT OF COURSE. I HAVE SAID, YE ARE GODS; AND ALL OF YOU ARE CHILDREN OF THE MOST HIGH. BUT YE SHALL DIE LIKE MEN, AND FALL LIKE ONE OF THE PRINCES.*" GOD gifted *and* created us is in his own bicameral image; to include both our forceful creative conscious soul, adjoined to a physical existence. John 10:34-36, "*JESUS ANSWERED THEM, IS IT NOT WRITTEN IN YOUR LAW, I SAID, YE ARE GODS? IF HE CALLED THEM GODS, UNTO WHOM THE WORD OF GOD CAME, AND THE SCRIPTURE CANNOT BE BROKEN; SAY YE OF HIM, WHOM THE FATHER HATH SANCTIFIED, AND SENT INTO THE WORLD, THOU BLASPHEMEST; BECAUSE I SAID, I*

*AM THE SON OF GOD?*" Phil. 2:5-8, "*LET THIS MIND BE IN YOU, WHICH WAS ALSO IN CHRIST JESUS: WHO, BEING IN THE FORM OF GOD, THOUGHT IT NOT ROBBERY TO BE EQUAL WITH GOD: BUT MADE HIMSELF OF NO REPUTATION, AND TOOK UPON HIM THE FORM OF A SERVANT, AND WAS MADE IN THE LIKENESS OF MEN: AND BEING FOUND IN FASHION AS A MAN, HE HUMBLED HIMSELF, AND BECAME OBEDIENT UNTO DEATH, EVEN THE DEATH OF THE CROSS.*" Strongly understand, that our being minute forceful creative godlike creatures, capable of creating ethereal particle-wave-based perceptions; is nothing compared to the fullness of the ever so great GOD who created everything and life. These Scriptures declare our empirical understanding of all reality; and otherwise, would be most confusing indeterminable things for GOD to say. We rightly define children or offspring, as directly descending out from their parent's qualities. Being new small growing representations, formed from not the same quantities, but the same qualities and inherent attributes and appearance as the parents. Children also exist in the same ways and means as their parents. FATHER said the Angels and Mankind are all children, originally born of the most-high. And CHRIST said that *before* the universe was formed, he too, being the only begotten SON, was the first gifted free-will soul and creation, to come forth out of FATHER. So, what then *is* conscious and physical reality? Is something real, because we strongly sense it in many ways, and because we believe we touch it and hold it? But we touch, hold, and feel things in our dreams; so, our soul and our created reality are there also. Is something more real, because no matter how many times we come to it, like a certain pen in a drawer, it is always there? Is this the standard fallen mankind chose to convince themselves of what is real, and what is not? But we can dream about the same things many times. Are some things simply *believed* to be more real, and others less real? Does our own subjective created reality have levels of belief? Are all things we see and feel real, but some more so than others by belief? How often and in what clarity do we need to see or feel something, before we are convinced it is real? Our choices are: are *physical* complexities forceful conscious creations, or, are all our *conscious* complexities, physical creations. Materialism has for so long spent vast treasure on research to show our awareness is a product of physical complexity. Yet to this day they cannot begin to write the first letter of the first word of the first sentence, defining what is self-awareness. Or where it resides, or comes from, or where it goes. Or how and from what do we create our perceptions. Their failed centuries of searching, *confirms* that consciousness is not a physical biproduct. Einstein's proverbial definition of insanely trying to prove a falsity is seen over centuries, as each time they expect a different result. Every soul is an entity of sensing awareness, and a forceful projector of patterned physical complexity. Each soul is an eternal seat of understanding, logic, belief, and a source of all forceful desires to exist. And FATHER'S soul is all of these, and the creator of all external reality, including physical life. It is GOD in their great creative forces above, and our lesser godlike creative forces below, partitioned and gifted out from FATHER. We are aware of all our neural senses, from sensing the body, up to our higher senses located in the head area. Yet we do not feel, see, hear, taste, or smell, through these. The information we sense in our fingers and toes is interpreted by our soul as being in those areas, and not the brain. Neurology focuses on the information running through our neural system, and its correlation to our experience. Every neuron and cell in the body is part of a two-way interfacing relay system, created to sustain our soul's awareness of existing. Our sending and receiving information, allows our soul to interact with GOD'S created reality. When studying the activity throughout the entire neural system, moving between the physical world and our soul. We see this central and extended neural system is *designed* to enable our rich blended and extended sensing of existing in this world. But observed neural correlations, do not in any way shape or form, show any true cause for our self-awareness. We were taught in elementary science, to *never* blur correlation with cause, each having distinct meanings. "Cause" has a true and distinct quality, standing among all other qualities; while "correlation" refers to all the lesser peripheral aspects, that occur in a loose association to the unfolding of any caused event in question. But now, regarding consciousness, we see correlations are used in the media to move people to believe neural correlations *cause* conscious awareness. All my life I have found that it is only the very beginning, and the very end of anything, like say a line, which can fully define its interior aspects. It's clear that in mankind's ever-changing, zero beginning or ending understanding, that should they ever see the true beginning and end of reality, which they can't. They again, must find everything in between to be something they had not believed, or predicted; and surely something they did not want, look for, or expect. I have focused my life on answering the big first and last metaphysical questions that when answered, form a never-changing center out model of reality. I exist in one ardent unchanging understanding; in opposition to mankind's ever-changing understanding. My knowing expands, but my body of understanding has never once needed to be corrected in any way. So, in this approach alone, I find immutable confidence. Our forceful souls exist only in our work of creating perceivable 3D perceptions out of that which is not our conscious force. But minus our work of perceived creations, our soul waits in an unaware uncertain state, akin to maybe an uncertain conscious entity wave. Now science, can't just coin two hollow terms for our lack of awareness, called sleep and death, and declare the question of where our self-aware reality goes, has been answered. If our received physical information becomes too weak, or completely shut down, our souls succumb to an unaware state, temporarily having no created existence, and reverting to a conscious state of uncertainty. And in such a state, the illusions of time fail, minus the continuum of one's streaming perceptions. Our expanding spherical *now* moment, is all we will ever be; which is so important to understand in living for eternity. After a hundred, or a hundred million conscious years, if asked how long our life feels. We must admit it feels like only a short moment, as all we ever are is our current moment. We say the aging body by increased disorder, feels slow, pained, and adversarial; but our soul feels new. Now if time were (it's not) a woven fourth-dimensional aspect of reality, then below our perceptive ability, time *could* be slowing and speeding up, it *could* be wobbling up down and side to side within the other three-dimensions. Yet all we can ever know is our next and next moments perception, giving us a steady forward linear feel. So, the after, and after, and after, can never equate too truly only forward. We observe and remember difference, so most proffer things beyond us. This adds to the illusion of a forward futuristic direction. But this all occurs on the surface of our current spherical conscious soul. If mankind needs a physical arrow of time for comfort, they can see one. And adding measurements to the unfolding courses of GOD'S created continuum, allows fallen mankind to try to manage and minimize the worldly chaos and destruction they set upon themselves. We exist on the surface of our current spherical expanding moment, while reality around us, including the body, deteriorates. Now we can choose to be *individuals*, or move towards our perfect *oneness* in GOD. For understanding, when we write about *individuals* or *oneness*, we never mean physically; but one's belief, understanding, and path of conscious existence. FATHER divided *out* from himself, vast numbers of souls of varying size, and created all physical reality and life to exist in his perfect desired ways. But we now see endless contrary understandings, desires, beliefs, and ways of living, called individuality. I find it an insane paradox that liberals say it makes life better when people are of different beliefs and ways; yet the same souls gravitate to those who agree with them in their personal associations. They know differing cultures bring disagreeable confrontation into their life, as they denigrate all those who live contrary to their beliefs. Most souls disagree with GOD; so, in defense of self, they lie to self. When souls say all cultures are equal in their eyes; they only mean their own culture is good. I fail to see the peace of Heaven surviving, filled with individual ways and beliefs. Now we consciously exist and then don't exist, over and over. And quantum particles move back and forth between our certainty and uncertainty of their existence all the time. They disappear to us, but they're not annihilated. They simply break down into aspects below mankind's comprehension. Now, consider that if we list all the requirements for the entire central extended neural system of the human body, to act as a *designed* interface to format and relay complex patterns of information back and forth between two independent qualities. In this we find a perfect fit. And all our observed neural correlations, reveal the back-and-forth formatting and channeling of sensed experience and desire between our soul and body. The central and extended neural system and body are an amazing created interface. If the neural body *works* in every aspect like an interfacing relay, if it *acts* like an interfacing relay, and *quacks* like an interfacing relay, it is in every aspect, just that. This brain-body interface is a created wonder; but saying our neural correlations are our awareness, allows materialists to make wild assumptions. By all things known, in no way is this physical interface our seat of conscious awareness. The eternal spiritual, *and* temporary physical perspectives, are antithetical as to our beliefs. For me it is so deceitful that the godless neural sciences continue to infer, and imply they expect to find more than these interfacing relay correlations. The eternal souls of mankind and angels are gifted measures or spaces of an other-worldly eternal creative projector of forceful desire. And we find ourselves among multitudes of other free-will souls, as each perceives FATHER'S one *shared* external physical creation. We see seven requirements must exist for this to be so, and those seven aspects do exist: one, there must be, and are a multitude of defined free-will conscious entities; two, there must be a prepared physical existence filled with sensing mobile interfacing bodies for each soul; three, there must be a background medium or aether, allowing souls to forcefully create their own particle-wave-based perceptions; four; there must be conveyable light, and other information traveling through the aether to describe *to* us, GOD'S great firm physical creation; these four, allow us to create number five, being our perceptions of reality within our soul; six, there must be a perfect reason and purpose for the great cost of pain and suffering in it all, being GOD finding a loving eternal family; and seven, there must be the one far greater loving conscious soul to design perceive and gift the complexity, truth, and understanding that allow us to then perceive and know his work. And for now, all created perceptions of each soul's existence are stored in the aether within our spherical soul as memories; being a book of us, if you will. We see by communicated revelations of HOLYSPIRIT, that non-verbal forceful communication between souls is a certainty under proper conditions. Like if one can play a song on the piano; other souls under proper conditions could overlay or connect to their soul and indwell that stored knowledge, and share that ability. Being fallen, this is no longer a shared conscious/physical attribute; yet we see savants show signs of this. Mankind lost the shared sense of one connected mind in the fall, but we should regain it in Heaven. Now I know I exist in my wakeful perceptions of the external world; and I *know* I also exist in *all* my created conscious perceptions of dreams and fantasy also. In my dreams, I am self-aware that I am me; so, I do consciously exist in dreams. The only difference between our wakeful perceptions, and dream perceptions, other than strength and clarity, is belief. In being awake and most vividly aware; most think they know where they are, when they actually don't. And in dreams, most believe they don't know where they are, when they should. In our dream and wakeful awareness, we consciously exist *only* within the ethereal perceptions we create. The constant vivid clarity we find in our firmer *wakeful* perceptions of

GOD'S created physical existence, cause most to draw a false distinction. Most see our awareness of perceptions as some being real, and some not being real. Consider how conscious reverie shows we are not limited in what we see and touch by what is physically there. There is no empirical line found in our souls between our sleep and wakeful reality; they are fluidly one in each of us. Yet each soul's core desires have influence over interpreting and perceiving the information they sense. And within each soul's size awareness, we build the ongoing particle/wave perceptions of our current moment. Our forceful process builds to reach its peak of our realized 3D certainty. And as we exist upon the top, or now, of our rolling streaming wave/particle creation as it rises and speeds up through our desires to meet us, the passing crest of our prior moment then slows and lowers into the aether as memory. We do not create a frame-by-frame existence; such proffers are not even possible for created conscious awareness. Our streaming awareness is like opening or closing a valve to allow a flowing rising streaming wave of information. Think of your awareness, as a surfer riding the crest of a flowing ever-changing wave of conscious created biased certainty. Yes, a flowing wave example, calls up the image of linear time; but please remember, this wave crest field always rises from within, to exist and cover our soul's spherical surface. This process creates the certainty of our moment, and being awake is just our process of strongly perceiving the external physical existence, while sleep or day dreaming awareness, is our process of perceiving ethereal information from memory and or the quantum realm. We can forcefully play with the created information we have stored within our soul. There is a true difference between our conscious awareness, and our consciously not existing. But a true line between being aware while awake, or awareness in a dream, does not exist. We sense strength and vividness in all stable or fractured information, but this is the only difference. We all dream create our reality, even those who hold a belief in the *primacy* of this firmer external reality. But no one can deny we forcefully and effortlessly create our complex, 3D, 5-sense perceptions, followed by our biased awareness of them. Can you believe your perceptions physically create themselves, without a projected forceful three-dimensional pattern or template to align their wave-particle structures; that being physically impossible to describe. Our perceptions of our differing conscious qualia, allow us to strongly sense as we interact with the created realms. So, in all our choices, we all exist within GOD'S perception. If one grows to see one's self existing in a strong physical way, one becomes a physical sided creature, enslaved by desired physical pleasures and experiences. Now many souls see the act of sleep, or any loss of awareness as meaning a physical conscious brain and body must rest or repair. Most godless studies on consciousness include sleep dreams, and other forms of weaker conscious existence. Yet to this day, for either the state of sleep, or being awake, science cannot define in any *physical* way, what is self-awareness, where it resides, or arises from, or what perceptions are, or from what are they made. They have *never* found one physical aspect, beyond those interface correlations of blood flow and chemical electromagnetic activity occurring within their map of GOD'S designed central/extended interface. This is because, that is all that is physically there to find. A switch turns on a light, is our awareness of brightness caused by that switch? A light switch is just a correlation that can be replaced. Now it is known, the brain does not have the programming capacity and ability to control all the requirements to orchestrate and sustain life, let alone the ability to create, record, store, and recreate all the facets to form our current and memorialized perceptions. One compressed moment of a multi-sensory perception of one complete occurrence, being interpreted through our understanding and belief would require at minimum, many millions of data storage points. So, to physically function, move, sense, create, compress, and store the memories and emotion of experiencing the three-dimensional existence of our five senses of awareness for 24 hrs., would require trillions of data points for a rare soul with an eidetic mind. Neural science has no clue what our miraculous awareness and emotions are; so how can they know their miraculous requirements? They say they know so much, yet an honest few speak about their profound limits of understanding. By adding all required work to consciously and physically exist, and memorialize every moment, a brain runs out of analog storage and operating capacity in just one day's existence. This inadequacy shows there must be a quantum place to store and retrieve quantized information beyond the brain-body-interface; that being the quantum aether hard-drive filling our conscious soul. By all these things, we see our souls exist and operate within the quantum realm. So, our awareness could never be closer to where scientists are being forced to look; yet by denying it exists, it could never be farther away. Now as to our loss of awareness in death or sleep, what does this truly reveal? Physical death is just the complete stop of GOD'S forceful patterns of life in our current interface. So, our soul enters and passes through a non-experiential non-existent state, lacking all physical information to enable our awareness. So, our eternal soul waits in that state of non-experience for a new platform. Our fallen body platform, encounters a corrupted external existence. It endures a harsh process in growing, only to then deteriorate like all physical life, straining against the hostile external forces, dangerous choices, and chance injury. Our eternal soul, can desire the respite of sleep from this world's confused tempting struggle, and sensing the weariness of a tired body. But our soul being eternal, requires no rest in its existence. In a state of continuous blissful physical information and GOD'S love, and a new eternal angelic body, our soul will never want or need the escape of sleep. Our nightly loss of awareness, teaches us certain lessons. One being to understand our impending physical mortality. If we did not know our nightly loss of existence in sleep, we could not begin to grasp the long state of unawareness in physical death. And our nightly loss of existence shows we can exist again after that loss. And dreams show our awareness can exist outside this body, so, its transferable. It has been said by many “life is but a dream” yes, truly a conscious dream. We are consciously aware in dreams, by seeing, hearing, and feeling, just as when awake. What are those minor differences in wakeful and dream awareness, and what is exactly the same? We see only degrees of firmness, richness of detail, and duration of information we receive to create our perceptions. This giving rise to the *terms* of wakefulness, the sporadic vague dream, and the vivid lucid dreams that are so real, in which we know we are dreaming. We consciously create our perceived reality in all of these variables. Whether awake and creating our representations of the external creation of GOD, or day or night dreaming, our existence rests solely within our created perceptions. The only poor example of the soul and body connection in this world, is computing software and hardware. We can use these, because their purpose requires both together; they individually have no existing purposeful outcome. We are truly no more than our awareness of our created perceptions, known through our current desires and beliefs, in being, and experiencing existence. Our soul's-awareness exists completely within this process and outcome. If this truth alarms or frightens you, I am sorry. I none the less welcome you to the truth of your eternal reality. If your soul is separated from all quantum and physical patterns of information, there can be no created third elements of visuals, sounds, or any form of experienced reality. Every soul to body, separation, leaves us needing a new sensing body/interface to restore our third element we call our perceived created reality. Please see and know how GOD shows us we can lose the awareness of our existence, yet he allows our reality to be restored to us day after day, showing our soul at rest, exists in an uncertain state in the quantum realm. The fallen body is now subject to reasoned physical disconnections, and a body-brain-interface-relay, being inanimate hardware, is interchangeable and easily replaced by GOD. Now please *deeply* consider, that an original source or cause of actual *force* itself, has never been *physically* found or proffered; because conscious creative force can never be physically found by mankind. Force is completely misunderstood by science, as they try to theoretically quantize force as being a particle, without knowing or ever defining what is force, or how to source and attribute force to that particle. They try to blur the distinctions between all force, energy, and mass at every turn. To create his aether, FATHER spoke with both his force and points, or breath. In all his firmer greater creations, FATHER again spoke with both. But FATHER can speak or just project his desires of force to sustain, or convey commands of understanding. Our soul, being gifted to us as a quantum realm entity, is immune to all physical laws. It is truly eternal and was once part of FATHER'S Soul. Even if we do not want it, it can never be given back. If we ever truly do not, and cannot, value and honor our *eternal* conscious gift any longer, there is then no good eternal love left to be desired, or anything eternal for GOD to know, love, or have compassion upon. There is only the ever-diminishing essence, of wanton self and all mutual destruction. But for *we* who desire eternal love and pure peace. 2nd Cor. 6:18, "*AND WILL BE A FATHER UNTO YOU, AND YE SHALL BE MY SONS AND DAUGHTERS, SAITH THE LORD ALMIGHTY.*" John 1:12, "*BUT AS MANY AS RECEIVED HIM, TO THEM GAVE HE POWER TO BECOME THE SONS OF GOD,*" Now GOD can slow a body, causing a soul to faint; and summon that soul in a temporary or translated body in the quantum realm of Heaven. Since this would not be of this world, this would feel like being in a most vivid lucid dreamscape reality. We might wonder is it just a sleep dream reality, as these lines get hard to see when searching for deeper meaning. If one is sitting, and said by neural science to be physically awake, but that soul is set adrift in reverie. Can that soul, be said to consciously exist in that chair? Or are they *literally* consciously existing elsewhere? If startled from daydreams, souls *instinctively* say sorry, I was somewhere else. Whether aware in sleep, or awake, or in between, we exist within the aether and perceptions of our current expanding moment of reality. We share a wakeful external existence with others; *but* our dreamscapes may also be shared. All dreamscapes are substantive quantum creations, so, these can allow souls to overlay and experience moments of shared existence, once there is a sufficient bridge to envelope and ground those souls together in one shared perception. Souls can share their perceived existence in all their senses, through any substantive environment. If GOD perceives a room on the moon, being filled with the aether, a table, chairs, air, and several mobile sensing interfacing bodies. This would suffice nicely for a true visit between several transported souls, even if one or all had died long ago. GOD'S shared ways, beliefs, a new created Heaven and Earth, and our new Angelic bodies, will allow us the bliss of eternal life among many. We see mankind making all manner of virtual reality headgear that allows them to be in the center of, and see and hear an alternate reality; even being shared with others in every direction. Mankind in weakness and confusion, is capable of this, so, what is GOD capable of doing. Mankind can only poorly imitate what GOD has already done. We have these created facets of our internal existence, and we have our shared facets of a vast external existence. Now most of us have awoken to say I just had a dream that was so real, yet I knew I was dreaming. Now how real can a dream get? Once in a *very* long *lucid* and most vivid dream. I was feeling the texture and warmth of a woman's hand I did not know. Feeling and seeing her hand and my surroundings so clear in such minute bright detail. I knew I was dreaming, yet I was in my most true and vivid awareness. I so strongly sensed the feelings of my presence and the warmth of the sun on one side, as the wind coming from behind, moved my hair. All near and distant surroundings were detailed in all exacting wakeful brightness, color, understanding, and continuity of weight, texture, smells and sounds. It was so real in every detail and way that I had to test it. I could see and feel the reality of my fingers gently brushing the minute blonde shiny hairs on her hand; and using pressure to slightly depress the brightly lit flesh of her warm smooth detailed skin. Yet I tried hard, and could not get my foot to depress the ground. I could not distinguish one single difference in all my senses, emotions, and comprehension, from my usual wakeful reality, which endured for quite a while. I awoke in bed with a full wakeful experiential recollection. We most often have a common vague confused sporadic dream awareness. But if being in that strong vivid dream lucidity, I see, hear, feel, taste, and smell, and cannot tell my dream is not real; how can I know being awake is not just a vivid lucid dream. Why did I receive my strongest of dreams, right when I was writing this part of this book. I have since had other such dreams, that were spiritual in nature. Now, what would have happened to us if you and I found ourselves sharing a vivid lucid dream that equaled our wakeful awareness, which continued days and weeks in the same surroundings. What happened, was GOD had perceived a new physical environment, and moved our bodies, or connected our souls to new different bodies. We must then accept that all shared reality and our existence, come to us only by our created perceptions of it all. And it is only the longer duration of our stronger more detailed perceptions of an external reality, that convince many there is a difference between a dream and wakeful reality. Our disconnection from the body in sleep, is not often complete. So, vague lingering connections, allow vague dreams to be created through faint confused information from just the body and our quantum memory. But a dream that is as real as being awake, this *may*, but *not* always be a vision of communicated purpose, or meaning from the other realm to be considered. These are of course rare. We read in Scripture how Prophets fell on their faces as if dead, meaning they fainted from their physical existence as the body fell. And they were lucid and vividly translated into the quantum realm of Heaven to receive Prophecy. And when returned and awakened again to this physical world, they held a most detailed wakeful memory of a true experience. And quite often they awoke in a far distant location, being GOD'S way to assure them it was a real experience of communication and revelation. Often, but not always, we see a soul's sporadic dream activities bleed back through the body, causing activity and movements of the physical body. In rare cases to the point of walking and acting out in sleep. We see a certainty exists that the last sleep we call physical death, can never preclude our soul from post-death episodes of awareness. Scripture says JESUS preached to the dead for three days between his death and his resurrection. There are so many records of spiritual connections, written about throughout the ages. Of course, not all were actual communications, but some were. We see GOD created dreamscape experiences, like the account of Lazarus the beggar in paradise, and the rich man being lost in his confusion of hell and death, sharing a short *multi*-sensory experience of awareness between them. Now Scripture warns many times, against seeking visitations from the dead; warning we will probably find demonic impersonators seeking to access, assume, and possess the powerful sensations of our firmer physical existence; as all angelic ethereal physical senses are far, far less than our firmer physical senses. I have witnessed the two different sides of one face in one body, fought over by two souls. There is in GOD'S design, a quantum air-gap connection of our soul to the body. Most *dis*believe that conscious desires and forces of fallen angelic souls, can by manipulation and prolonged effort, work between this quantum air-gap connection to hijack another body's rich physical information, enabling them to have strong physical experiences, and even gain control of that body in this physical realm. There exist many signs and medical records of this, outside all Scriptural references to demon possession. Struggles between two or more quantum souls over the senses and use of one interfacing body, present as multiple personalities, hallucinations, schizophrenia, sleep walking, turrets, hearing voices, and violence. But as more imperfect neural deformation and chemical imbalances also now exist in most fallen bodies, and by the denials by science of our soul. Understanding the use and control of one body by a second soul is hardly believed anymore, even in most churches. There are diseases, imbalances, injuries, and many drugs that alter, scramble, and impair neural transmission and reception of sensed information. Yet total possession, and even lesser demonic influence of weak unguarded souls, is a real true phenomenon. Some prescribed and illegal psychotropic drugs help demons take control. We must know that by souls being closely aligned, yet not hard-wired to any neural structures of our current interface, this can allow a fallen angel to work between and push back a weak soul enough to assume all or partial controls and feelings. If diagnosed with *certain* psychotic issues, it can be difficult to rule out possession.

Possession can form perceptual anomalies that blend the sensed and projected desires of two or more souls through one body interface; this creates the overlayed visual perceptions formed from two visual reference points, forming blended out of place images and voices, called hallucinations. Scripture says those who turn from all evil, and are indwelled by HOLYSPIRIT, are an impenetrable fortress against this. Lesser weaker demons, seek souls who are spiritually weak and sinful. The strong demons seek those with power, privilege, and wealth. Now most souls derive comfort from plans and the repetition of daily events and routines in their lives that bring an immersion that strengthens a soul's delusions towards seeing a purely physical reality. This great external complex creation from GOD, gives us true question upon question, needing true answer upon answer. These true questions, have only one possible answer, being the eternal loving GOD wants to share with us, the ways of eternal life with them. And our answers are seen in our desires that lead us to, or away from GOD. Now about that one great question, is there one true belief. Secularist declare there is no such thing as one universal perfect truth or way of life to live, saying there are only individual conscious preferences, and an ever-changing governing consensus. I emphatically disagree. GOD'S omniscient created ways and our true adherence, equals our knowing obedience to that one perfect complete information that formed GOD'S creation, and sustains its existence. As we spoke earlier of an omniscient quantum system or particle, and our projected patterns of conscious desire, creating *perfect* mirrored particle arrays. All our projections of force, are patterned and logical in projecting our desires that create our *perfectly* mirrored perceptions. But only our mirroring and emulating the beliefs and ways of CHRIST, sustains eternal life. Please see that if GOD were to remove every projected conscious pattern of his force from all reality, mathematics and all complexity cease to exist. Our projected forceful patterns of desire, *and* created perceptions, have perfect symmetry within our soul. But if our conscious force patterns of desire are contrary to GOD'S, and go beyond our soul in our words, or actions; those force patterns conflict and interfere with the perfect sustaining conscious force patterns of GOD, resulting in the divisions of hell, conflict, and death. Lost souls may compromise with each other, but the creations sustaining mind of GOD cannot. All *individuals* live in their degrees of confusion, delusion, and insanity, forming divisive unholy groups. Created reality's initial state, was the pure projected patterns found in GOD'S two realms of perfect eternal life. Until Satan, then the fallen angels, then Eve and Adam in the Earth, and then most souls who were born after them, rebelled to interject their contrary force patterns into GOD'S once perfect creation. Now, endless contrary force patterns are being projected at GOD'S once perfect order. We must understand, that the only way for souls to not destroy our environment, created specifically by GOD to support perfect eternal peace and love, requires we all adopt FATHER'S perfect belief and ways of life that align and reinforce his design and purpose. Sin, is all the contrary forceful ways, souls project against GOD'S created ways. And sin's contrary forces have caused a runaway dissolution effect within this universe. In every sinful dream, each souls projected forceful desires and fulfillment, resides in enjoying their contrary created dream perception. But imagine *you* are the only great soul to exist, and being so lonely that you must gift and create a perfect eternal family. You must first perceive an aether and a perfect external environment, and gift a multitude of minute free-will souls, who could consciously perceive their own awareness of your firmer created perception as their eternal environment. But what if not all, but *most* rejected the required perfect loving forceful conscious ways that would eternally sustain your perfect conscious forceful creation. And they lived by endless forceful contrary desired ways, projecting constant contrary forces into your creation. *Your* desired environment for eternal love, could not be sustained minus the endless burden of your constant forceful repairs. But since you are *vastly* superior in force and size than all other souls combined. You can *temporarily* allow these souls to multiply as long as possible, and save the souls along the way, who of their own desire, share in your love and ways. Then after this process of diminishing returns, renders all it can, you can perceive a new Heaven and Earth for your found eternal family. I don't understand prideful envious resistance to living by GOD'S created omniscient ways that sustain the loving eternal good for all. The forceful patterns of desired belief in all souls, must to a great degree, align in eternal peace and love to perpetuate our eternal environment. We see beyond the projected force of GOD and all souls, that the external physical reality has no *inherent* eternal sustaining force or knowledge of complexity on its own. So, the dissolution of GOD'S Creation will leave us all adrift in a sea of *pure entropy*, and in a state of eternal unawareness and non-existence. Science sees GOD'S four levels of force, he projects to perceive the firmer universe; but as to the conscious *source* of those forces, science has no desire to know the truth. So, a peaceful symmetry can be seen to exist in between a soul's contrary desires, and their own created perception alone. But the eternal desired perfection of GOD outside our souls, is real. We, and all we love, reside in our accountability and eternal need of GOD'S spiritual ways and physical provision. Individual beliefs, and mankind's changing consensus beliefs, have no eternal power of life. They have only the alure of hell and death for those who hate GOD'S ways. The universe was perceived to fulfill FATHERS desires, and everything he *creates* is physical, but everything that he *gifts,* is conscious and spiritual. Our souls were *not* created by FATHER, they were divided out and gifted in love, hope, and sacrifice from his soul. Now *many* shared aspects of FATHER'S omniscience can be learned; but his many personal experiential things cannot. Think of your own soul as a sphere of conscious relativity, connectivity, continuity, and desire, that you need to fill with GOD'S omniscient truth. But every *contrary* desired belief within your soul projects entropy and insanity. Most souls build individual structures of belief that can't sustain any eternal connections to other souls. We need to be eternally known together, or we wind up eternally separate and alone. If we replace our false individual knowledge, with GOD'S true knowledge, we will find *his* eternal peaceful love among many. In our expanding moment, we are known to GOD as the congealed belief we have come to be. Now if we were to suddenly erase the complete structure of our every memory, though we do not consider it all in each moment, we would lose all our believed understanding. It firmly appears, that our built-up layered understanding of belief, enable us to relate and consciously continue in the current moment of who we are. Each soul is a spiritual body of congealed true or false understandings. A perfect Godly understanding must be built piece by piece, and our evil, must be erased piece by piece; which requires our fervent loving determination. And since we are all eternal, this task should please us. The gift of eternal love and joy are of unimaginable value, and not thrown into the crowd like candy. All our imperfect knowledge and evil memories, are for the moment, stored in our aether. But we can soon see a perfect force coming down from GOD, cleansing our soul's aether of every past evil perception, in preparation for the New Heaven and Earth. Leaving only our chosen loving perfect understanding and belief of love, pain, and loss, for us to remember. Scripture says all the forgiven sins of GOD'S Children in Heaven, can be remembered no more. So, their created patterned information, must no longer exist in our aether. Once GOD'S family has freely learned and chosen to love all things good, our past created shame, we wish would cease to haunt and taunt us, will be gone from our aether; we will be sealed and free from all shame. GOD'S saving process is found in our emulating CHRIST, and turning from individualism, to be filled and taught by HOLYSPIRIT to reach our soul's sealing point of adequate omniscient understanding. We are then a sealed eternal Child with FATHER; there by our free-will loving desire. Perfect Godly belief of understanding, can never eternally coexist with contrary beliefs. Mankind can temporarily play a god with all they see in their small slow boundaries of perception; but in this, their beliefs of all understanding must forever change. Until they see and accept the realities that truly rule beyond their perceptive boundary, they must wander from error to error forever. The physical body and vehicle we are provided with, was formed and aligned to our soul in GOD'S hope of our experiencing, wanting, and finding our desire for GOD'S created ways. The first thing FATHER did, before creating a physical reality, or gifting us our souls, was to plan his desired eternal outcome. And FATHER has only *every soul's* eternal spiritual happiness with him, guiding all he does; but *not* any one soul's selfish physical desires. Professed believers may become defensive at all of this; but all components of eternal life must be considered to have an informed opinion. Can or do we desire eternal life in an eternal peaceful joy and love among many souls, minus every imperfect disagreement. Where no one ever dies, or is hurt, and there is no *new* pain ever caused by those there. If so, our basic goal aligns with GOD. But do our desires still align or depart in how we all get there? Do we desire to lead ourself or others in a self-determined way that does not align with GOD'S desired designed created ways and purpose? Do many desire an eternal life where individual belief, disagreement, and imperfections of sin are allowed? Eve and Adam chose this. If *any* disorder enters perfection, it spreads its separation, destruction, hell, and death, just as we see in this fallen world. If GOD'S original projected equation for eternal physical life were written down for mankind; they still could only find the perfection of life by aligning their desires and ways with GOD'S; as it is FATHER'S desired perceived creation in which we dwell. We have no power to re-apply that perfect formula to all physical things; but only to learn and apply it to our souls, and trust a loving FATHER for the rest. Yet many believe in a physical beginning, and think self-awareness emerged. "Strong emergence" is a modern scientific term, meaning that *unpredictably,* at a pure and random point, a new unexplainable essence and quality appears out from within a far less-complex and completely different essence and quality. Coincidentally, that is the same definition for a miracle of creation. In GOD'S true model, all physical complexity is desired and forcefully projected, prior to its creation, or strong emergence, as projected patterned force aligns its complex entangled particle/wave properties into all larger greater realities. The dual essence of the soul and body of GOD, never emerged one out from the other, they have both always been. And all perceived physical reality springs forth from the Dualist union of consciously generated patterned desire, projected into the aether, being of GOD'S body. Even the realm of abstract thought, exists in its created particle-wave perceptions. Now the planned creation of GOD, used only prime number patterns at a foundation; being less divisible orders of mathematical logic. The aether itself is formed from the strongest prime number of one as a structure formed of single points. The first created orders of GOD reside in 1, 3, then 7 points or layers…; but now by introduced entropy, the structures of GOD'S original creation, that reside in all values above 1,3,5, and 7, are dissolving. FATHER said in Isai. 55:8-9, "*FOR MY THOUGHTS ARE NOT YOUR THOUGHTS, NEITHER ARE YOUR WAYS MY WAYS, SAITH THE* ***LORD****. FOR AS THE HEAVENS ARE HIGHER THAN THE EARTH, SO ARE MY WAYS HIGHER THAN YOUR WAYS, AND MY THOUGHTS THAN YOUR THOUGHTS.*" There is a far greater quantum realm existence, filled with immense ethereal bodies, living within the quantum firmament called Heaven. All reality on the quantum side was created first; being the foundation from which the firmer reality drops down secondarily. Our souls exist within the quantum realm consistencies and laws, and where we are loosely aligned to every quantum aspect in each cell of the body/vehicle we currently use. Our soul has its up and down sides; and only our *down* side faces the body to receive the information that describes GOD'S physical realm, but why? Because our *up*-side is meant to face and align to HOLYSPIRIT. Now in physics, if a physical body nears the speed of photons, its mass inflates. And as our speed and size outruns our relative use of photons, we require a faster light particle. And moving at that faster quantum conscious frame of reference, we would realize the firmer physical reality is the far more unstable realm; so, it will dissolve first once FATHER loses all desire for the fallen things. Mankind's flesh platform is at the mercy of entropy and fission, and the deteriorating illusion of time. These bodies also suffer from a "Loss Differential Principle". These bodies we use, are inflating relative to the universe nearest to us. Yet we actually shrink in value, as the far reaches of the universe inflate faster than these nearer aspects. So, we suffer in a growing, yet relatively shrinking physical reality, that like entropy, affects entanglements. Now consider that if your believed understanding of reality, produces unanswerable existential questions? This shows there must be layers of reality unknown to you, below and above your belief of understanding. Materialists deny the primacy of conscious forceful projected physical certainty we all can see in the

Observer's Effect; even as their denials raise paradoxes, ambiguities, and uncertainties, in their physical belief. But they desire and truly need all such mis-directions. The *Conscious* Principles of Proportionality, Cause and Effect, and Conservation, say the Universe, being an *affect*, could only be brought into existence and sustained by a *causal* conscious force and body of mass, at least 7 orders of magnitude greater than the observed force and mass of the universe. The primacy of conscious forceful projection of desired intent, shows the one straight path to the one true interpretation and understanding of all things conscious and physical, being minus all ambiguity and paradoxes. Now, we see only two physical occurrences: one being creative occurrence; the other is destructive occurrence. In their simple strongest forms, these occurrences are called fusion and fission. But in seeing fusion and fission at their root, they are both only witnessed effects; and neither is seen as a root cause. So, they are just abstract conscious force terms. Fusion and fission are occurrences, and not projected forces, and they have no discernable values or attributes prior to their occurrence; yet they appear as antithetical enemies. From fallen mankind's perspective, the four known forces appear to exist only in between the parameters of fusion in creation, and fission in destruction. And these four forces are not found outside the boundaries of physical creation and or its destruction. So, these four, are *not* the governing forces in Heaven, being GOD alone; in fact, these four are products. And the entropy residing between the concepts of creation and destruction, say these forces are diminishing. We first see FATHER'S original projected sufficient forces that created all reality in its stability, that now by inflation grow more insufficient. The stretching dissolution of this universe started the moment forces of imperfection were introduced at its fall, bringing disorder, and later fission, havoc, hell, and death. Now science finds the two nuclear forces have weakened to a greater extent than electromagnetism and gravity. We know FATHER sees reality from two ways; from outside the physical realm as a whole, where electromagnetism and gravity work. And all quantum structure near the nuclear forces, where he consciously works for all things to consist and live. Cosmic inflation brings a rising need for additional force and work on the quantum side; but FATHER will not bow to sin and increase these. Science believes this weakening of the nuclear forces occurred suddenly, at a long past point. But there were actually two sudden dispensational weakening events, followed by a third continual weakening. All projected force, has its part *over* creation or destruction. And the ratio of GOD'S *unchanging* initial sustaining force, over all the destructive forces, brings an inflating creation that brings its foretold dissolution. Mankind brings destruction in their forceful desires against GOD'S ways. Living in contrary evil ways, pushes FATHER away as his hands pull the edges of the universe with him, leaving behind an increasing force insufficiency. FATHER has no desire to increase his work of ongoing corrective forces to sustain evil. The failing condition of FATHER'S Created universe, of which there is only one, is the result of all contrary forceful desires. And as minus the Quantum Zeno Affect, we see the failure of souls to project the required sustaining forces that align with GOD'S original design. FATHER says he no longer sees *all* things as he turns away; and at the complete loss of his forceful desire, we will see the loss of all reality's consistency. GOD'S hope for mankind is weakening, as the creation's purpose dies with every new contrary desire. Confused rebellious souls stand and scream at the world to say the one pure truth and way of life does not exist; and other confused rebelling souls stand up and say the one pure truth and way of life do not exist. So many agree, and in comfort of a consensus delusion, they exist in their continued self-destruction. The one true creative purpose underlying *all* reality, is to bring to FATHER his eternal loving Family. Minus this truth, that vacuum of purpose within each soul, must be filled by a false individual purpose. False purposes are now endless and constrained only by the limits of individual desires. How could the miracle and mystery of our conscious awareness, exist in, and only by a fleshly brain meat. And at death, like blowing a fuse, lose that awareness to the dust, and never again exist. Death is a sleep GOD says we *all* will awake from in a new body; recognizable, but of differing properties, consistencies, and needs. We must see how all physical wonders are dwarfed by the wonders of GOD and all forceful conscious souls, who together create or destroy all these physical wonders. Consider the firm anomaly of trying to talk to someone of your native tongue, yet they are so contrary in belief of understanding, that any true communication is impossible, just the disagreement and misunderstandings that arise between two antithetical perspectives. We need to see language as having many properties; but it is *not* the actual words, phonics, or inflections, it is one's *belief* of understanding that is the far greater part of our language. We all speak *and* hear in our own belief and understanding of a language. And now all consciously created things in existence, act in this same confused way. From particles to galaxies, to souls, all things now dwell together in a mess of projected divisive desires. GOD has their perfect omniscient eternal belief, so they too, are also cut off from true communication with most all of mankind in this same way. Individual souls, hear through their changing individual beliefs, so as GOD says this, they hear something different than intended. To truly speak to, and hear GOD, requires we share GOD'S belief and understanding. We must learn and adopt GOD'S eternal perspective and truth, and discard all our individual understanding. We must learn a new Godly language, or we will undoubtedly misinterpret all GOD spiritually says. Why do lost souls hear so many differing meanings when reading the Bible; because they read their own desires into it. Imagine an *individual* trying to talk to GOD; then asking GOD, do you know what I'm talking about. And GOD replying, child, you don't even know what you are talking about; and you are so defiant and so far from hearing what's real. Why is this phrase "do you know what I'm talking about" become so common and necessary in these last days. We all need GOD, but part GOD, is not GOD. No *individual* can ever dispel the paradox, being does our belief govern our understanding; or does ones understanding govern their belief. All false beliefs, must bring false understanding. All *individuals* acquire a false foundation on which they build their false beliefs. All misunderstandings and misinterpretations of every individual opinion and belief, are all the same thing in GOD'S eyes. These are all the individualized conscious effects, that come from denying GOD'S one truth. Now in that all relatable information, being quantum information, exist in or moves through the aether. Let's consider deeply where the brightness, colors, and dimensions, being the visual perceptions of our awareness come from? Is our created particle-wave perception itself bright? No. It is *only* our awareness *of* visual perceptions that is bright. There is only one inherently aware source of brightness that perceives all otherwise dark moving information. The entire physical universe and even physical light itself, are utterly dark entangled patterns of waves of spinning particles, and rotating, orbiting mass. Dark photons of physical light bring no bright observable quality to the universe. The waves/particles of light give no luminous quality to our perceptions. It is only our soul's attributed awareness that brings brightness to our created perception. All stars are utterly dark; and when two physical things touch, neither physical aspect is aware of it. There is no awareness of touch between the physical body and the world. There is only our soul's inherent awareness of the feelings of touching or being touched in both wakeful and sleep dreams. A soul surely has more inherent conscious abilities and properties than the five physical senses we currently exercise. But the body we currently use, can't sense the higher forms of information required to exercise those higher aspects of awareness. But we do have non-verbal communication between souls if *we* silently pray to GOD, and *are* heard and inspired. If one chooses to use the acronym ESP, I see no need to argue. But as to certain *lost* souls, physically exercising this attribute over others without a stronger larger conscious help, is impossible. What additional senses await us in Heaven? Now why do souls believe light is responsible for the brightness of their perception? It is because science, the media, and a godless education, *imply* and perpetuate this lie. Science also knows that a fat meaty analog brain storing physical analog memory is impossible, but again, they infer and imply otherwise. Consider how under physical light, if we study a sealed-closet's contents, then turn off the light. We find that in our focus through wide-opened physical eyes, we can't see anything. But if we close our eyes, within our quantum memory, the bright closet's contents appear. We pick up a familiar pen, and hold it in front of closed eyes, yet see the writing on it in our created bright awareness. If not ocularly blind, we perceive *patterns of information* from two ways: from the external physical reality through the eyes of a body first, and sent to our soul for added brightness; the other way is through sources of quantized memory, or from a more powerful source, followed by a bright visualization by our soul. Yet many with their physical eyes open, are blind to the fact of conscious brightness. When we dream in sleep, eyes closed, in a dark room. The information that enables our bright dream awareness, must come from the quantum conscious realm alone. Consider how persons born blind, touch distinct shapes, textures, and sense lines, turns, and intersections. They must form abstract semblances of understanding that align to their conclusions in some dark, yet 3D visual form. Consider how all physically stored records of information on a hard drive, a photo, or film; or for the two unprovable examples, we include our point-based created perceptions stored in the aether, and the secular theories of recall through neural activity alone. Now, if physical light were a part of *bright* awareness, *all* of these types of media would require new light to be re-illuminated back into our awareness, like a photograph. But for memory, applying new light is impossible, yet we see brightness. How can the information of five-senses, be stored in a brain, when brains only process electro-chemical analog information? Are touch, smell, taste, sound, and sight, just 3D analog information? And so they don't have to define it, secular science says awareness is an illusion and not real. So, how do we store an illusion and how do we find and restore an illusion? And how do we recreate all these illusions in new altered ways? How do we recall sights, smells, tastes, sounds, and feelings without a soul, these being *abstract* experiences of qualia? How can a chemical-electrical based interface, made of nerves and fat, possibly create, store, and recreate the analog information of our 3D, five-sense perception? Yet we see this occurs within in us, so, we must accept how this truly happens; being it is stored in the aether within our soul. And for recall, it is forcefully pulled up through the body relay interface to be interpreted through our current array of desired beliefs. Then the old and new representations are both stored in the aether. Our recall is *always* a new version formed under current beliefs and desires. People live in new changing emotions, and fantasize about changing past events; and in creating many new versions, the original often becomes lost. Yet GOD sees the original and all subsequent versions in their proper order. If any believe the fat meat of a brain, has the godlike creative power to build particle wave-based perceptions, using a desired 3D pattern of force. They should show the world in some detail, how a piece of wet brain meat might do this. Tell us about this brain meats creative working medium, and its brain meat logic, and this brain meats projected patterns of forceful desire. If you cannot offer all of these in a relatable brain meat way, you must stop teaching your godless subjective religious faith of belief to naïve children as being objective facts. Consider how strong physical senses and impressions, minus any true understanding, can enable a belief that we are only physical, and in a physical world. Please try to see that the phenomena of our awareness of created visual brightness and other sensed qualia, could never exist now, unless they have always existed. Modern research now forces neuroscientists to admit consciousness must be a quantum quality; being the only place left to search, and the last place they want to look. But though our souls exist and project within the quantum realm, our soul have no particle/wave structure. We are only our eternal forceful desire to exist. And by most all souls being individuals, we see the near certainty of everything's antithesis is coming into existence. In this lost world, all souls find antithetical opposition, even those who doubt or deny everything we say; yet *we* face more opposition of understanding and belief than most. We see antithetical light and darkness, good and evil, right and left, up and down, existence and non-existence; we see the antithetical conscious states of Heaven and Hell, love and hate, pain and euphoria, all being created antithetical concepts and sensations. And it is only by many fallen free-will souls, that this *Conscious* Antithetical Principle has permeated all reality. Nothing physical has any arbitrary good or evil attribute, except the good or evil that GOD, or lesser souls, attribute to it. So, we all exist within the bounds of our current core desires and choices of belief. And here we see the one true reason and purpose for all creation, in FATHER *needing* to give and receive omniscient eternal love. We see that the more we live in GOD'S omniscient created ways, the more we love life in an eternal faithful responsible way for the good of all; and the more we hate the destruction brought upon it all by every great and tiny sin. FATHER is the only soul who can create the ways and provisions of eternal physical life; so, he alone defines sin, hell, and death. An actual disconnection from the eternal ways of life, is enacted in every soul not following GOD. Do you know what being an individual means. It means creating growing degrees of separation between who you are, and who all others are. If you are a believer in GOD, do you understand what the introduction of omniscient perfection and eternity into the equations of life means and requires. Do you see the only way to live different from GOD'S created ways, is to be antithetical. Do FATHER, JESUS, or HOLYSPIRIT, ever disagree about the omniscient ways of eternal life? Have they ever *once* wanted to be different or separated from each other in any belief? Have they ever desired to be an individual with a growing separate understanding? Do the *Holy* Angels ever disagree with GOD or each other? Have FATHER'S true eternal children, ever once, having the mind of CHRIST, wanted to be separated in belief from each other? No. They have always agreed on every perfect thing, all being one in eternal understanding. Now the fallen angels and fallen mankind, *do* disagree with GOD'S omniscient eternal perfection, and with each other. And these *do* want to be different and separate in belief. These want to be a god over their own life. Even most *believing* souls, make GOD a lesser part of their life, wanting GOD to approve and provide for their individual imperfect lives. But all who see and want GOD'S common perfect plan for all souls, are on the true path. All believers in any false god, are individuals; this is the righteous definition of an individual. All the patterns of information we receive, are there in hope of our learning the true cause of life and death, and our turning to that perfect way that perpetuates life. So, we see that science needs to develop spherical equations, and not linear equations. And we see why we are given such limited control over the physicality of the body; being just experiencing the richness of reality and our free-will choice mobility. We see having ownership over every particle of a body would give us eternal life, and the ability to rearrange ourselves into new forms. So, we are not granted this level of control. Now most see an ever-growing computational power in super computer arrays; and most see conscious influence over reality to some degree. But most cannot grasp the vast conscious force, FATHER spoke forth, to create all the complexity we perceive to externally exist. Yet many *can* believe it just occurred on its own, and came from nothing. All materialist theories rely on a miraculous impossibility of *nothing* knowingly turning its non-existence, into a most forceful, immensely complex self-awareness, by physical miracles against undefinable odds. How is that easier to accept? These souls resist every idea that challenges their beliefs. All they believe in are wave-particle patterns that hit their retina, or manmade detectors. Now certain peer reviewed materialist theories, pose that the universe is a holographic illusion; to include us. Materialist's limited perspective and physical senses, are all they have to study with, so, they wonder aimlessly. Now we know if we see, hear, touch, taste, and smell a thing, it must be physically real in some way; but we can't equally trust, just our sight alone. What we see far out there, could be no more than created patterns of photons, projected by greater conscious souls. The Heavens are filled with vast forceful souls, capable of such things. 2nd Thess. 2:9-12, "*EVEN HIM, WHOSE COMING IS AFTER THE WORKING OF*

*SATAN WITH ALL POWER AND SIGNS AND LYING WONDERS, AND WITH ALL DECEIVABLENESS OF UNRIGHTEOUSNESS IN THEM THAT PERISH; BECAUSE THEY RECEIVED NOT THE LOVE OF THE TRUTH, THAT THEY MIGHT BE SAVED. AND FOR THIS CAUSE GOD SHALL SEND THEM STRONG DELUSION, THAT THEY SHOULD BELIEVE A LIE: THAT THEY ALL MIGHT BE DAMNED WHO BELIEVED NOT THE TRUTH, BUT HAD PLEASURE IN UNRIGHTEOUSNESS.*" By relying on five senses and technology, mankind knows our moon, sun, and planets exist. But light, out from beyond the first several hundred million stars, declared to exist in Scripture, could all be a vast created trick of the light. Scripture says if souls look for and *strongly* desire to see a certain thing beyond their horizon, GOD says he allows their delusions. I look up at night and perceive the same points of light we all do. But in knowing GOD is at *minimum,* seven times the sum of all that physically exists, and reading how the 300,000,000 angels have a gifted/created relationship to their own star, we can know how reality truly appears. And yes, a few fictional portrayals of reality are closer to the truth, than what we physically sense. If our ability was enhanced to see the fastest particles of light, and see the full spectrum of all mass, energy, and force patterns. We would see a union of two distinct aspects everywhere in an echoed overlaying field of force upon each physical particle creation, being a spherical halo of projected desire. Two qualities bound together, being the two halves of reality. Projected desires to exist, seen as our opaque forceful patterned guidance, causing all physicality to align into all desired ways and projected dimensions. In *our* innate conscious brightness, we could see waves of patterned forceful frequencies upon swirling points, as they align new larger creations. We would see the warring effects between the aligned *creative* brightness of truth, and all misaligned antithetical *destructive* darkness of lies. If we moved out beyond it all, using the fastest light to see the entire universe, we would be afraid in seeing a once condensed perfect light filled universe, now having been stretched, has lost around ninety four percent of its light to a darkness that can be felt. We would see the remaining created physical complexity is also dissolving, to be drawn back into the single point grid of the aether. Now, in our moving back to the conscious center of reality on earth, we could make out the larger distinct hazy forms of mass, such as human shapes, buildings, and oceans. The smaller and more common the shape, the harder things would be for us to distinguish. Using that fastest light, we would receive so much information, we would see things defined across a spectrum of light to darkness, and lacking color and finer physical textures. We would see how mankind's sensing limitations, are what bring color and detail to all firmer things. By the fastest light, we might even see the wave function, as all larger dimensions of certainty appear in being created. *All* sound frequencies would compress and become garbled, like a lot of sounds traveling through water. We may see a spectrum of projected emotion and desire, and their countenance in all forceful projections between souls. At one end of this spectrum of emotional countenance, we see oneness of the omniscient truth, bringing in us the brightness of eternal life. At the other end, pure godless dark individualism and disorder. We might see each soul's projected entangled streaming perception of desire as they rise out of the aether upon their surface and return to the aether to be stored. We could learn to know each other's moods, and what others think, without speech or hearing. This is a possible heavenly sense we cannot now experience through a fallen body.

We would see a varying countenance (brightness) in all souls *and* their perceptions. We would see disorders of darkness attacking the bright creative and binding healing forces that remain. If you can visualize this, then know this is what the Trinity sees, and to lesser extent what the Holy Angels see. For many Scriptural, Spiritual, and Physical reasons, we see the New Heaven and Earth will be firmer than the current Heavenly realm, but less firm than our current physical realm. There may be only one realm in eternity, no longer a physical and quantum realm; just one being the best of both together. Now let's turn to the Holy standard, or banner, stood up in the Earth by GOD; being Israel and Jerusalem, which at its best moments, held tightly to GOD'S commandments. And to whom the SON came with his forceful physical manifestations of healing and altering physical realities. CHRIST brought with him the original created spiritual example of eternal life, and for a few decades, CHRIST worked to realign many misaligned things. His corrections of spiritual and physical confusion, had a profound shaming effect upon the world that altered the course of mankind. But his affirmation of the perfect truth, has again been mostly buried under the evil desires and lies of individual religious beliefs. Most professed believers, wish GOD would just make us all do what is right. But our soul, under any such control or restriction would not be us. If GOD forced his will upon us all, we could never be self-aware. Only GOD would be aware of what he caused and forced these bodies to do. GOD *cannot* make us love *them,* or anybody else. And finding a loving family is the only purpose for the whole of creation. After our being gifted to exist, if we allow ourselves to be mired in desires to experience physical worldly pleasures, this leads us away from the spiritual ways of eternal life. Now the physical laws *within* GOD'S person and self, must differ from GOD'S restrained laws applied to his *ex*ternal created reality, shared by many new gifted souls. So, FATHER'S forceful desires were once applied to all particles and mass filling his creation, but not over our souls. FATHER'S projected varied forces of desire upon the firmer physical reality, are opaque yet real; and are literally the intensities of his focused desire. If we took the far-reaching yet weak forceful entanglement we call gravity, and focused it down into a certain local area of space, it would be the equivalence of electromagnetism. And greatly focusing that force and area again, we would find the weak and then strong nuclear forces. Each of

FATHER'S broad to narrow fields of patterned conscious force, do exactly and only what he desires them to do. This is not GOD controlling a pre-existing force, these are the forceful projected desires of FATHER for all things to be what and where, and bent or curved, to attract or repel in how he once perfectly perceived them to be. Do you see how in *our* dreams, all the physical perceptions we create, move, and appear according to all *our* creative projected forceful desires. We see varied strengths of patterned force projections form our entangled particle-wave dreams in their alignment and consistency? How do we walk on the ground in our dreams? Is it gravity? Or is it our forceful created aligned entanglement? How do we float in the air in our dreams? If we jump up in our dream, or fall, we usually come down to our created ground, but why? It is our projected patterned forceful desires, that align and direct all these things created within us. And we have been consciously loosed, deep within FATHER'S far greater created perception. This is hard for most to accept, seeing their every projected contrary desire, has an undesired evil effect upon the quantum fabric of GOD'S creation. Again, there are only the *two* true opposing force polarities: being attraction and repulsion, or convergence and interference.

These are by omniscience, created to fulfill FATHER'S desire to find *his* Family. By sharing the loving polarity of the mind of CHRIST, we perpetuate a reality of eternal life and love among all. The only other polarity is individual and divisive. So, we find only two perspectives in souls: those who see life as temporary, to be individually and physically experienced and enjoyed; and those who see life is eternal, to be spiritually cherished, understood, and perpetuated by all through GOD'S shared truth. Every one of every soul's projected force patterns are of one of these two polarities. Now the fallen angels want to be more physical like mankind; and among mankind there are the few who want to rise out of this disaster to be like the Holy Ethereal Angels. What do Angel's ethereal body feel? We can't exactly know, but if an angel's arm is touched by another, they must sense it, but not the way we do. They don't experience our greater sense of taste or our physical euphoria. But their conscious emotional senses of love, hate, and such, are far more refined and intense than ours. FATHER'S faithful responsible belief, forcefully creates all physical life to exist eternal in pure peace and love. But individuals seek a way and truth they physically desire; so, they become divisive, seeking out physical pleasure and sustenance, projecting forceful disorder towards universal dissolution. Every projection and act of contrary desire, adds to the destruction of physical life and reality. Every free-will soul was gifted in FATHER'S hope of everyone wanting eternal life and love as his Family. Simply put, we see FATHER'S forceful loving ways and creations; and we see his allowing all the rebellious to destroy it all and bring their self-destruction. Conscious loving fusion, or allowing complete fission as FATHER loses all loving desire for this fallen reality. The most common effective attribute of evil, is its relentless blurring and conflating its individual liberal desires with the love of GOD. This happens in all souls who deny GOD'S command for us to be perfect and righteous as he is. Now, all current godless physical theories on reality, are founded upon unreasoned physical miracles, and that initial miracle is required to enable their desired interpretations of the limited things they sense. It is their first false physical miracle that enable the leaps of false faith within the mind of all individuals. Ask them to discard their first required miracle, and see what they can theorize. All now broadly accepted first miracles, were at first dismissed as fiction, nonsense, and non-science. And to this day, every one of those founding miracles, is argued for and against. Fallen mankind has reached a point where all physical theories, are admittedly beyond proving; so, having every aspect of pure religious faith. Now for fun, we proffer our own *false,* far more plausible physical theory, we call "Theory of Physical Perfection Lost". The great fatal flaw of materialism, is it can never explain the miraculous *origins* of the universe, physical life, or s soul's awareness of it. Such complexities and the phenomena of life and our awareness, could never come by chance out of random physical chaos. If you somehow *accept* a first physical miracle, like an exploding singularity, and accept the prerequisites for force, energy, and mass existed within that singular quality, that being a stated contradiction. That belief may get you to clumping of matter, but never the phenomenon or life or awareness. So, *our* false physical only theory, is based on our sensing what remains of a once near perfect universe, having no prior reason or purpose, being a near physical perfection that has been lost. Now, in trying to know anything about the remnants of a *designed* creation, we know there were both conscious and physical reasons applied to its design. And we know science is disadvantaged in their efforts, by not knowing a remnants reason and purpose. Consider there was once an imbalance, called a Quantum Physical *Certainty* Principle, that overwhelmed uncertainty and chance occurrence to point of forced complexity, which eventually formed a complex *perfect* environment for physical life to form. But this only gets us part way. We must also allow for a mysterious Abiogenesis Principle, a Biogenesis Principle, and an enigmatic Conscious Force Awareness Principle. We must have *all* of these higher unbalanced complexity Principles, as the creation of the complexity and mystery we see, is beyond all balanced physical chance. Envision the physical universe forming a perfect physical environment, at least starting on earth outward. A universe being greatly settled, minus most all expansion and fission, attaining a near zero diminishing state of entropy. And so, under these four once dominant Physical Certainty Principles, a great quantum stability and equilibrium began its reign for eons, allowing these unimaginable complexities of life's orders, to be forced into existence. But after near perfect physical life formed, a neural forceful free-will awareness and choice appeared and grew to over-rule all Physical Certainty Principles; so, a Free-Will Forceful Disorder Principle was introduced by the differing fluctuating forceful choices being made. So, then Physical Disorder gained control, and that once ruling Physical Certainty Principle faded. Reality then turned to a ruling physical Uncertainty Principle, under a godless force determinism. Which spread as entropy does, and started the universal expansion we see today. Now, this false physical theory makes far more physical sense to me, than all other godless theories. But seeing FATHER'S projected forceful perfect creation from the beginning, followed by all free-will projected contrary force; we see not evolution, but ever-increasing devolution. Remember the illusion of time is relative to each conscious entity's size, speed, and understanding perspective of GOD. So, the entire moment of this universe of life, from GOD'S perspective has been at most around twelve thousand earth orbits around the sun. Time is such a malleable conscious perspective; but eternity is under the fixed perspective of GOD. What is the spectrum for defining "all eternal souls" is it never having a beginning or ending at one end for FATHER, and just never having an ending for all gifted souls? Our soul never had a beginning as it was part of GOD. Now the force of will in GOD'S peace and love, can bring a whispered gentle caress to a small single particle; yet at great provocation in his total loss of desire for it to exist, he can bring fission to all physical reality, dissolving it back up through the quantum realm and into the aether. FATHER could also dissolve and pull the aether back within his person. But for his great work and found love being his eternal promise and reward, FATHER will never do that. His found eternal souls and the aether are where FATHER'S found mutual eternal love now exist, and where his new Heaven and Earth will exist. Mankind thinks they see different external forces, but these are all of one reasoned purposed conscious force. Romans 13:1, "*LET EVERY SOUL BE SUBJECT UNTO THE HIGHER POWERS. FOR THERE IS NO POWER BUT OF GOD: THE POWERS THAT BE ARE ORDAINED OF* GOD." If not correctly understood as one conscious source for all layers of force, the four forces of GOD'S projected intent, take on an invalid appearance. Now as mankind sees this physical reality in so many inferior ways, all their many theories have discrediting things in common, such as *none* has a relatable beginning. Their delusions stem from their blindness to the conscious truth, and their proverbial delusions of grandeur arise in those infamous tricks of the light. This blind confusion to GOD'S truth can be acute or chronic. And using the word confusion correctly, we mean all conscious states of errant understanding about reality. Most people define a confused soul, as someone holding an unaccepted amount of confusion, outside the bounds of any accepted confused consensus. If one thinks they know the current state of something, yet don't know its origins. They *must* be confused as to that current state. If a revered consensus fervently agrees on something not true, this is called the confusion of a mass consensus delusion, and not the truth. People play with meanings of words in their minds and speech as coping and propagation mechanisms. A true effort is lacking in most souls to see the growing confusion that arises from all intentional and unintentional misuse of language. But the strange thing is, *we* can overlay every false physical theory of reality that excludes GOD, and then superimpose every false religious belief upon it all. And though it is all still wrong, after cancelling out all duplications and irreconcilables, we see something closer to the truth, than with any single religion, political belief, or scientific theory. By removing FATHER and *his* Kingdom, from the top and bottom of created reality, what they see as the remaining highest and lowest levels of reality, must set and form their false beliefs. *We* see how our patterned forceful desires projected into our aether's spherical field of possibilities, force larger probabilities to break free and form certainties of perception within us; yet some in this world reverse this cause and effect, saying the energetic frequencies in the collapse of a wave phase, may be the source of raising conscious awareness. But only the truth of our entangled quantum conscious force projection, explains all the affects and effects required to form our streaming multi-sensory perceptions. All purely physical scientific views, leave questions as to the cause of a wave collapse unanswered. Or how collapse could give rise to the complex multi-sensory particle-wave perceptions that enable our awareness. The physical sciences are caught in a perpetual revolving door, going round and round for centuries, while refusing the only conscious creation exit, looking for a physical creation exit that does not exist. Now it is impossible for GOD to communicate and reach souls spiritually, who want GOD to reach them physically. It is only in our moments of doubt towards this physical world, that GOD can best be heard. And the more deeply a soul is entrenched in the beliefs of physical primacy, the more difficult their rescue becomes. GOD does not want to physically fulfill us in this fallen physical realm; but to lead us *out,* and *into* his perfect eternal spiritual realm. In mankind's godless attempt to physically explain self-awareness, minus our soul, they are forced to define our awareness as an illusion, and deny our forceful created substantive perceptions. We can never be aware, minus a real perception of that which we are aware. Why do they skip these most important required steps? Well, because these steps destroy all they believe. Their views are truly unscientific in believing awareness is an illusory product of fat brain meat; they see awareness as an inexplicable physical *miracle* they don't like to talk about. Materialism can never find a physical way to explain created particle-wave based multi-sensory perceptions, or our awareness of them. Our awareness is an experiential phenomenon of qualia, requiring a generated forceful logical projected and sensed patterned information as its language. If any can proffer another start to finish process, that creates real quantum ethereal multi-sensory 3D perceptions, please share it with me. Now two renowned minds proffered a theory for an energetic path to conscious awareness, called OOR, or Objective Orchestrated Reduction, which focuses on frequencies microtubules emit in all cell structures. Now while all frequencies are interfacing informational correlations, as is neural activity; this proffer is still set in a narrow materialist struggle. The resonating microtubule structures in every cellular system of life, are a part of the whole interface. *Every* energetic frequency in the body interface, is relayed information to and from the soul, and that is all. Their proffer *replaces* two-way communication, using a body-interface between two separate qualities, being our soul, and all physical reality; with one physical system that can't be explained without the word illusory. Now there are three basic forms of religion in this world: a *few* hold the loving logical pure beliefs of the one highest soul of GOD, as a growing number hold a loveless false belief in materialism alone; and between these two, there are endless blended emotional illogical divisive beliefs called worldly religion and politicized science. All souls have faith and bias, but most individuals cannot accept that *their* beliefs control how they falsely see *all* things, at least to the degree they do. They inversely believe that all they see, controls their beliefs. We see the staggering mysteries of our conscious comprehension and influence that should stop us in awe. And these truths exist without any *credible* opposition or doubt. Our soul obviously creates particle-wave perceptions, yet we exist, move, and dwell as a quantum shadow of the body. Our quantum realm soul is forcefully there; yet we can't show it, as it eludes all physical discovery in its own way and place. The only possible reason for why our awareness can never be found to be physical, is because it is an inherent aspect of our desirous-force-based-soul, residing within the quantum realm. If we eliminate every physical possibility, whatever remains *must* be true. The brain/body neural system is akin to a miraculous complex virtual existence gaming console *and* vehicle, which to certain extents, we control with our soul. But fallen mankind grasps for lies to replace GOD'S truth they deny without reason. Our leaps of enlightened awareness of truth, can never be attributed to physical reactions in neurons; just as computers can never make any true unprogrammed or entangled leaps, using layered AI algorithms. There are all our eternal souls, there are individual delusions, and there is the one created truth of understanding, flowing down from GOD. Rocks and mud are essentially the same elements, as brain meat, and if our awareness understanding and memory resided in neurons of a brain, it should be a most evident thing. But no such evidence is ever found. Some endure Hemispherectomies without reciprocal memory loss, or change in personality; yet their sensed information needed to control their body is far more affected, unbalanced and incomplete. These patients experience physical processing problems and diminished controls in losing the true bilateral aspect of their physical interface, designed to be mobile. There is no great reason for a brain to be a bilateral physical system, other than a soul's perceived physical sensed mobility and balance. There is zero understanding, belief, or awareness, within a brain, these and all good or evil desires rest within our soul, to form our conscious vantage point. Materialists see our awareness as a strange act of physical mysticism, that eludes empirical understanding. Their *unsupported* theoretical proffers show they have no desire to know or study the truth of the matter. They know the world sees inexplicable unaccountable conscious strangeness every day and everywhere. By all the conscious strangeness *you* see and know, we ask *you* to answer a question where no one can hear; is it an honest scientific principle to say without a single reason or proof, that as to our forceful will of awareness, you see only *two* credible probabilities to be allowed in the scientific study of conscious awareness, and not *three*. Not just *one,* being that by some rising physical complexity alone, the physical aspect must be the mystic cause of our awareness. And not just *two*, being that our projected forceful consciousness inexplicably emerged, and evolved to now have influence over many physical certainties. But there are *three* probabilities, and of the three, only one is highly credible and sound; being that all projected force in every soul's patterned desire, and the non-conscious physical aspects, have both always been. And that by each soul's awareness of their own projected created perceptions, we see how the dualism of conscious force over every physical creation, defines the beginning and fulness of all reality. That physical complexity, nor experiential forceful desire, at any point, existed, minus the other. Which again, says GOD *must exist necessarily*. In physical science, the term *necessarily* means that by everything observed and know, that a certain aspect is *required* to be there for everything else around it to exist, as it is seen to exist. Let's also discuss the word miracle, and accurately define what is truly miraculous. Not the narrow definition used by materialist; but one acceptable to all souls. According to materialists, a miracle is any occurrence that physically manifests, which not only cannot be explained by their believed physical laws, it must violate those laws. This is a false vague definition, as we see they don't *possess* the required full understanding of reality, required to define the true parameters of what is, or is *not* a miracle. For materialist, consciousness is a mystery, and actually an immense problem as they try to define a miracle. To truly define what a miracle is, requires knowing and understanding all that lies beyond fallen mankind's highly interpretable, politicized, and limited physical senses. To define a miracle, requires a complete understanding of conscious and physical reality from its source and origin, to its current fullness and even beyond. I trust only in a search for the one great truth of life and reality; I trust no purely physical definitions that arbitrarily exclude or deny a single article of conscious or physical evidence. In our gifted authority of now holding a far higher knowledge and understanding of reality, we declare that *every* substantive quantum particle wave perception we each consciously create and perceive, occurs beyond, and stands contrary to mankind's physical laws, constants, and understanding of reality. So, by *their* definition, our forcefully created particle/wave-based perceptions, are *miraculous* manifestation. All our consciously created perceptions, occur not only outside of mankind's physical laws, they violate them. Yet a soul's created perceptions don't violate the far greater laws of GOD'S reality and so, are not miraculous to us. They are a lawful most common created occurrence. The word miracle, is then a term used by, and needed only by this confused world to describe the laws of GOD. To GOD'S children, it becomes a slang term that has no real meaning. Why is it that all godless science, hold inexplicable *physical* miracles as the foundation for their belief structure, and then tells the world we can trust them to interpret and define reality *for us* through their non-empirical, exclusionary, biased miracle-based beliefs. I say all souls *are* truly freed by GOD to believe such things with all their hearts and minds. They have their GOD given free-will choice to be falsely convinced beyond all reconsideration; but that changes not one thing for me. By reading, hearing, believing, and now knowing our souls were each gifted out of FATHER'S creative soul of logical emotional awareness, and that he perceived these bodies for us in his image; and in hearing all true Scriptural interpretation from HOLYSPIRIT. I now see and know the truth of GOD'S reality. There has been no real effort by secular science to prove, or as importantly, disprove GOD'S truth. They just espouse their false slow small perspective beliefs, and deny particle-wave-based perceptions exist, saying they can't be physically proven, but wait, neither can all their theories. Are they not saying at their first *biased* glance, they see no *desired* evidence, and so they dismiss this case without investigation. Where was their long honest search, required to say created particle-wave perceptions are disproven, and non-existent. But we do see the *endless* invested political lies and efforts to get all souls to believe that our awareness is a physical emergent of wet brain meat. So many decades of scientific deception, and so much money spent on indoctrinating all to believe in this mystical brain meat. Will they ever admit a physical consciousness can never be found or proven; so, like GOD, their materialist religious doctrine, must be removed from public schools. Anyone who says our soul can never be proven to exist, has never tried to prove or disprove it. They just blurt out their bias, and demean all believers. Did certain darkened souls at a certain long past point of ignorant bias, simply declare and accept this by nothing more than their desire to not be subject to GOD? And for their comfort of belief, do they join together to fill positions of power, and ridicule any mention of a soul in their halls of study? The world is looking at the proof of GOD all around them in a clarity, they refuse to see. If science in all honesty says we are so far from all that can be known of the physical reality, let alone the far greater conscious mystery. How can science then say they know what can never be proven. It is a crime of a most sinister bias; not against certain beliefs, but against the honest scientific search for the full and final truth of reality itself. Why say you search for the truth of reality, and do all you can to narrow, restrict, and limit, that search. There is no good reason to steer all souls away from the truth; and only devils do this. A true demise of what materialist refer to as the unreasoned foolish notion of an eternal soul, nor their proof of physical awareness, can ever come from their current approach; which is why they choose this approach. They know a physical awareness can never be proven, and our eternal soul can never be disproven; in this they find comfort. It was believed in science 100± years ago, that a proffer that cannot be physically or mathematically tested *in that moment*, is not part of any true *current* science. But then materialist saw and could not deny the contrary quantum realm that taunts and haunts them, over-ruling and shattering their pure physical beliefs. They looked into that ghostly realm, and lost their firm standing and reason of purpose. They are now lost, confused, and like zombies, meandering about in a singular focus, as their once held objective empirical standard became subjective. For many decades, their three secular pillars of a Big Bang, Earth-bound Abiogenesis (life from non-life), and the Evolution of Species, have all encountered insurmountable quantum problems and odds. So, in a panic, they began to scramble for anything that even might appear to hold their desired delusions together. This resulted in their growing number of unreasoned delusional theories, each being further and farther from the truth, and being beyond proof. They lost all scholarly constraint, and grasp at any radical possibility in their dark confusion and fear. Their former accepted defining line of true science, has moved from only testable probabilities, to un-testable impossibilities of Multi-verse, Inflation theory, Quantum gravity, Abiogenesis, and Evolution of Species. These form the vacuous leading edge of scientific research, wasting billions of dollars. They know in the solitude of their mind; these fantasies are all beyond the bounds and odds of empirical proof. Yet they need them in their desperation to counter the fine-tuning and mind body problems, and their so many other problems. Conscious force projection explains the fine tuning of the universe, and all other things; while their theories are no more than slight-of-hand, bait and switch tricks of the light, designed to deny GOD'S existence. How are these unrealistic physical proffers accepted and taught as truth by no more than blind faith. These people have invested much of their life and belief in a physical only direction, it is a religious part of them that is hard to impossible to let go. Just as with all other false religious doctrines, they patch and brace the flaws of their beliefs with anything that appears to prevent its collapse. I see their created secular puzzle of patchwork fantasies and unsecure braces. I see a purely faith-based assumption in their founding physical miracles, that allow almost any physical theory to be built upon them. These souls do not see the eternal defeat they inescapably run right towards. Many ask, how can our conscious awareness be qualified and quantified and added to any equation. Yet we know our conscious force projections exist, and that our created particle-wave-based perceptions are our sound basis for such a belief. So, *we* ask; how can *you* leave all conscious awareness out of any equation and find any true answers about reality. FATHER'S U.E.D. model, that supports all conscious creative belief and understanding, has a base of true and sound certainty; and there are many areas of study to be pursued in this field of conscious force determinism over particle-wave creation. Whereas, we see a true zero possibility for multi-verse theory. Now philosophically we ask you, what real conscious and physical implications would arise in souls who search for, find, and hold, just a basic yet sufficient measure of FATHER'S all-knowing love and truth? Will such souls thrive eternally? I*s* that our wondrous spiritual connection of faithful responsible eternal love among many under GOD? Yes. A conscious bond which brings a euphoria of truly shared love, that will endure eternal, being minus all things of division. Do feelings of faithful responsible love exist in much higher forms and measures than a fallen world can comprehend? Yes. Will loving peace, endure in a perfect conscious and physical environment? Yes. Is there a higher spiritual love we can never tire of, and never get enough? Yes. There is so much to FATHER'S higher spiritual existence, belief, and understanding that fallen mankind cannot imagine, while dwelling more and more in their slowing small desired physical senses. They cling so tightly, as all their physical senses fade and diminish, and their soul grows weaker; and as it all physically pleases them less and less. They become trapped in longing for all their strong former past physical pleasures. Now our projected free-will force of desire is known to have great affects in the quantum realm, beyond our created perceptions. The Observer Affect is one of the most documented and argued affects in modern science. We also see many societal, emotional, and physical forced affects, which are all true correlations in observing a world filling with more and more conscious confusion. Which lends more proof and support to *our* belief of understanding. Now in our current experience, we can sense certain future probabilities, like if someone is about to be angry or happy; we sense many such things just prior to their actual occurrence. So, if those sensed probabilities become a reality, what does that mean? Did we technically and actually see into the future just a little? Yes. Do we see farther into the future in many other ways of prediction? Yes. So, can we through quantum realm connections, tap into a true ability to see much further? Can we gain great foresight and valued knowledge from being one with the Prophetic mind of CHRIST through the indwelling HOLYSPIRIT? And is this a gained aspect of our eternal perspective? Yes. Now Philosophy of Science, reaches out *past* that pure physical inquiry that materialist believe has the only firm profitable value. Philosophy looks to the unknown foundations and horizons, seeking inspiration rather than working on the current internals and minutia of physics or technology. Philosophy is exercising one's spatial reasoned visualization in all unrestricted undefined ways. So, what requirements must exist within reality to allow all conscious perspectives, choices, and physical chance and free-will beliefs to exist at once? And have all these requirements been met? But what if one of these entangled requirements were removed, and put back repeatedly; would our conscious free-will choice of belief, and all physical chance itself, flicker off and on? And by all things known; is a belief in GOD, not the most natural inherent thing? And is that natural inherency in all naïve young children, being hijacked by an indoctrination from the most-evil of fallen godless souls? And for what cause and for whose purpose is this evil done? Who sits in those higher and highest seat*s* of opposition to GOD, working to directly challenge and destroy his authority? It is Satan and his fallen angels, followed by every lesser soul on earth who denies even one true word of GOD. Now, seeing our that the angel's and mankind's conscious estates exist minutely under GOD, we ask, once souls fall and continue to fall from GOD's ways, does GOD only desire to reveal themselves to these souls, only after seeing a true loving desire to return to GOD? Yes. Please see that *if* all fields of science would accept the real probability that in every projected forceful intention of every soul, there arises a new created wave-particle perception, the scientific world would explode in public awe. But to introduce such a physical heresy into public domain, would require souls with meritorious standing, and a courage to do so. Dualism of a ruling conscious force over the non-conscious aspect, was long ago vaguely considered among certain sciences. But quickly cast off as a bastard by the parental physical sciences, discarding and aborting this unwanted child that did not fit in their family portraits. But how can the orders, complexity, and fashions of all *life* we see devolving due to disorder, ever be first formed of chance from within that disorder. Have laws, life's complexities, and math itself, been created by randomness? Or is everything created by GOD'S patterned desired intent. Such vast and ubiquitous entangled purposed complexities could never arise by chance out of nothing. Mathematics reveals the existence of a prior logical pursuit and planning, and how projected patterned force created all complex reality. The symmetry of perfection, and the growing asymmetry of imperfection, exist by all projected agreeing and disagreeing patterns of force; the latter being the cause of disorder. The quantum Heavenly realm's vast ethereal forms of life, can, but rarely pass through the firmer things playing out in the world of mankind, as it all moves through the aether. Now materialist concede the universe must have been designed and created by the same mathematical laws it now obeys. So, these *conceptual* abstract laws of reality existed prior to all creations formed under those laws. These concepts of law existed in the only place they could, being the mind of FATHER; needing only his spoken forth force patterns of desire, and his body's points to create all these external realities. Materialist's admission that these laws must have existed first, demands a conscious arbiter of those laws must have also existed to both hold and apply those laws. The laws of physics are based on no more than the lawful forces of GOD'S desire, aligning his conceptual creation. And again, all knowledge reasons and purpose, exist only in each soul who knows them. Knowledge reason and purpose, cannot exist if they are unknown. The definition of knowledge, is that which is known; ergo, knowledge can't exist without at least one conscious knower of it. GOD'S omniscient mathematical awareness, is *relative* to all aspects and appearances, from the initial *origins* of the universe, on to as it appears now. We see the strength of science's physical model is immensely over stated; and professors without tenure are admonished if they share this fact. We have no desire to deny *true* data, but only to add the proper perspective and interpretation; and add our far greater structure and support to build far higher. We see mankind's limited physical data, and we see a few secular interpretations of that data, in limited parts and ways, supports our true and far greater conscious force guidance perspective. We accept that the observed constants, are the *current* scientific currency; we also see an ever-changing, too slow and small currency exchange rate. Mankind has not known any values or appearances for long enough to know if anything is a constant; and I assure you that none are. Reality once resided in conscious stability, but it now resides in its conscious variables. We are *not* surprised the value of many constants were not correctly predicted by mankind's standard model. Certain values coming in at many magnitudes off from their model's predictions; like the cosmological constant and dark energy. Even the value of what is now accepted as the Higgs Boson, fell between the Standard Model's two most predicted values. For us, the cosmological constant is the current force value of the minimally excited aether, occupying a cubic meter of space. We see beauty as the *a*lpha constant nourishes the universe with light. This fine structure *variable*, once had its perfect prime value of exactly 137, which is now changing. Feynman said it was a number "written by the hand of GOD" and yet "we don't know how he pushed the pencil" Gen. 1:3, "*AND GOD SAID, LET THERE BE LIGHT: AND THERE WAS LIGHT.*" This number 137, stands alone in that it is not predicted by, or connected to anything else in physics. GOD perceived a vast broad spectrum of light waves, with at least three light particles, as one of his first acts upon his aether. This, so all vast to minute souls, can sense the relative information that to them, represents FATHER'S creation. The smaller the light particle, the easier and far faster it moves through the aether. A photon is the far largest slowest particle of light. Please try to grasp the drastic changes to scientific beliefs, understandings, and theories, that two more far faster light particles brings. Once you see GOD is real and defined; meaning they have their shared qualities, yet their variable parameters and boundaries. This says FATHER has his ultimate internal and external limitations, being the only limits on reality. In a true understanding of GOD, we find no contradictions, paradoxes, or unknow variables. So, once you see and accept the conscious/physical aspects of FATHER, and his created reality, you can know your own projected force of desire is the creative aspect of your own internal reality. Now we see Photon levels of $C^3$, and the two far smaller lighter, faster, particles of light, being $C^2$ and $C^1$, and maybe more, all have a singularity point count. And all that mankind calls energy, is just the resonating wavelengths of the fields of patterned force surrounding and binding physical singularities to congeal in a desired designed complexity. Believing nothing can go faster than a photon, is a false belief born of a biased purely physical approach, and the incomplete information of mankind's too small, too slow, fallen perspective of reality. Classical observations and quantum measurements from a slow small materialist perspective, in inconsistent problematic ways, leads to mankind's standard model of reality. Yet those same observations and measurements, interpreted through conscious force primacy, lead you to FATHER'S consistent problem free, UED-WPDT model of reality. Now FATHER'S UED-WPDT model, and mankind's Standard Model, lightly agree in some areas and aspects of formalism, phenomenology, epistemology, and ontology; yet FATHER'S model also has many unique conclusions and aspects in phenomenology, epistemology, and ontology. Now strangely, GOD'S WPDT quantum mechanical interpretation, *viewed through mankind's too slow too small perspective*, allows for the same values and measurements as all other interpretation; yet from GOD'S *faster larger perspective it does not.* So, from mankind's slow small timely perspective, GOD'S WPDT violates no rules to qualify as a valid QM interpretation. Yet from GOD'S perspective they are violated. But again, from GOD'S perspective, those rules are *in*valid. GOD'S UED-WPDT spherical model of all created reality, defines and explains all particle, elemental, and complex construction, and our sustained existence; and answers all questions as to reality. Now as to mankind's use of unknown hidden variables and perturbations that they require to allow and support their too slow too small theories on reality, we have zero need for these. UED and WPDT, reveal all the influences which cannot be seen from a fallen slow small perspective. Yet are seen and known through a larger faster conscious size, Godly-perspective. UED is a true physical reductionist model, consciously building reality from fundamental spherical singularities; or looped strings if it helps you grasp. We find beyond doubt, that each forceful observer's size, momentum, and their subjective biased conscious perspective, are central and relative in defining their own experiential normalcy. For WPDT, we take mankind's over-all slow small ideas of physical determinism, and add all *prior* consciously determined guidance. Now as with some QM Interpretations, WPDT has *literal* fields and patterned waves made of points being forcefully collapsed or congealed into larger quantum spherical systems. WPDT shares in the certain conscious affects with Von Neumann-Wigner, and Copenhagen; seeing conscious force causes all wave collapse. Our great departure from these two, and others is that in UED and WPDT, conscious force creates both the waves *and* all resulting particle systems. WPDT agrees in ways with all interpretations that hold observer deterministic aspects. And by UED's expanded Conscious Uncertainty Principle and Observer Affect, WPDT and all Many Worlds interpretations, share in one sense, that new physical realities, or worlds if you will, *are* created within each spherical soul. UED and WPDT require each pure Dualist sees both an *objective* outer, and a *subjective* inner wave-particle based reality, or idealism. WPDT shares with the

De Broglie-Bohm Interpretation, or Pilot Wave Theory, in that all ejected lesser particles, such as light, can be carried or hindered in moving through all pre-existing waves. WPDT shares a few aspects with Relational Quantum Mechanics. WPDT shares a few aspects with Quantum Information Theory along its central quantum process. Time Symmetric theories and interpretations share a loose connection in that all *current* conscious determinism, applies its additional conscious force desires upon GOD'S *past* conscious force desires, which creates a perceived event within the aether, that affects the *approaching* experiential occurrence in each soul's next moment. And though we see a past and future moment, in its ways touching our every present expanding spherical moment, we see no arrow, dimension, or directions of physical time. All relative and relevant reality, rests *only* in each soul's consciously deterministic present; being a formal type of created present*ism*. Now super-luminary particles violate some laws through a too slow small physical perspective and model. But in GOD'S UED/WPDT, super-luminary particles are predicted and violate nothing once you move your mind out beyond mankind's slow small perspective. And as with every QM interpretation, WPDT does bend and violate to differing degrees, certain aspects of the secular standard model. Yet in a profound way, UED and WPDT being consciously deterministic, when properly known, completely depart from all other models, theories, and QM interpretations. Now consider that if mankind could know, qualify, quantify, calculate, and formulate, all *prior* conscious and physical occurrences and realities together, they still could not predict in any near certainty, the free-willed conscious physical influences of GOD, the angels, or mankind to come upon reality. If GOD ceased all salvation-efforts, mankind might predict physical probabilities slightly better; but by the realities of Chance, Chaos, Butterfly Effect, Interference, Convergence, and their hidden variables and perturbations, mankind could still not in any way predict nature, or life, out past a few days. Now as an abstract and near meaningless conscious exercise, we offer this most basic creation equation, that defines the act of FATHER creating his Aether. So, the value of *one internal* "physical singularity of FATHER'S Physical Being", is noted as (1(psFpb). And each singularity that was spoken forth, was projected under the value of "1 equal quantum of FATHER'S conscious forceful desire". And noted as (1(eqFcfd). Yes, the actual force value is an unknown, yet (1(psFpb) and (1eqFcfd) spoken forth = (1(sras), or 1 standard-resonating aether singularity". If it helps, call each of the aether's singularity a closed 3D loop or string. And as creation rose up in complexity, this simple formula just expands in its orders of magnitude to allow all greater and greater conscious creations. Yes, currently it is a meaningless gesture, yet it is the equation that started the creation of all physical reality. Now a strange question arises in this physical reality, being, are we big or small? When we speak about FATHER, we know he is far greater in every way than this universe he has perceived. And within the Milky Way Galaxy, our solar system is unimaginably small, and the Earth is unimaginably small in our solar system, and atoms are unimaginably small next to the earth. But the countless singularities of the aether are the smallest physical aspect of reality, which give mass to the layered spherical fields we call quarks, that form the larger particles that form the atoms, which form all larger aspect. And FATHER'S perspective sees all that exists from each tiny singularity, to the whole of his created reality. So, only he sees the outer-most bounds or shell or edge of his created reality. Now all FATHER'S children are meant to exist together and comprehend his created reality; but fallen individuals have chosen a self-limiting perspective. FATHER'S size and reality are beyond fallen mankind's conceptual ability, as the universe exists between his outstretched pulling hands, as fallen mankind chases those physical sensations in their small slow lost way. And from their *fallen* temporary perspective, mankind will never know the true reality. We know FATHER is consciously and physically real, being an immense and forcefully defined personage. But according to FATHER'S own Scriptural expressions, he is not infinite in every way; but *towards us* and his creations, he is seen as infinite wherever, and in whatever he desires and perceives. All reality comes by FATHER with an eternal purpose; and our accepting this in all its fullness, shatters our former fallen mind, so our new mind in CHRIST may begin. We know there is one FATHER above us all, for he has searched in his ways and found no other. Isaiah 44:8, "*IS THERE A GOD BESIDE ME? YEA, THERE IS NO GOD; I KNOW NOT ANY.*" Before gifting anyone, or creating anything external, what did FATHER see beyond himself, trying to sense and know if anything or anyone was there, searching in loneliness for another. Can FATHER describe to us, what does not exist beyond his awareness, or how it feels to be utterly alone. These unknown real boundaries of FATHER, make and show him to be so real to me; yet for many souls, having boundaries, raises their doubt in him. For many, the only acceptable concept of GOD is an infinite impossible God. We must understand and accept that in FATHER being real, he is a truly *defined* entity, or personage. And he is the image or person we are meant to eternally follow in the perfect reflection of his SON, CHRIST JESUS. Now FATHER'S Children, all reside in partial omniscience, and they all strive to live within his ways. And each Godly souls expanding boundary of spiritual relativity will continue without end, as a loving unfolding part of FATHER'S eternity. We can only experience and understand FATHER'S perfect reality, by our having the mind of CHRIST; that being the obedient eternal perspective of FATHER. Yet the grown and growing measure of our own Godly conscious awareness our FATHER sees, must for now remain an unknow to us. We wish with letters and words we could convey the rare joyous sentiment, and also the pained sorrow each soul holds in knowing the reality of GOD. But instead, most souls lack any desired ability to hear or comprehend such wonders. We see most of mankind's fallen physical search leads them to hell and death. *Your* adopted perfect eternal perspective of HOLYSPIRIT, corrects all misconceptions and all misinterpretation. The vast conscious and physical eternal perspective of GOD, can be hard to accept; yet until one adopts it, one can never truly or fully understand anything. Imagine FATHER looking out at one side of the universe being held by his one hand, then gazing across to the other side held by his other hand. What would fallen mankind's sense of passing be during FATHER'S short visual moment. Was fallen mankind's minute perspective the same, or did millions of years go by, or did mankind's moment stand still. Again, time does not exist, there is only each soul's created perceptions of experience. So, the speed of our experiential moment of reality unfolding was the same as FATHER'S moment. Yet, by a small slow false physical perspective, mankind could misinterpret reality and falsely see it could be very different. If GOD zooms in more and more until he sees only the earth, has his eternal perspective altered down to mankind's slow small timely perspective? No. And when JESUS put off his immense eternal ethereal body in Heaven, for his soul to come down to be consciously born into, and walk in the tiny flesh of mankind, did that alter CHRIST'S mind to a small slow timely perspective, causing him to no longer see the true reality? No. So, as we are adopted by GOD, we adopt and hold to their far larger faster perspective of reality, where *we* now see and find all pure consonance between GOD'S true science and his one true religion. By their ruling physical desires, secular science, and worldly religion, can never find or desire or see this consonance. Consider how physical behaviors become engrained by a soul's desired nature. Free-will souls desire and repeatedly commit rebellious physical acts. And like clay, each soul's attributes are willfully molded to need more such physical sensations. Our souls sense the physical body in a strong experiential and exercised way, which *can* result in strong continual desires for the same. Habitual rebellious sinful acts, enhance desired physical sensations, and increase a soul's expectation and anticipation for the information that stimulates their soul's senses. If one forms a habit, it can affect a body's DNA, so a proclivity towards such things can be passed down to their offspring. Num. 14:18, "*THE LORD IS LONGSUFFERING, AND OF GREAT MERCY, FORGIVING INIQUITY AND TRANSGRESSION, AND BY NO MEANS CLEARING THE GUILTY, VISITING*

*THE INIQUITY OF THE FATHERS UPON THE CHILDREN UNTO THE THIRD AND FOURTH GENERATION.*" Now science says they see the physical wiring of our interface, grows a new neural connection with each new strongly desired experience. If so, this is a newly wired interface path to that same-moments created experiential information stored in the aether. Science also finds long term avoidance of a memory withers that connection; yet that information remains in the aether just the same. So, in turning back to follow GOD in all our exercised corrected ways and desires, we see all our hardwired experiential desires must wither. And in our *enduring* repentance, GOD heals our DNA. Now science rarely talks about the inability and inadequacy of DNA to align and guide *all* the physical systems required for a functioning physical body; or that DNA is most often ruled by physical chance in aligning the basic proteins that define inheritable distinctions and attributes. DNA once held only the original programmed physical proclivities required for eternal life, but all choices for sin, enhance or diminish certain proclivities. The inheritable proclivities of each parent's interface, or body, are blended and passed down after their kind. Yet throughout the embryonic, and following phases, science sees there must be more guidance than DNA and electrochemical reactions offer. Now again, please observe any physical object within your purview, and know that upon every single point in every smallest to largest quantum system within that object, we see all the projected patterns of lesser to greater force quanta of FATHER'S projected desires that aligned that object's elemental consistencies. Now in all the wonders of reality, how do *we* know *we* hold the one true science and religion. We know, because we see all physicality, all awareness, all desire, all logic, and all purpose, falling seamlessly, effortlessly, perfectly, and wantonly, into pure alignment. And once we remove all worldly science, all worldly religion, all fallen perspectives, and all fallen desires from the equation, and ask our science and religion any existential question, HOLYSPIRIT perfectly answers *every* question we can imagine. If your understanding of reality does not do this, something huge is missing, and confused imperfections exist within the foundations of your science, religion, and even in your questions. If one thinks every question is legitimate, they have no true foundation. The one true understanding is learned from its foundation up, there is no instant gift of our knowing these things from GOD. The HOLYSPIRIT, the KJV Bible, and our four books, are a path to know GOD. If you hear and interpret anything about reality from a slow small physical perspective, and not a Godly perspective, you must and will misinterpret it. Remember, GOD does not limit free-will questions, to true questions. And though GOD will never answer an untrue question, beware that many others will offer you their false answers. If you go down a wrong path, you will have only wrong questions, till your path is corrected. Fallen mankind's love and understanding are so fractured and flawed, and they ask most questions in error. This precludes *them* from seeing that they now experience the remnants of something that was once of a far higher faster and perfect reality. Now certain physical theories postulate there are continuous linear physical-time-frames of the entire universe, existing all at once as a segmented continuum of space/time. But FATHER'S aether, and the forceful patterned realities of every creation, thought, and experience of every eternal soul's spiritual and physical existence, are the only real physical aspect and record of reality's continuum. And for the now and present flash of reality, there is only our ever-current conscious moment of it all, and all ongoing new creations of each desirous eternal soul. The physical aether is both our medium, and book of record for all created and perceived existence. We use the old term aether correctly, to mean a superior prior background foundational substance or medium, that facilitates and allows all the realities of perceived physical creations, forced movements, forced interactions, and every aspect of forceful desire. All desired conscious physical movements, and the creation of all things physical, move through the aether's grid, or dwell within the aether that fills each eternal soul. The primacy of all new generated projected desirous intentioned force, rules supreme over *all* physical creations, without reservation. These conscious forces *and* physical aspects, are not placed and rested in the angels or mankind at any level of self-sufficiency; but they are supremely set in the whole Trinity of GOD. Now again, we see no Newtonian gravity, or Einstein's warping of space/time. These are two similar false perspectives, interpreted through mankind's slow small understandings of entangled attraction, asking does force of mass pull us down, or does mass warp space/time, and push us down. These two physical proffers, see the effects of gravity from two differing physical ways of seeing FATHER'S desired forceful entanglement. Yes, all three describe and predict the exact same affects and results. We see FATHER'S weaker, broader, far reaching desired forceful entanglement called gravity; and we see his three gradually stronger more narrowly focused entanglements. And then there are those we all project to create perceptions. Can mankind's science even ask, let alone answer why and how force acts in these varied, determined, required, and exacting ways. We see beyond doubt, that the quantum realm can exist alone, with our consciously existing through a quantum ethereal body. But all physical realities and life, existing without the quantum realm is impossible. The physical is obviously a secondary lower realm. Now we see that the closer to each quantum realm soul's attributes we get, the more all conscious forceful influence creates and affects all reality; but our conscious forceful affects quickly diminish in the other direction. And there, GOD'S forceful influence rules alone. Materialist have their contrary perspective, yet we see the same fatal flaw in *all* their theories, being they all start with a required *physical* miracle. They can't see why such theories can never be true, being founded on such unknown and unknowable beginning. They have *so many* false assumptions that enable endless false interpretations.

Yet beyond all their confusion, I am aghast at their great evil of forcing the indoctrination of their false beliefs into naïve children, as being truth. Which seduces, blinds, and steals them early and quickly from GOD. We see there is, and must always be an assigned purpose and motive to life. We ask how and why we have come through what was, to what is; and why we go on to what will be? It is only and always our purposed motivations of life. I have always loved the one true GOD, *and* the one scientific truthful understanding of GOD'S gifted souls and his Creation of physical life. But the definition of *true* science, is asking in the true and proper order of importance, these two great questions. First and most important, is *why* are we here; and then secondly, where and what are we? This *order* of understanding is a vital part of creations design, as no soul can ever answer just one of these questions correctly. All lesser to greater things must be answered in a full and relative six-dimensional way, and in their proper order and way, and at the proper speed, to instill faith in knowledge and remove doubt from *all* souls. I am an absolutist in all ways, and these great questions of why, do exist. So, I find that to know any single or everything *absolutely*, we must first have a relative answer to every possible question that can be raised around and within those same things; and not just the questions raised from a small slow temporary purely physical perspective alone. I remember something most don't about a 1960's elementary education, coming to that early fork in the road of being told in fourth grade science lessons and grades thereafter, that science applies only to things we can test and prove; so, we must disregard any and all questions of why are we here, and all thoughts of GOD in our scientific studies, as these are un-testable, unknowable un-scientific matters. And each naïve child that accepted this lie of hell, had a biased curve set upon their thought process that pulls them down, to drown them in physical obscurity. At that tender age of not knowing, by a teacher's statement, the understanding of our soul's conscious force projection is severed from most soul's future scientific study and belief. Such instructions are a great evil to innocent children, and even adults; yet this seems to be a minor thing to most souls. But to *us*, these few words are filled with a murderous intent, horror, and the spiritual disfiguring of hell. My eternal perspective that interprets my understanding of observations, comes by holding onto all known conscious spiritual variables, and seeing their endless waring secular beliefs, failing again and again. So, from my youth, I held my firm suspicion toward all purely physical scientific approaches. I could never be turned by any unsubstantiated statement from a teacher to disregard, or for that matter, accept any limited avenues of existential study without being convinced by solid reasoning. But no reason was even offered, just a bare false statement. So, to accept that, would be a blatant assumption on my part. And after nearly sixty more years, I still have not heard one good or true reason offered for such separations in any scientific study. The one common factor I find in all souls who espouse this exclusion, is they all have no desire for GOD'S ruling ways. Pure Dualist see the complete truth of every evident aspect, correlation, and fact; and these reveal our creative projected conscious force. These truths set our clarity and guidance for seeing all things congealed properly; which then declare reality's highest reasoned purpose. In that pure physical theories are all based on unfounded physical miracles, this shows us why all their models raise many profoundly missed predictions, and create a need to surmise and accept unknown variables and perturbation theory as a safety net. They can't see that what they call unknown variables and perturbations, are the higher matters of GOD'S and lesser souls forced influences, waring among all the physical aspects of raw chance and entropy. Their theories are *forever* unprovable, missing any true foundation. In their unprovable godless theories, they see *only* the aspects and part of reality they look for, and want to see. Their picture of reality appears to us as a slowly yet constantly changing peripheral area of a picture on a TV screen. But they lack the defining heart or center of the full picture. They search for a hollowed-out cause and belief without real reason, meaning, or purpose. While in UED's loving conscious primacy, we see a non-changing ardent picture, with the center being clear and complete, lacking only the tiny periphery at the very edges of our TV screen. And this perfect omniscient picture, eternally draws us into GOD. We see both pictures physically overlap here and there in certain vague ways, but our interpretations of all those overlapping aspects differ. If you can desire to see it, our picture is bright and quite complete, and far more hopeful, beautiful, and full of all eternal answers and the meaning of life, love, and truth. For the truth to be told to, and heard by a secular audience, we must remind them often of the intractable problems, weaknesses, paradoxes, mysteries, and physical miracles, that haunt, taunt, and fill materialist beliefs. And a secular audience must desire to hear these things through our expanded shared frame of reference. This being a rare wondrous occurrence indeed. Consider please that at this moment, there exists a line out there for us all, being a line of attaining our own set understanding of reality, and our own existential meaning of it all. So, let's look to GOD'S *Omniscience,* and ask what it actually is, and does. Omniscience or *Omni-science*, means all *true* science, or *true* knowledge. And FATHER'S pure *true* omniscience, includes no errant confused understandings, or any free-will soul's idle errant evil thoughts, or any evil free-will acts, motions, or desires, or each occurrence of chance and choice. GOD'S omniscience endures every perfect reasoned consideration, and holds every relative formula, equation, and factor of true knowledge pertaining to all that firmly, ethereally, and forcefully exists. Also knowing by what means and for what purpose, all things came to consist and live. And lastly, knowing all FATHER will physically bring to pass. FATHER'S omniscience holds the complete true understanding of *GOD'S* reality, reason and purpose; but it is not in his currently knowing the sum of each confused individual. GOD'S omniscience does not include all fallen mankind's misunderstanding or misinterpretation of all the data they see through their slow small confused perspectives. FATHER by choice, does not want or choose to witness every evil thing everyone does or thinks; yet he sees enough to have a true understanding of it all. And all such evils are recorded within the aether for the day GOD does need to see all and know all to judge all. GOD'S omniscience is the sum of his truthful love and science with*out* error or deceit. And the very essence of the conscious term "one pure truth" means one fully formed, *errorless,* conscious conceptual understanding of every created existential aspect, *as* GOD has made them to be; and seen *solely* through FATHER'S original created intent. This conscious truth is nothing born of the physical, or of mankind. Most souls, including most scientist and preachers, exist within layers of confused belief, moving from one to the next. All such confusions are good for nothing eternal, and are discarded and forgotten by all lost souls as they accept each new equally wrong replacement; examples are Quantum Gravity and Multi-verse Theories. Omniscience is complete in what it is within GOD, and it is perfect and unchanging. Our souls were each gifted in the adequate measure and capacity to learn, grow, reach, and hold a near full omniscient truthful understanding of

GOD, as it is then laid out before us by HOLYSPIRIT. But if any soul desires to fill their soul with partial and false understandings that can never eternally satisfy; this each soul can easily do. Finding the basic foundation of GOD'S omniscience to build upon, is in first denying *all* physical primacy. Then you listen, learn, and grow into your eternal image of a child of GOD. For it is only by and for the reason of FATHER'S love and need of an *eternal* family, that *we* were gifted our free-will soul. If FATHER wanted, he could view every soul's confused thought, belief, and action just after they create them. But this FATHER *says,* he has no desire to do. For FATHER would not find his understanding or belief in any of it, so what would be the purpose of that. FATHER'S avoidance of witnessing most rebellion and error does not subtract from his omniscience. FATHER *does see* the overall acceptance or rejection of him, by the strength of our visible countenance; being the spiritual brightness in all our projected desires. So, nothing is hidden and none can ever say to FATHER, things are not as they physically appear. If a soul is dark, FATHER need not see every reason why. If a soul is a radiant brightness of truth, GOD looks deeply into this child's life. GOD'S original perfect knowledge eternally exists; but all false and damaged things and ways can only temporarily exist. FATHER'S once perfect external physical Creation, is now thoroughly corrupted, yet his omniscience remains untouched in having the answer to every *true* existential question concerning all that souls are and do. By FATHER holding the omniscient knowledge of his created reality now, means he held the exact same knowledge before he created it, unchanging and fixed. In this world, mankind can sense a feeling of creating new knowledge that never exist before in this world. And if one desires, thinks of, and believes anything outside of and beyond the boundary of FATHER'S omniscience, they may have created a new, yet false individual knowledge. The sum of FATHER'S knowledge, and his perfect understanding of all created realms, and all life, is there in its fullness right now. Not ever to be added to in any manner, or taken from or altered in any way by any lesser authority. The die has been cast, the sum of all the true knowledge that was, and is now, and the sum of all the true knowledge that will ever be, as pertaining to our very existence and FATHER'S eternal purpose, are exactly the same. Fallen mankind's too slow too small physical temporary perspective and their false interpretations, must change continually; but FATHER'S knowledge never changes. It is now only the unfolding, seeking, and finding, of each souls chosen acquired true or false sealed belief of understanding that remains. Surely as a logical rational argument this can't be denied. But this is only one abstract side of the coin of knowledge. And we must accept in accordance with all these facts, that all true knowledge says beyond doubt, that minus an associated consciously created substantive perception of it, there is no *knowing*; being the meaning of knowledge. Knowledge can only exist in a soul's knowing awareness of it. For FATHER alone, every act of desired dualism is set in his omniscience. For lesser gifted souls, we consciously and spiritually and physically sense, desire, create, interpret, and then we believe, or deny. This is the only definition of knowledge or knowing we have, or can ever have. What is our own founding source for knowing, understanding, and creating our desires and perceptions that form our beliefs? It is our soul. And it is FATHER'S purpose in gifting us our souls to know the mysterious-experiential-qualia of his vast physical creation, that raises our existence? Now our firm UED/WPDT model of particle-wave based, patterned forcefully created perceptions, is the firm beginning and Biblical foundation of all true science and religion; so, search no further and speculate no more. We all know there is no real brightness or sounds outside our soul, there is only our created awareness of brightness or sound. The experiential qualia of sound, exists only within our quantum conscious soul. Beyond our soul's awareness, there are just moving utterly silent waves of utterly dark forced frequencies traveling through the physical medium of the air. There is no external brightness colors or shades to reality beyond our soul's own created and interpreted experience of these qualia. There are only the moving patterns of dark silent particles and dark silent fields of wavelengths and amplitudes we call physical light and dark silent moving mass. The Sun, and all stars, and all physical light itself, are actually darker than our soul can ever imagine. Outside our soul's awareness, reality is only moving utter darkness and silence. There is also no real physical feeling of anything or anyone, there are only our created *perceived* sensations of a touching experience, felt within our soul. Beyond this, there are only unknowing, unfeeling, unaware forms of created mass, coming into the proximity of, to collide and affect one another. All of this sensed physical information, triggers our soul's creative forceful awareness process. Please remember, we also see, hear, feel, taste, and smell things in our dreams. Please open your eyes wide to *your* bright reality, and do not stay deceived in the dark. All we sense and perceive in our own awareness, is in our minute godlike manner, forcefully created within the sphere of our soul's conscious estate. And we will all eternally experience a joyful or anguished reality in this way. The last *true* physical correlation, just prior to our soul's indwelled awareness of it, is our forcefully created, dark patterned particle-wave-based streaming perceptions, that mirror all the physical and quantum and all other consciously sourced forceful information, sensed through our physical interface, and then sensed by our soul. This is the sum and process of who we all are, actually, and eternally. This process allows our desired interpretation and believed understanding of our experiences. It forms our evolving or devolving over-all core concepts of belief. The term "conscious concept" cannot mean anything more or less than one's knowing awareness of a defined relatable process or aspect of our desired understanding and belief. And we cannot define any aspect or concept of knowledge, minus our knowing awareness of it. *All* patterned intended forced information, can equal knowledge, only at a soul's perceived knowing interpreted awareness of it. So, this Knowledge Paradox arises, being that if all the omniscient information and rules needed for creation, existed prior to all creation. Then all this information and all these rules had to exist somehow somewhere in a knowing organized useful way. Meaning knowledge must have existed, prior to this reality created using that knowledge. And once again, knowledge must be first perceived and known, and only then can it be forcefully projected in its *known* patterns to then create. So, if not FATHER, who perceived and knew this required knowledge, prior to creation. Could mankind throughout history, be collectively forcefully creating and sustaining this one vast shared universe through our incomplete knowledge and all our differing desires and beliefs. That theory fails instantly under mankind's endless waring understandings. So, again, we see FATHER GOD our creator, must exist *necessarily*, by both knowing and forcefully perceiving all the complex physical realities into existence that we now secondarily perceive; there is no other rational option. This is the age-old quandary of circular, chicken or the egg reasoning, that confronts and divides the rebellious souls of the angels and mankind. I hope you all see, that once you accept that a supreme eternal omniscient creative conscious force projector exists. All such circular reasoning is answered by GOD'S projected created perceptions around us; and all things become most clear at this acceptance. If you cannot accept the pure truth, then quandaries, enigmas, and paradoxes will forever drag your soul down into false dark endless unanswerable questions. Only one path answers every true question for all eternity, and all other paths question every true answer for all time. Let's take several more conscious steps further away from the false small slow physical perspectives, and towards the far larger faster omniscient conscious reason and purpose for it all. In pondering our comprehensive capabilities and capacities to change and grow our believed understanding of reality. We must ask, is our knowing capacity finite, or infinite? And in our knowing an unimaginable Supreme Conscious Soul is out there, who first *perfectly* perceived the physical realities that most souls now perceive *im*perfectly. Do you see that though *our* souls are eternal, our required physical half of reality is dependent upon FATHER. Please consider what amazing wondrous results would come upon a fallen soul's awareness, at acquiring a sufficient sum of GOD'S omniscient understanding. Imagine not knowing anything true; then learning and knowing the true foundations of all reality. In this alone, comes something very different from learning little false pieces of physical knowledge, which in adding each piece, never proves anything beyond doubt. This only cause souls to forever in error speculate further. But in accepting all basic aspects of GOD'S omniscience, a wondrous new property appears as the pure truth of all things. At attaining just this basic level of true knowledge, we are awakened to a subsequent disproving of all prior false beliefs and understanding. And when souls attain a certain adequate understanding, it eliminates all individual beliefs, and all things speculative and divisive. There we find and share the mind of CHRIST, in the one shared truth of life and love and experience a profound alteration to our soul's conscious properties. Again, you must disavow *all* physical supposition and conjecture, as finding and adding to your found measure of GOD'S omniscience brings a true metamorphosis to your belief system. And this pure understanding is what enables your *true* communication with the mind of GOD; being on the same page if you will. Now our desired and projected forceful belief and intent, define every perception created by us, the angels, and all the vast firm perceptions of GOD. All consciously created systems of information, are to be considered complete creations. So, at a perceptions creation, it *can* be said it is a minute form of mathematical omniscience, or a complete perfect information that defines that perception. So, even the created systems of each evil perception, *can* be viewed as a *mathematical* omniscient system by itself. Yet GOD'S faithful have a desire to learn, choose, and align our belief and understanding to the full and ruling omniscience of FATHER that creates and sustains all reality. Now we ask, at the created concepts of up and right, did down and left come simultaneously, or did they follow by a conscious certainty, or an unimaginable conscious probability. In conscious creation, they must follow as a soul conceives of one *first*, and then the other. So, in GOD'S creation, filled with billions of minute conscious entities of vastly different size, did the original perfect creation fall due to the Antithetical Principle? No. Did it fall because of all the created differing size relative perspectives? No. Neither of these two aspects of choice, previously existed in any soul's conscious practice before the fall. These both were surely secondary correlations to the fall. But reality fell by the unpredictable free-will choice probability of all the gifted lesser souls, within it all. And after that very first contrary anomaly of Lucifer, rebellion became a free-willed contagious certainty, being pushed and spread by him. But by the omniscience of FATHER'S existence, being a final control over all physical creations, including Satan's required body, FATHER remains the ruling soul over all creation's equations. We are each a spiritual over physical desire probability equation, having two possible consciously determined outcomes; being ours, or GOD'S. Can minute conscious entities such as we are, have a true desire to yield and adopt the understanding and belief of a vastly superior entity? And then exist happy, deep within GOD'S created physical perception? Yes. But for reasons of confused pride and envy, most hold no loving desire for such a life. They must rebel in their ever-falling confusion, despair, and anguish, reliant on their inescapable physical surroundings. Which again, are created controlled and sustained by FATHER. *Individuals*, are each a physically driven antithesis to GOD'S purpose. Now let's visualize a free-will soul's conscious space, filled with their false understandings. It being a spatial sphere, filled with many aspects of dim light and aspects of pure darkness. These representing an individual imperfect partial truth and false understanding of GOD, life, love, and reality. Then see this soul deny all those false things, and begin learning only things true from GOD. And continue on until that soul adds that piece that substantiates an adequate portion of GOD'S omniscience, and immediately starting a chain reaction of true connections of enlightened understanding. Picture now the properties of the one truth of all that GOD has perceived, spreading throughout this conscious sphere that *was* once filled with their imperfections of grasping at darkness and speculation. And like the birth of a spiritual star, they now alite and burn in a blinding bright epiphany of the pure truthful understanding of all things. 1st John 1:20, "*BUT YE HAVE AN UNCTION FROM THE HOLY ONE, AND YE KNOW ALL THINGS.*" This speaks of our finding the properties and understanding and truth of GOD. And this is our only true path into the knowledge, understanding, and our true communication with GOD. 1st Cor. 2:15-16, "*BUT HE THAT IS SPIRITUAL JUDGETH ALL THINGS, YET HE HIMSELF IS JUDGED OF NO MAN. FOR WHO HATH KNOWN THE MIND OF THE LORD, THAT HE MAY INSTRUCT HIM? BUT WE HAVE THE MIND OF CHRIST.*" John 15:15, "*HENCEFORTH I CALL YOU NOT SERVANTS; FOR THE SERVANT KNOWETH NOT WHAT HIS LORD DOETH: BUT I*

*HAVE CALLED YOU FRIENDS; FOR ALL THINGS THAT I HAVE HEARD OF MY FATHER I HAVE MADE KNOWN UNTO YOU.*" Hear this promise of gifted truth and wisdom from MESSIAH. Please see how holding just an adequate relatable portion of omniscience, minus *all* suppositions, reverses one's former devolving fall into an *individual* understanding of belief, born of all worldly desires. All such evils, by our desire, can disappear from our conscious abilities, taking forever all confusion and doubt. Our religion and science will then fall together in consonance, as ardent and everlasting. *Our* search for all existential answers, now no longer exists. We who desire and hold such knowledge, fear and strive to fulfill GOD'S hope in us. And all who don't want this truth, will feel trapped by a oneness of mind in CHRIST. The fallen desire and need to exist in a reality that is an open-ended question. They need their desires to run wild and change, unfettered by any final created truth. Yet they cannot see the final loneliest horror of where this eternally leads them. So sorrowful to watch, and more-so when we try to show them, and they call us religious fools. FATHER'S omniscience knows what enables and perpetuates eternal life and love, and what destroys it all. If we will see the

Omniscient loving logic in FATHER'S original desired created ways and designs, we see they are the defining sustain force in all external things and ways, created for our eternal loving existence. Do you want to know what it feels like to never again doubt in GOD, or be confused as to life. And to never *not* know a physical fact of creation. For me, it feels loving and calmly settled, even while here. But for our far greater final fullness of peace, we must wait as we still exist in this now horrid world. We know the omniscient considerations, desires, and plans of FATHER, are all born in a fashion and purpose to serve and exist for eternity in perfection, and never to sustain anything being imperfect or temporary. Consider how all the destructive intentional and unintentional errors of mankind have been consciously born of all divisive desires and subsequent beliefs. All divisive misunderstandings and disagreements come by all contrary warring understandings and spiritual weakness born of fear, pride, envy, and all considerations of self-preservation. In opposite fashion, we see FATHER holds all power above, and has no worries for his life to consider, he has only an eternal purposed desire to find an eternal family of love. FATHER *can't* have any imperfect understanding or considerations for self-preservation; so, unlike mankind, he is *incapable* of confused destructive desires, having no place of error, weakness, fear, envy, or self-preservation from which all imperfect evil thoughts are born. But there is in GOD, his eternal desire to separate and protect those who eternally love him in omniscience, from those who don't. And we see the most-evil desires to dissuade and force their own fallen beliefs upon those who love GOD, come from all dark souls, like those of Sodom who demanded that Lott send out those two Angels he was lodging for the night, so they could have forced vile physical relations with them. The same angels GOD sent to remove Lott from Sodom, before destroying that *entire* abominable reprobate city. Evil cannot know or accept boundaries. Now "truth" means a conscious conception, and nothing physical; ergo, *two* who disagree on the true foundations of eternal love, can never converse or walk together in eternal peace. We must learn and hold GOD'S omniscience to find our eternal peace with the loving mind of CHRIST. All communication between GOD and lesser souls, requires both hold a common truth and belief to eliminate all misinterpretations and disagreements and misunderstandings. Souls who believe life is finite, hold an individual temporary belief about life, bringing endless conflict between them and GOD and others. Consider that most do *not* want an eternally defined truth to be bound by and accountable to; so, these must then *den*y their eternal soul exists. Eternal loving joyous life among many, can only exist among souls holding FATHER'S omniscient loving truth. All individual or personal beliefs, do conflict to differing degrees, which do weaken and destroy *eternal* peace and life. No one can objectively deny that *all* knowledge and the one true reason and purpose for this reality and all life, are out there and known to at least that one very first supreme soul; while all the fallen of mankind search their confused minds, looking for their own desired individual versions of it. The proverb is true; they can't see the truth, starring them right in the face. Can anyone *knowingly* deny that we forcefully align the quantum substance of our perceptions, enabling our awareness of existence. The pure truth shows only false knowledge and beliefs, dwell within all individual souls. *Our* true reality is a spiritual path of eternal faithful steps. And the first eternal position we take, is that before all that now exists, there was only the soul and body of FATHER GOD; being all his forceful truth and love, and the source of all the physical uncertainties, possibilities, probabilities, and certainties, that have ever been or shall ever be perceived to exist. FATHER created the truth of eternal life, being the founder and only source of all such things. Let's all use *our* soul's godlike 3D force projections to create within our mind's eye, a simplistic vision of two very long trains traveling through space. Let's call these, the two trains of conscious existence. Now we see one train going towards eternal omniscient love, life, and truth, with all its souls, who by their own loving desire, have received the mind of CHRIST, and are wantonly and obediently attached as this train's cars. And out in front, we see the most powerful conscious source of reality, being the creative FATHER, pulling us all souls along towards the oneness of providing a perfect loving eternal existence. But going in the other direction, we see no guiding engine out front, just billions of individual unconnected cars, caught in the dissolution of rebellion. And in the unfolding of reality, individualism grows within them, dividing them onto an ever-increasing number of tracks, with fewer and fewer cars on each new track. Please see that in the forever, this leads to the final loneliest separation of single cars on single tracks; being a final never-ending destination of singular desired belief. Rejecting FATHER, who is the only anchoring truth that can eternally bind us together, leads to pure loneliness and insanity of singular belief. All individuals will devolve into their singular understanding, destroying all ability to communicate with anyone else forever. Individual souls must forever sink deeper into a bottomless abyss of being confined in a loneliness, once known only to FATHER. But unlike FATHER, these were formed in FATHER'S wisdom, having zero control or possession over the points of a body of their own. And being only sufficient in conscious size to learn of and love or reject GOD. So these have no possible future of dividing *their* soul to gift other souls to escape loneliness, or be a god over others. Now how godlike and effortless was it, to *desire*, create, and see your particle-wave perceptions of those trains. Please see beyond doubt, that wet brain meat neurons could never do this. Only the soul of FATHER and his gifted eternal offspring can create and know their existence. A growing separation between all fallen individuals that fills this entire world, is fearful and destructive. We so love our bonding sustaining understanding of sharing the joy and peace found in being one, now in the mind of CHRIST. The sum of knowledge defining all external creations, exists only with FATHER, being beyond fallen mankind's horizons. Now being desirous learning spiritual creatures, we *strive* to satisfy our desire to know GOD'S truth. And there is the pure truth of GOD alone, or our own desired false truth of GOD we choose. We are all unknowingly or admittedly bound in a search for either our interpretation of what true knowledge is to us, or the true knowledge of GOD. And we who seek and find GOD, learn through the eternal perspective that there are eternal consequences that come in finding an adequate omniscient understanding of GOD. Like finding a single pure truth that we must know and live. I hope all are honest enough to see that from a small slow godless perspective, one can perceive the many aspects of this firm reality in many ways that allow one to assume a false miraculous physical beginning, and build a belief upon that error. Yet, if you dwell in such *im*perfect knowledge, you may sense the eternal perspective begin to disavow your prior beliefs. No one should lean out over a deep canyon of false belief on any brace of assumption, which can quickly give way. If we hold to the one eternal perspective and receive the one initial understanding of life eternal, and the Scriptural truth of creation. By having this, and discarding *all* our former beliefs, disavowing *all* physical assumption, we are then on a path of hearing

GOD'S perfect truth; to become one in the believing mind of CHRIST. 2nd Cor. 13:11, "*FINALLY, BRETHREN, FAREWELL. BE PERFECT, BE OF GOOD COMFORT, BE OF ONE MIND, LIVE IN PEACE; AND THE GOD OF LOVE AND PEACE SHALL BE WITH YOU.*" If you and I share the eternal perspective and truth of GOD in our adequate portion, rejecting all conjecture; we can never disagree on any fact. Personal expressions of belief and morality, as they pertain to the created ways that propagate eternal life, love and peace, can no longer exist between us. GOD'S true Children, understand and agree with GOD on all created aspects. Many can't see why their individual desires are declared imperfect and divisive by GOD, or what makes GOD'S desires perfect above theirs. This is actually the same question as asking what is eternity. But eternity can't be defined using anything of substance in this fallen existential theatre as an example. Eternity can only pertain to things perfect; so, as for eternal physical things, they no longer exist in this fallen creation. Eternity means there is no markable beginning or possible end; as those are two defining parts and elements of time. Eternity is not endless time, but the absence of all the destructive elements that are part of time. Eternity shares no common aspects with time. Death, entropy, and fission, are all a part of time, but not part of eternity. Eternity and eternal life, have no physical example for us in this world. There is however a physical question, which if asked, the answer to, and example of it, are a great part eternity. Being, what defines the perfect physical body of FATHER GOD. I hear souls question GOD'S perfection, but always in moral and theological terms in relation to *their* own desired beliefs. But we see the ruling perfection of GOD exists beyond his conscious boundaries of truth. Let's remember that this fallen creation is now permeated with physical confusion, disorder and fission; yet we know that eliminating all that corruption, would return us all to an eternal physical existence. Mankind tries desperately to correct tiny physical errors, trying to extend physical life. This fact itself, declares in removing every *physical* imperfection, physical life must continue. We see GOD lives in their omniscience and truth, being immovable and immutable; but do you see their *physical* perfection, and FATHER having all ultimate conscious control over all the physical things we see. By these facts, GOD is physically perfect and eternal, without any physical imperfection, weakness, or limitation that pertain to us. There exists nothing that has a physical or conscious capability, to physically or consciously come against, or penetrate the conscious or physical perfection of GOD. All perfect physical things, are eternal physical things, so, all eternal physical things, must be perfect. And these we see are only found in being one with GOD and his ways. All that ever exists, dwells within the concepts of eternity, or time. And one of these concepts, defines our perspective and belief in our desired continuum of our ever-expanding conscious moment. The feelings of eternal life, come only by a soul's perfection, having no error of belief. The feeling of time comes by errors of belief, and sensing one's physical imperfections. Human bodies, and all physical creations are now forgone to FATHER, as he allows time to consume them. The two concepts of eternity and time, both exist, and are there in the *now*, being between all that once existed, and all that will exist. Eternity is inexplicably comforting, peaceful, and loving, as there is no looming end, or any further loss of loved ones. There is no *new* pain or sorrow coming to souls in eternity. There are no pressures or constraints as with time. Most souls could embrace a joyous eternity except for one thing; being eternity is only found in being one in the perfect love and ways of GOD. There is one

GOD, and our acceptance of their sacrificial love makes our life at peace. Nothing temporary can know eternity. If anything has a temporary nature, that nature came by desires against GOD'S ways being forcefully projected into all FATHER'S initial perfection, which caused its fall to a timely existence. We see the perfection of GOD *rests* in his omnipotent supreme properties of being eternal. These can only be attained by us, in our loving alignment with GOD'S created truth, light, and way of life. All souls have been lovingly asked to join GOD'S Children to reach our common perfection in GOD'S ways. Yet mankind loves and worships their fallen bodies and environment. But a perfect physical body and environment were once a created platform for mankind for us to learn and embrace GOD'S spiritual world. But most souls have turned 180°, believing the soul is their platform for learning of and enjoying the physical world. A Pure Dualist is a *spiritual* environmentalist, who allows GOD be our true physical environmentalist. This means *we* cannot destroy GOD'S Creation in any way, except by sin. We know that GOD'S initial perfect physical environment ensured Adam and Eve nothing eternal, because, as we see, they were sinful spiritual environmentalist. A perfect physical environment was created to eternally support his loving family; and we read how CHRIST taught them his spiritual omniscience daily in the Garden. GOD sees us all in a gifted potential; but now we see endless contrary ways that cannot coexist in peace. At a certain point of slowed spiritual growth, a pride and envy appeared in the angel (Lucifer). And from that, he spread all the imperfect individualism we now see. To recap: there are inarguable laws that define a real far greater conscious size relativity, and laws defining eternal perfection. We see laws defining our forceful conscious creations; and we see our perceptions form our desired and relative, yet subjective reality. We have vast evidence that shows we could not altogether, perceive this universe. So, again; GOD *must exist* by necessity. GOD'S Children run to him in love, stumbling less and less in their passion towards his open arms. But there are far more who stand still, or walk to him in their belief. Or run from GOD in pride and envy, falling blindly into their lonely abyss, as individuals. And across this spectrum, we see many try to have it both ways. Running to-and-fro in a futile attempt to straddle and define GOD'S spiritual and physical ways for themselves. These are caught in an individual confused belief of a personalized god, imprisoned by their physical desires. We present as evidence of this, *all* false scientific and theological beliefs. And it strangely appears as *we* run towards the truth in our fervent love and desire, that most of the world en-masse, go in circles of confusion, running from the truth in a hate for GOD. For all souls, there are only two possible forms of believed existence: one is finite and physical, as with fallen Mankind, the other is infinite, ethereal, and spiritual, as the Holy Angels. This is where understanding, beliefs, desires, and all good and evil souls part ways. We are all gifted in GOD'S hope we would desire to be *one* in his eternal majestic family. At a soul's gifting, we are dimensionally and forcefully defined, one from another. Yet meant to be spiritually one, in one true belief and love. Mankind sets their societal laws and boundaries in trying to restrain their evil ways in their *temporary* life, and maintain peace. And if a soul violently crosses a line to hurt, it requires incarceration to preserve life and peace. But initially, we had FATHER'S perfect eternal boundaries for souls to live by *eternally* in love and peace. FATHER'S creation is not a democracy or in any way, politically correct. And Heaven is an Eternal Holy Monarchy, a Holy Kingdom, existing as FATHER desired and created it to be. If anyone tells you GOD is liberal, or mutable, or inclusive, turn and run. All perfect eternal *sustainable* boundaries are set by GOD'S eternal truth and omniscient love. So only a harmonic eternal love, *not* in conflict with GOD'S eternal truth, can perpetuate peace and life there. And only souls having this common love, are searched for, found, and eternally loved by GOD. If one feels it unfair, that GOD does not love them because they are contrary to him and his created ways. They should consider that by all their choices, when GOD did first love them at their gifted creation; they did *not* love GOD in return. And GOD gave all souls more chances and reasons to want to change than they will ever know. And after all his many efforts, FATHER then sent his SON to suffer a brutal death to show his love and build a bridge. Yet that is not enough for most; and there is no more FATHER can righteously do. Such souls are cold and dark to GOD, and beyond his loving reach. Eternal peaceful life for many, can only exist in passionate fervent faithful responsible love for all souls within Heaven's boundaries created by GOD. We fear losing those we love, so in our great joyful desire, we protect all from the introduction of any contrary divisive ravages of time, hell, and death. Some say this fear of loss, means self-preservation must permeate and define Heaven, as the only way to preserve one's eternal life is to in fear, live to preserve the eternal life of others. This is a subtle untruth, and is how Satan speaks. Although a discernable part of all love, is a fear of losing that love; the love *itself* is *first born*, and then a fear of losing that love. So love is what truly sustains. GOD sees the loving and fearful emotions in all souls; and saves no souls having only a legalistic self-preserving fearful obedience alone. In life, the only way one can ever believe a lie, is to first want that lie to be true. In this way alone, souls reject the truth of GOD. This world is full of liars and lovers of lies, all by their desires for those lies to be true. But in *our* understanding of the omniscient truth and loving desires of GOD, we find eternal life has far higher meaning. The flesh that was originally created to live eternal, being now corrupted, must acquire energy from an external source. But GOD has taken his perfect provision of life away, leaving us in a rising confusion of predation. Gen. 3:24, "*SO HE DROVE OUT THE MAN; AND HE PLACED AT THE EAST OF THE GARDEN OF EDEN CHERUBIMS, AND A FLAMING SWORD WHICH TURNED EVERY WAY, TO KEEP THE WAY OF THE TREE OF LIFE.*" All that physically lives is now a potential source of temporary physical and emotional energy for others. We can't in any way expect a conscious spiritual relationship with GOD, to be like this worldly physical relationship between fallen souls. To properly define this spiritual relationship, we will borrow four terms that define biological relationships, and use them to describe a true Godly relationship. These terms used by mankind, define the desired and undesired effects of all inter and extra species physical relationships. The first being physical Mutualism, meaning two differing physical entities, physically benefit from a relationship and neither are physically hurt. The second is physical Commensalism, meaning one entity physically benefits, and the other is physically unaffected. Three is physical Amensalism, meaning one entity is physically hurt and the other entity is physically unaffected. And four is physical Parasitism, where one physically benefits, and the other is physically hurt. We discard the fifth category of mutual physical destruction called Predation, that cannot apply to any relationship between us and GOD. Now does GOD need or benefit from us in any physical way? No. Our bodies, the earth, and the universe are GOD'S perceived property. *We* benefit both consciously *and* physically from GOD'S eternal gift, and ongoing physical work. But GOD in no way, physically benefits from us. Can there be any physical parasitic relationship between us and GOD? Can we *physically* hurt GOD? No. GOD does physically give of himself, but no lesser soul can physically take from FATHER what is not given, or *allowed.* So, there is only our Spiritual relationship to understand and seek. John 4:23-24, "*BUT THE HOUR COMETH, AND NOW IS, WHEN THE TRUE WORSHIPPERS SHALL WORSHIP THE FATHER IN SPIRIT AND IN TRUTH: FOR THE FATHER SEEKETH SUCH TO WORSHIP HIM. GOD IS A SPIRIT: AND THEY THAT WORSHIP HIM MUST WORSHIP HIM IN SPIRIT AND IN TRUTH.*" The term Spirit here as always, means a soul's attributes, and not the soul itself. A soul's spirit is an accounting of their desired understanding and belief. Let's use these 4 terms to define a loving relationship with GOD, where we see Spiritual Mutualism, and just one specific Spiritual Commensalism describe a relationship desired and sought by GOD. We start this discussion with the last of the four terms, being a parasitic spiritual relationship. Are there souls who have a parasitic spiritual relationship with GOD, who benefit from GOD spiritually, yet *hurt* GOD, spiritually? Yes. Seeing at our creation, GOD did sacrifice to *benefit us* in gifting our lesser eternal spirited soul. Scripture also says, GOD is grieved (*hurt*) in his soul, by all sin and rejection of his loving ways. All souls benefit, yet most return a spiritual conscious hurt; so, yes, we see parasitic spiritual relationships between souls and GOD. But can there ever be spiritual Amensalism between gifted souls and GOD; where only GOD is spiritually *hurt*, and the other soul is *unaffected*? No. GOD has ordained his judgment of all spiritual rebellion. A soul may move and act through their temporary continuum of an earthly existence, not seeing or caring how they hurt GOD, yet at their judgment, both will have suffered great spiritual hurt. Now conscious Commensalism is where one soul spiritually benefits, and one is spiritually unaffected. Which we see is *only* relevant between us and GOD'S omniscience, in that we are called to come out of all our individual imperfect understanding of belief, and come into the shared omniscience of GOD. We were all gifted in the hope we would spiritually *benefit* from GOD'S omniscient perfection, to hear, learn, follow, and receive eternal life. And though GOD feels spiritual hurt, at each soul's rejection of his gifts, his omniscience remains spiritually *unaffected* by all rebellion to the same. So, we can benefit from it, but his omniscience is never hurt by any soul rejecting it. Lastly, we see conscious Mutualism, meaning both souls *benefit* spiritually, and neither are ever spiritually *hurt*. This is the greatest desire coming from FATHER, which rests in a balance of all souls living in his ways of loving created perfection; a perfection not found in give and *take*, but in giving and receiving. This is called the eternal flowing love from, and towards GOD. We see spiritual mutualism requires that all higher and greater eternal needs from, and to each other, be fully met. And in eternal love, GOD wants to share all they have, as do we. If you're a parent, or want to be, you may better understand this desire. But in our perfect loving desire to give all we are and have to our children, we also desire and need everything loving from our child in return. So, we see FATHER teaches us for our own good, to surrender all responsibility for everything we eternally require to him. And in return, he wants to give to us, as we grow to understand and receive it, all GOD is and has; being unimaginable to us. If you are given a true choice that must be made in the depths of your desires, between living the created truth of eternal life, that eliminates all new pain, loss, and suffering, and brings us eternal joyous existence; *or* a short self-reign, leading to death and total loss for you and all you love. Which do *you* choose? Desiring to live by your own insufficient imperfect beliefs, against the truth of creation, is a confused desire for life to be temporary and filled with pain, loss, and finding endless conscious hell and death. Strangely, even most believers in JESUS, will never desire the ways of mutual eternal sacrifice. It is by their pride and envy, anathema to their core desires. Most weak believers in GOD, try to have a mutualistic relationship with others in this world. But as to GOD, their relationship is narcissistic, seeing only how GOD can help and affect them, rarely seeing or caring how they pain the heart of GOD with sin. Most believers spend most of their prayers asking for physical things; and their gifts to GOD are mostly physical and mechanical, like church attendance and physical acts of charity in this world. And though these are good sanctioned things, they are secondary. Such souls can still lack faithful reverent spiritual obedience to FATHER'S perfect design. One of the many defects found in having faith in the spectrum of physical evidence alone, is by itself it is so broadly mis-interpretable. And this is why the many will cling to it. Its unanswerable questioning frees souls to hold their ever-changing differing beliefs, saying no one can know the truth. All *individual* souls in *denial* of the truth of creation, forever search for any aspect of highly interpretive data that helps allow their individual beliefs. Now our saying that our perception of it all, equals our existence is nothing new. We have Renee Descartes's (1596-1650) "I think therefore I am", George Berkeley's (1685-1753) "If a tree falls in the woods, and no one is there to hear it, does it make a sound?". Now in respect for some of Berkeley's pros and flourishes, his question properly phrased should read "if a tree falls in the woods, does it ever make a sound period"? Because it makes no sound if someone is there or not. A falling tree, causes waves of outward emanating dark silent energy through a dark silent medium. But if those silent dark waves find and stimulate a created ear, those wavelengths of silent dark resonance are formatted and sent via the neural interface to a quantum soul. Where they will be forcefully projected, *created*, and perceived in the awareness as our intrinsic inherent audible experience we all call an audible sound. It is scientifically argued whether the woods are actually there, minus someone seeing them. We know all physical realities of FATHER'S creation are there, at and beyond our sensing boundary. Yet our created visual perceptions of the woods, and sound of a falling tree, and our awareness of them, are all very different things. The dark firm physical properties of the woods, and those dark silent waves radiated outward by a tree falling, are there if no soul on earth perceives them. For the good reason that FATHER perceived to create this world to exist in all its ways, thousands of years ago. No physical complexity exists, until at least one conscious force entity, *necessarily* perceives its entangled existence; and we have no better, fuller, or more adequate definition. Our second-hand, hearsay, created, brightly seen and heard awareness of reality, are our only proof for the existence of reality we can ever have, or offer. And I find these to be the most valid, honest, illuminating, scientific answers, which give me my complete and undeniable understanding. Materialists say questions as to a soul are distractions beyond answer. But why say this, seeing *we* can answer these questions now? Please see the simplicity of the thing; questions as to the soul are never rejected for being *beyond* answer. But because these questions, no matter how long and hard they try, they know they can never find a *purely* physical answer. But see how all *their* as yet unanswered physical questions, *are* answered here and now in GOD'S truth of all creative forceful souls. I so love GOD'S conscious pure truth, as our souls are truly freed and raised in love by our admission of its existence; and we are raised more, by our acceptance of it; and raised ever higher by our adherence to it. The initial infancy of acceptance in each soul, occurs suddenly between two conceptual boundaries: one, is in being beyond disproving, and two, being our first new loving desire for it to be true. If you just now, feel any *new* desire for these things to be true, then a new infancy of the highest love for FATHER is born just now within you. In you, a new brightness of polarity just came to exist that GOD now sees. This conscious concept called faith, is required in everyone to accept any and all beliefs in anything and everything. *Everything* believed, has been accepted by faith within all people. One plus one equals two, is a very strong simple faith held by most. Yet by faith in certain emergence theories of physics, one plus one, as an abstract concept, can strangely, yet mathematically be believed to equal three. So, even math like all other conscious constructs and concepts can be bent and deceitfully woven, and used against the truth, like other tricks of the light. As imperfect understandings arise more and more, diverse faiths and beliefs appear more and more. There is truly all that is dark, silent, moving, and external to us, being the complex creations of FATHER. Yet it is only *within* our creative perspective of desired biased belief, that we truly witness and recreate it all as our perceptions, to then live and exist within it all. Yet we do not consciously exist, in that which is *without*, we live, exist, and remain within our created mind. The way in which we perceive ourselves to exist in a relative way, is affected and interpreted by what we desire and perceive to be within and without. And what of the wondrous experiential external existence promised to GOD'S own; and the never-ending-falling experiential anguished external existence promised to the lost. Such coming good and bad things, that we must recreate and perceive internally to be experienced in our awareness. All things, understandings, feelings, and desires, come to exist to us, only *within* our soul. Luke 17:20, "*THE KINGDOM OF* ***GOD*** *COMETH NOT WITH OBSERVATION: NEITHER SHALL THEY SAY, LO HERE! OR, LO THERE! FOR, BEHOLD, THE KINGDOM OF* ***GOD*** *IS WITHIN YOU.*" We can never directly observe the Kingdom of anything that is external, as everything physical and every spiritual understanding and feeling and belief, exist for us only in our internally created awareness.

This can raise the question of, is physicality real? In Dualism, the physical in all its forms, is as real as anything can possibly be. But the big question is; can many souls *eternally* share in a coming new perfect spiritually created physical Heaven, or will *one* soul start a rebellion like Lucifer did, to destroy it again. Life is not about any physical understanding most souls expect, or want, or believe. Life is born and consists of multiple complex layers and parts; yet we exist only in our subjective re-creations of all things in the biased way we desire them to be. One sees justice, as another sees a tragic miscarriage of justice. We see FATHER'S great conscious work going before us, bringing its many facets that form and allow all the possibilities of *our* own created interacting reality. Our soul of conscious quantum force projection, exists in its three-dimensional size and form of space and reach. And when in a dreamless sleep, our soul might remain within the size of the body we use. But when awake or dreaming, our soul envelopes the shape of each current expanded perception on the surface of our soul. Our body to soul sensing, and our perceived creations together, bring us a strong feeling of being one with our three-dimensional experiential qualia. Picture your soul being transferred into a futuristic robot that looks and feels fleshly like the body. Having all the same sensory systems, sending all information to your soul through a wireless interface, and synced for you to create your perceptions. And you control this robot through the same wireless interface by all your projected forces of willful desired thoughts. You would know this robot is not you; yet this is exactly how we exist now. Many hope and believe that a conscious transfer to an android will be possible in the future; yet they deny FATHER did this at our beginning, because this is not what they desire to believe. FATHER created and shared with us the perceivable dimensions of his vast physical reality in such wondrous strengths of perception, that many do not care to see GOD any more. Most only want to experience their physical senses in their illusion, and abandon of their Creator seeking only more ever-stronger perceptions of every physical desire. They push past his wondrous created reality, flocking to amplified concerts, thrill rides, and 3D movies with CG graphics, electrified colors, and booming sounds. We see the abandonment of GOD'S eternal morals, giving many the strong sensations of their more individualized physical choices and controls, beyond FATHER'S created physical ways. Those who truly find and know GOD, seek and have our different enhanced feelings; being shared eternal peaceful love for GOD and family. Having known and experienced both sides, I find my feelings with GOD to be far richer and enduring and rewarding. While every temporary physical sensation, quickly and forever fades; even as our desire for more of the same grows. Now GOD'S conscious primacy is merciful, and may be reasoned with, being consciously deterministic; whereas physical determinism is not. Isaiah 1:18, "*COME NOW, AND LET US REASON TOGETHER, SAITH THE* ***LORD****: THOUGH YOUR SINS BE AS SCARLET, THEY SHALL BE AS WHITE AS SNOW; THOUGH THEY BE RED LIKE CRIMSON, THEY SHALL BE AS WOOL. IF YE BE WILLING AND OBEDIENT, YE SHALL EAT THE GOOD OF THE LAND: BUT IF YE REFUSE AND REBEL, YE SHALL BE DEVOURED WITH THE SWORD: FOR THE MOUTH OF THE LORD HATH SPOKEN IT*." FATHER hopes all would desire to come to find his uniting omniscient truth, love, and the provision of eternal life only he can provide. FATHER says in coming in what he defines and sees as a true loving faithful repentant responsible desire, none are turned away. Our eternal love for GOD is seen, defined, and judged to be in us, by GOD alone. Beyond all reasoned doubt, the truth of eternal life exists, and it brings peace, comfort, reason, purpose, and every true answer to every true question. We see an immeasurable immovable stone in GOD'S truth; being that FATHER'S understanding, belief and love, has no changing evolving or devolving form, like we see in lesser souls. Yet even for FATHER, there are all the unknowable choices of free-will, now set within all gifted and created creatures that allow love and hate to exist. FATHER has his perfect knowing desire and eternal wisdom, having great patience, mercy, and love, towards all turning to search in desire and hope for his eternal promise. And again, please know that FATHER'S omniscience does not include other soul's thought, or acts of free-will choice, or each instance of chance, FATHER created reality in enough complexity to be beyond any and all firm knowing or prediction. FATHER created reality, and let go and looked away to a certain extent to allow for all lesser forceful free-willed chanceful things. Only when FATHER is not focused on forming and controlling all aspects of his creation, are our free-will choices, physical chance, and true love within each soul, possible. FATHER has to turn his view, to stop the desired force of his own creative perception process, that may influence or control our choices of physical acts within his creation. And by rebellion, we see chanceful and chosen disorder increases, and lives conflict more and more as individual beliefs grow and thrive. We need FATHER to not let go of us eternally; yet most don't want his *conditional* provision. Now by super-computers arrays, many view omniscience as an emotionless logical program. And it is beyond doubt that FATHER qualifies and quantifies all physical things perfectly. But GOD is no machine, they are the perfection of all needs, set in 3 great loving souls, bound in one *perfect* logical understanding of FATHER'S creation and ways. GOD is all perfect logic, combined with perfect love and mercy. These born of FATHER'S loneliness, being far beyond *fallen* mankind's comprehension. What is pure omniscient love? We are shown, it is the forever circular sharing seen in CHRIST'S example of giving his very force of life *to* all others, in a hope of receiving the same back *from* all others. It is an eternal flow of the forces of life and love from and back to FATHER. We see love and life must both flow, or die; these cannot be held onto in anyway. This can be hard at first, but easier with growth. For proper understanding, our greatest love is our greatest anchor, and we must see that in both love and life, we must hold our greatest love for the very source and definition of life and love, being GOD. Or we *must* fall by creations unavoidable design to find and separate an *eternal* family for FATHER. All we each are and have, are gifted and created by FATHER. What can it mean to love any gift above the giver? Many say those who love GOD'S ways above self, do so out of fear and weakness. But it is impossible to *love* out of fear or weakness, as love is our *strongest* force. One can obey GOD in fear, yet never love them; but one can never truly know and love GOD, and not fearfully obey them. Fearful belief alone, brings souls nothing. The many cannot love GOD or others more than themselves; so, for comfort they project their weakness upon those who love GOD in their eternal strength. This truth is too harsh for most fallen souls to accept or admit. The faithful responsible love for GOD'S omniscient truth in these last days of great confusion, is now rare. FATHER can see every strength and form of force we project in our desires. And in this, he qualifies and quantifies our desires. There is an aligned, or a divisive polarity of force in our every want and desire towards GOD. In all judgments, GOD is always right by their perfection and omniscience being the source of life. GOD is loving, compassionate, and merciful to the point of saving all who repent and want to be pulled out from among those which do not. The size and dimensions of every creative soul, is gifted in enough measure to allow all higher *loving* eternal understanding; yet small enough in total, that FATHER retains control over his creation. GOD'S faithful responsibilities towards life, are pure in their eternal ways. Knowing love and life from a physical perspective is impossible, and by adding eternal souls to their equation, their theoretical confusion grows. We see mankind's ruling triumvirate of individual *power, privilege,* and *possession* that FATHER did not conceive of; he conceived of all top-down faithful loving responsibility. The desired conscious forces of angels and mankind are real, yet we see a law, being the far greater size, force, and reaching soul of FATHER, holds supreme sway over all physical reality. The sizes and force of the Angels are far greater than ours, and their lively existence is different through their vast ethereal bodies. Yet they too find only non-existence, at FATHER taking their ethereal body away. We see the angel's immense souls and bodies, vary in measure and size, far more than for mankind. Yet the conscious force within FATHER should be seen as limitless for all intents and purposes, as he resides above us all. We cannot define or understand the boundaries that define the realities of FATHER. But we know we do not own or control the physical points in this body, yet by it, we forcefully act, react, know, and influence the external creations of GOD. Yet there is no way round the limitation set in us all. We see that within FATHER'S universe, there are no conscious or physical limitations upon him, other than those set upon him, by him; being the boundaries of who FATHER eternally desires and knows himself to be. Now our awareness of sight, sound, smells, and taste, are sensed only in the head area, where those interface sensors are. But sensing these strongly in the head area, should not mislead anyone as to where our seat of awareness resides. Again, our awareness and sensation of an injury, is sensed at the injury, not in our head. And again, the absence of all information, received through the entire body, we call dreamless sleep, or physical death, are the same thing. One is short and recurring, the other is singular and prolonged. Now interestingly, Scripture (KJV) uses the religious term "Resurrection" no more often than it uses the scientific term "Regeneration". Most worldly Christians believe at physical death, their soul is then aware and transported to the current Heaven, but this is false. Scripture says at GOD'S discretion this does occur to Martyred Saints and a few others being given early resurrection into an angelic body. But most saved and all lost souls, find a long sleep until the Millennial Reign or their Final Judgment. For an example of seeing death, and nightly sleep as the same, JESUS said a man called Lazarus was asleep, who was in his tomb for four days (four, for important reasons). Then JESUS calls him to awake again. How can we reconcile, that JESUS said Lazarus is asleep *and* dead, yet believe he was also alive in Heaven; meaning Lazarus was returned to his harsh earthly life, after he had received his eternal reward; and it is appointed unto men, just *once* to die. Also, Lazarus upon being resurrected, gave no account of any awareness of those four days to anyone. It is shown that our soul's free-will desire, controls the motions of our physical bodies. Mankind takes this, and claims a free-will right to physically live and believe however they desire. Do you see that to argue for or desire, what are in every way, non-scriptural *individual* rights, is to hold a belief that mankind has a primacy of self-rule. GOD gave us a choice to obey; but no way to give rights to ourselves or others. if we add GOD and eternal life to our belief, we see we have zero individual rights. We have only our ordained responsibility within GOD'S perceived ordered creation to perpetuate eternal love and life for the Kingdom and the Family of GOD. Which defines *all* rebellion as a rejection of the same. A soul, by their every conscious avoidance of GOD, declares their conscious freedom; and by extension declares rights never truly given. All individual physical rights, are privileged considerations for certain broader lively allowances. It is impossible to show an individual right that comes to anyone, except by an *earthly* authority. No one was ever born with innate rights. We have been given free-will, which does not translate into any GOD given right for any-thing. Rights originate not from matter or energy, they are conscious malleable concepts. If life is gifted and granted on the basis of the ruling desires of GOD. Where, how, and why, can one claim more than that which is given and intended. Rights only appear between souls in a fallen world. Individual *rights,* mean individual control that is not to be infringed by any greater governing force. We see that if physical life flows from GOD, any self-declared contrary rights, are taken in rebellion to that source. So, do I then disavow my enumerated guaranteed rights in the U.S. Constitution? *I do not*! In this now fallen world, that would force me to bow to its evil. My religious liberty to serve GOD is my necessity. And as stated, came first by a command from GOD, and now further enshrined by men as the first supreme article of our Constitution. It is a required earthly protection, as liberals want to deny all religious liberty. All responsible commandments towards life are *gifts* from GOD. And individual rights of power, privilege, and possession, are *taken* by mankind. Do not hear in this, any indorsement for any evil Godless Socialist or Communist Government, it is not there. I am governed by GOD, and no other. The more powerful and godless a government is, the less liberty to openly serve GOD I will have. We are all gifted souls under GOD'S supreme law, not worldly law; and we are called to eternal responsibility towards love and peace with GOD and Family. No one has a right to ask or take the provision of life from anyone, to then spend it on things of death. This is what this world sees as entitled physical provisions. This sentiment exists equally in both the lost rich, and lost poor. Most who have gained wealth under worldly laws, feel an entitlement to such. And most who have not found any wealth, feel the need for a new law that grants them those same things. The quandary is there is not near enough wealth or production in the world to give to everyone a median standard of living the world at large has attained; let alone the luxury of the rich, beautiful, talented, and ambitious. There are no rich or poor rights of worldly entitlement. These are all rebellion to GOD'S authority. In our world filled with imperfect divisions, every possible system of governance conceived of or instituted, has *never* begun to remove the class distinctions of power, privilege, possession, beauty, gender, age, religion, race, intellect, and endless cultural preferences. They just shift them around, as godless liberal authoritarians work to over-rule the beliefs and ways of others. Obeying and preaching all the Godly commands is now called bigotry and hate by evil liberals. Should a liberal government select all marriage partners, homes, jobs, and even children, to make all life equitable in *their* eyes and in their biased judgment. How can those who hate GOD, even know what is equitable. They can only create a broader, more unfair struggle, under a centralized far more imperfect rule. They only make everyone equally unhappy; except for the few in power who are not subject to all the rules they impose. In this increasingly confused world, accusations come from all souls at other souls. This world exists under the feeling of time, and not the eternal, which raises the insanity of urgency and unmet temporary expectations. In Scripture's described eternity, we see no desire for rights of privilege, power, or possession. We will receive all we need for eternal peace, joy, and life, as it all comes from GOD. We will have our desired faithful shared responsibility to preserve and perpetuate life as one. These are our endless treasures in eternity, where time does not exist. All lively created perceptions of reality that fall beyond GOD'S truth, are all individual illusions of grandeur. If one is grounded to their external physical illusion to a point of it being their reality, it is so difficult for them to see the true, far higher enlightened spiritual things beyond that desired illusion. Lost and losing souls, hold on to what they think they have in fear that it is all there is. But it is all only a dark silent vapor and wind that comes suddenly, and leaves them just as suddenly. These trying expanding moments of each soul's existence here in this fallen world, are teaching moments of a great gift, all to make our choice for or against GOD'S truth of love, and love of truth set before us. An offering made by GOD to all of a pure bright love to be wanted and treasured, or unwanted and rejected. Free-will choice cannot create truth. The free-will choice to inwardly believe what one *desires,* is just that, and nothing more. Please read that last statement carefully, as it contains the word "*desires*" and not "*demands*" as the one pure truth of love, cannot be demanded. If any soul says GOD has their truth, and I have mine, GOD says, let it be so; but all should consider what that truly means. Individuals may create their individual truth, but it rules nothing beyond their own mind, and it can't create or sustain their physical reality or eternally bond to other souls. The ability to believe what one wants has nothing to do with being true, pure, right, or just. It has only to do with revealing and declaring our core desires for what we want to be true. There are all the wrongful worldly pulls of physical desire, reaching into the quantum realm arena, pulling at every soul. These are temptations to become an individual with power, and in essence play a god. Most souls grasp at the air in this world for what; where does their search eternally lead. Humanists think if they convince us they have falsified the existence of GOD, that this world will find something wonderful. All souls are wrong who think I over state here, that there is an *epic* battle of desired cultural realities raging around us. Please describe for me the reason, purpose, and joy, and show me where is the answer to life and love that comes by a belief that life is temporary without eternal meaning, and leads only to eternal death. The biggest question of all, is does life win in the end, or does death win. Why should any fight and desire for death to win. Please, this is not to be confused with, or equated with Pascal's philosophical wager on believing in GOD, as we do truly *love* and live for GOD'S ways. FATHER holds his created universe and all life; but in CHRIST, HOLYSPIRIT, the Arch Angels, Seraphim, Cherubim, and all Holy agents, FATHER'S command to perfectly affect and influence the courses of his physical reality are implemented. And though GOD can heal the *physical* aspects of our personal lives, yet rarely do at this moment. We see FATHER *cannot* control a single choice made by any gifted free-will soul. GOD is about teaching and affecting our eternal path and picture of belief as one. And never about the *individual* physical desires most pray for in not seeing GOD'S grace is already sufficient. Imagine if someone had a small mysterious resonating glowing sphere that if held, gave one an unimaginable physical elation this world had never felt before. And they said they would let me hold it for a short moment. *I* must refuse; but many would pay to hold this sphere. Nothing eternal comes from temporary physical elation, and at it being taken back, one's physical state is only more sorrowful than before. Nothing temporary or physical can satisfy a soul that truly knows and desires eternal life. Materialist insist life is temporary, godless, and indoctrinate their belief into little naïve children who need GOD to find eternal life. Humanist seek to kill and drain all hope of the eternal perspective, and replace it with a lesser value of life, through beliefs in abortion, physical abiogenesis, an old-earth-new-species evolution theory, and assisted suicide to end suffering and confusion. And they pardon the godless vile confused souls who eye and lust for and rape or murder children. Under FATHER'S commands, I allow none of the above, or a death penalty administered by the state, being murder. If a soul is guilty of murder, and the surviving loved ones were allowed to stone them to death; I find this ungodly, but not murder. To find the pure truth, we must see why truth is only found in GOD'S loving eternal perspective of life, being strength for our soul to resist worldly sensations. In these last days, most seek only ever-more complex physical sensations to feed their desires, this is now their central purpose to exist. And as all evil physical desires become engrained, a soul's turn back to the ways of GOD becomes more difficult, as they feel disfigured in trying to deny those strong desires. Imagine if GOD took away your very sensitive body, and connected you to a far less sensitive body, and erased your memory of all strong physical sensations. You would be free from all curses of physical addiction. Has fallen life become mostly about the pursuit to experience the ever-changing qualities and quantities of physical pleasures, excitements, comforts, and luxuries. Is it these, and not a higher spiritual love for all brothers and sisters in CHRIST and eternal life itself, that most see as their great reason for life. How can one blinded to GOD'S truth, understand anything, being in ignorance of the fact that all external things exist *only* by GOD? And for us, all things only exist within our conscious awareness. I pray at this point, in just starting this book, that you are sensing unity and oneness of understanding. I pray for you to sense the many more relative aspects to come. Most souls, existing through an individual physical perspective, *can* say this book contains little science, and lots of theology, because they can't see that in GOD existing, science and theology are one. We can no longer separate them once we see how all physical reality is bound to all desired believed conscious forceful projection. Reality must be explained from both ways; and any denial of this fact is a formal rejection of GOD. The purposed creation of GOD, is impossible for *me* to discuss, minus its loving purpose. I hope you are understanding and gracious towards *my* literary quandary. I've struggled to find the best way to say these things for years. How can we physically show spiritual things, and declare there are the searchable higher reasons and purposes for all we see. Now confused believers in a god, allow for a physically conscious brain, *a* sub-conscious brain, and also an eternal spirit *or* soul. This confused description of our conscious awareness is desired by the fallen proclivities of all flesh. But in holding such blurred confusions, souls then reside in a self-made religion, worshipping a self-designed god, who happens to greatly reflect the desires and appearance of the worshipper. Do you know the term "*Pagan*" means a soul who creates a personal god. All have heard the terms: the soul, the spirit, the mind, the conscious, and the sub-conscious. But we are only one, being our eternal soul, fighting or enjoying our physical desires. Our *soul,* is our essence and agency, being our awareness of experience. And the term *spirit*, speaks to current descriptive beliefs, attributes, and proclaimed purpose. Such as a soul being of an angry confused spirit, or a soul having a spirit of joy and holiness. And all so-called sub-conscious activity, is FATHER'S pre-determined will in his perceived work to sustain physical life. We are a soul, having our acquired description (spirit) of what we have desired and chosen to be and believe. And we have only two *possible* directions of connection: one is between our quantum soul and all the physical information moving back and forth through our relaying interface; and the other, to the conscious spiritual world. But a confused belief in any awareness by and from within a brain, allows souls to believe many crazy things. Like their saved soul is not responsible for the evil things the brain and body enjoy. We must ponder our free-will control over this temporary vehicle, seen as our body. Will your hand to move and dance before your eyes, and it does. Weaving effortlessly back and forth up and down in and out, faster, then slower, adding and changing all rhythms. See your controlling force that flows from your soul; you desire it and it moves, you command it and it obeys. This forceful control is your desire to exist. Your soul generates forceful information patterns that cause these otherwise inanimate body parts to move. Disconnect the soul from the neural system, and the brain is just wet lifeless meat. How can one *believe* that wet grey brain fat, creates our bright 3D senses of visual, audio, and other qualia? Or believe it is responsible for our other-worldly experiences of thought, belief, contemplation, and emotional of want and loss. Or orchestrate and feel the body's commanded movements in any other-wise unreasoned faith. That is a most *unreasoned* and profound faith of belief to place in a piece of fatty meat; this is the greatest of cults. Our soul's wondrous influence is continually seen and felt, even as it is denied by science. Yet by all new evidence, science sees more and more that our awareness does not reside within a brain, nor in the quantum information arising from the entire body combined. They are forced to see there must be a quantum aspect or source of our conscious awareness. We see the physical correlations of sensing and relaying forms of chemical electromagnetic information; but wet neural meat can never perceive anything in awareness. And no soul has ever perceived or known anything, without three elements: being forceful desire, some sensed physical information, and our created perceptions within a physical medium we call the aether. Physical science fails to even start to explain how or why wet brain meat could ever know anything; but they admit dry brain meat cannot know anything. So, is water and chemical charge the secret sauce poured on a brain meat to trigger awareness. If so, maybe a wet type of computer can become aware? We see pure distinctions between the animated and inanimate, the lifeless and the living, and the aware and the unaware. We see the Dualism of conscious forceful patterned desire, projecting upon the inanimate, is the only researchable equation. If we truly see through GOD, we are shown many things that point us in the one true direction. The first question is, was there a conscious, or non-conscious beginning? But can you see that any answer or proof of a non-conscious beginning, can never be physically found in itself. It can only be found in disproving a conscious beginning. Yet the conscious force question is always just denied, leaving an endless futile search. But why? Because this conscious force question has that one feared answer they do not want to find. Being we are creative perceiving souls; that meaning there must be a supreme soul who perceived all things external to us. Trying to avoid the question of GOD, and our accountability, cannot make this question, or its only possible answer, or our accountability to it, disappear. No souls search for things they desperately don't want to find. I desperately search for these wondrous things, as GOD'S enduring love and immutable promise denies us nothing good and eternal. The only way to avoid the true *conscious* force answer, is to deceitfully reject, avoid, and change, all conscious force questions; which is what materialist do. We know asking any *physical-based* conscious question, is asking a false question. We all are gifted in our sufficient conscious measure and capability to grow to see all things clearly and truthfully, if we so desire. Our eternal physical half rest with GOD, who alone are self-sufficient, self-reliant, and responsible over all lesser life. Many say their mind and self would be lost in such a belief; and this can be true for them. But I know myself and who I am very clearly in seeking my oneness with GOD'S loving ways. I truly desire GOD'S ways; this is who I desire to be. I gain everything good, and lose nothing of myself. Minus GOD, a soul's desires and beliefs change constantly through life, yet they feel like their self. Yet most will feel diminished, in living as one mind in the eternal ways of GOD. These need divisions in themselves from others, before they sense themselves clearly existing. One says I am like this, and another says I am like that, and they both derive a stronger sense they exist in their distinctions. Now some ask, if we are created so infinitely small, how can GOD love souls such as us;

can we love an ameba? Such questions come from a confused temporary small slow physical perspective. A *created* ameba has a limited purely physical purpose, with no eternal understanding or soul, or possible love for us. Our far higher properties and purpose, being we are gifted an eternal soul in sufficiency to know and love GOD, that distinctly elevates us far above all lesser purely physical life. We need to brightly see that love is not a physical aspect having size; it is spiritual and conceptual. It is defined only by its existence, purity, and its responsible faithfulness. So, when mutual love is born, it disperses through and fills the giver and receiver in all its forces of sacrificial desire. Do we love our tiny child less because they are smaller? Of course not; pure love is pure love. Our tiny four-year-old may love us more than our big teenager. If we know and love who GOD *actually* is, we are then a soul of understanding and desire, making and projecting that true forceful loving choice. All GOD sees, is we desire and love *them* for who they truly are in our faithful emulation. In GOD'S great joy, they can only sacrificially love us in return as their dear child born of their love, truth, and belief. I wonder why most religious souls hold a view that GOD needs or desires little from us, by being GOD.

We are made in FATHER'S image and likeness, having all his emotional, creative awareness, and outer physical attributes. Scripture speaks of GOD'S pain in being rejected, and being grieved by the loveless rebellion against their created ways and loving attributes. We are told, we are given the greatest power that has ever, or could ever exist. Being the power to love GOD in our learning and knowing who GOD truly is. And this being who *we* truly become in our hope of many souls finding all good eternal desires together to be loved back. And we are created to know all things. John 1:12, declares that to us is given the power to become the Children of GOD. In this promise, we can in love, rise to the great heights of being in the very presence of all that is; or in rejecting *any* one true aspect of GOD, we fall into a bottomless realm and dark pit of increasing loss. I am amazed and wonder by what strange conscious powers, can and do souls deny the goodness of GOD and tie themselves to false things. In this book, we do *not* teach those created ways of eternal life, set in Scripture. We believe each soul must search the Scriptures for life, cleansed and led by HOLYSPIRIT. If we all receive one identical interpretation, we are taught by GOD; but if any receives an individual interpretation, they are being taught by themselves and the lost. We must peripherally touch upon certain doctrinal ways of eternal life for context; but we focus as best we can on all means and ways that bring all conscious forceful acts of creation, spoken of in Scripture. Scripture defines FATHER'S omniscience, omnipotence, omnipresence, his loving ways of life, his eternal perfection, and all his perceived works of purpose. Scripture says GOD hates mankind's *constant* evil efforts to water down or alter his Commands. We share *our* spiritual revelations, and all *we* know and affirm in the court of the spirit; being all the motivated opportunities and forces of GOD, over all created physical evidence. Most souls can't begin to see just how far this world, and all worldly churches have fallen from GOD'S true ways; being a difference we do strain to describe. Mankind revels in their physical false knowledge, but all for naught. Their only true physical knowledge, is to know to keep putting stuff in the top, so stuff keeps coming out the bottom. Every *true* description of *fallen* mankind is not pretty. Now later, we examine the Scriptural chronology of creation, being a scientifically required order in which all things were perceived and made manifest. This being relevant, as this ancient chronology beats immense odds in aligning to the required step-by-step process to create the complex reality of life we see. We pause to honestly define the term *Religion*. This definition I received from GOD, and my authority is to teach. The word or term "*Religion*" in all honesty, is rightfully ascribed to those things that each and every living desirous soul, believes, declares, and vows they must do, receive, give, and hold. Being each soul's definition of life lived to its highest ultimate true purpose, in honoring a spiritual, and or, physical life; under *the one* highest knowing source, or any knowing source, or honoring physical life by a physical source alone. Religion means each *soul's* strongest guiding core belief. Everyone's beliefs are their religion, and every firm belief system, is a religion. By this true universal definition, all the belief systems of humanism, materialism, secularism, and atheism, have always been and always will, in the *eyes* of GOD, be sects of free-will religious belief. If you deny this, and say that a belief of reality, minus a god, is not a religion; we find that to be an ignorant denial. Everything believed by every soul, is held as truth by their faith. Not one existential aspect of reality has been proven by or to anyone beyond all doubt. Search the W.W.W. and see the established Church of Atheism, the established Church of Humanism, and all the rest are all there in their practiced professed faith. Yet we see by biased liberal sources of public information, their constant forceful labeling together of *only* all beliefs in a god, as being religions. Often implying these are superstitious, problematic, and often extremist people. Secularists think they get to exclude *their* beliefs of pure faith from being a religion, by their own simple narcissistic declaration. They say their unproven, *un-provable* beliefs, are not matters of hope and faith, but of science. And that their search for their cause and reason for life, is not a search for their lifeless god of a meaningless religion. Surely, such souls are not the picture of mature intellectual integrity. Theirs is a typical form of insanity, and their views of GOD'S Children originate from their baser physical instincts to control others out of fear, envy, pride, and hate. So, all beliefs are forms of religious practice, including science and politics. Politics is no more than a conscious battle for control to preserve and propagate one's religious beliefs. Politics when exercised, is the art of defining and attacking the lines of least resistance in front of us all; while building greater lines of acceptable alliance behind. These front and rear lines are usually void of any pure truth, and formed by calculated compromised desires. A uniting voice is a forceful artist; and *world* leaders having differing beliefs of science, politics, and religion, are to differing degrees, of the same artist mold. We see when an *individual* seeks to prove anything to themselves or others, they are biased and hold to anything that appears, even in confusion, to support their desire. They are equally biased to deny and disbelieve all that even in a stronger way, does not. And as they do these things in attacking their opposition, not one truth endures. We offer as evidence, that not one country, government, worldly religion, or physical theory, has ever stood for one day in its original form. Since the fall of mankind, every belief mankind builds, they immediately begin to tear down, change, and replace. They think they are always improving but they are not, as this is their spirit of rebellion. Mankind's modern theories on reality, can no longer even stand up to then be torn down. For a *physical* theory to be taught in most schools, all that is required, is for it to be a godless possibility. Consider Multi-Verse, Abiogenesis, and Evolution of Species, all being religious doctrines of impossible unprovable *faith*. Each, *initially* shunned by a huge scientific consensus; yet now they are broadly worshipped. There is one fact that all must agree on, being not *one* thing has been proven beyond every doubt to everyone who has studied it. If we honestly apply this test to our own beliefs, we see the strength of our own biased faith. Physicists, Cosmologists, Neuroscientists, and others, say they *think* they are on the right track, but admit that at any moment a new finding could send them back to the drawing board. So, they offer no warranty to their follower, where we do. Our belief of understanding cannot ever change. Within their theories, we see a blend of two parts: one being biased mathematical probabilities; the other, a biased speculation of hope, faith, and desire. And their current theories are mostly biased speculation and hope, and near zero probability. We hear scientists say things like: we *hope* to find, and we *need* to know, or we *want* to find this or that, or we are *trying* to invent a way. But we know that hope, need, want, invent, are all terms that show great desire, and all desire is bias. Now worldly knowledge has two parts: one is mankind's limited sensed data and their slow small *desired* interpretation of it, that allows them to assume the things to *support* their desires. Two, is all they can't sense, and their interpretation of all they don't want that doesn't support their beliefs. If you can see any probability for a conscious force primacy over the physical, you must fervently pursue it. Consider that since a single omniscient truth must exist, it must exist within at least one supreme soul. And by this one truth's existence, it must be of an unknowable value above all that exists. So, if we desire to know this one pure truth, we know it can only be received from its one true source. But most don't want that, so by them, it is ridiculed, defamed, and rejected. Now the embodied pure truth came back to this world for about 33 years, around two thousand years ago. But this world only wanted the pure truth's physical healing and provision, without any spiritual corrections or adherence to its perfect ways of eternal life. So, they took the embodied loving truth of eternal life, and beat him beyond recognition, and nailed him high upon a cross of shame. So, we ask, how can the truth of omniscience be shared with all, when it cannot be forced upon free-will. Can it be gently cast among all peoples, blinded under the spells of their own physical desires, to be misinterpreted and never truly seen. Or can it *only* be held up high and greatly illuminated from afar. Drawing only those, who in themselves find a true wanton desire to look up and go to where it is. We know that climbing a ladder of knowledge built from purely physical evidence alone, is way too short a ladder, and no one can ever reach the higher answers in that way. *The truth must be seen to be heard.* But by the omission of conscious force in research, a true *inability* to do so exists, using only 6% of the physical half of all the evidence. Now we know that the left and right sides of a *conceptual* line can never be unified or divided, there are simply two conceptual sides with no true center. And the two sides of physical reality seen in the quantum classical realm distinctions, can never be unified or divided either. Once we see two distinct sides to anything, a created conscious division has occurred, and no physical theory can ever explain a consciously created divide. Now, under our "Antithetical Probability Principle" do these two real and conceptual sides of all realities rest in mutualism? Or do they exist in conflict? We see individuals constantly project divisive distinctions into the conscious, quantum, and classical fabric of reality. And our "Forceful Belief Division Principle" predicts this confused disorder we see. So, in every way, we see we need a "One Conscious Force Belief, over One perfect Physical Creation Principle". By his desire, all external conceptual antithetical aspects, such as day and night and male and female in FATHER'S initial creation, stopped at a perfect place. But further antithetical concepts desired and believed by all lesser souls are not perfect. And after GOD'S final Judgment, if the eternally lost should forever try to find the center of the line of that final conscious divide, to cross over to the eternally *found* side. They will only descend into their own smaller slower consciously perceived insanity. One may conceive of and believe in abstract forms of contrary concepts. But unless you can divide the *entire* underlying entangled reality that concept applies to, all it is, is a false belief. By GOD'S supreme aspects above ours, no one can ever find the truth, except through them. By seeing only the physical realities, one can feel as if they are a physical part of reality; but in pure Dualism, reality forms through us. Consider how by the existence of many souls, all their differing beliefs are not a search for the truth, but the entropy of their sanity. It is all a disillusioned struggle against the forceful eternal truth of life. The truth has become so obscured in this world, that if the eternal separation and protection of the truth came down, and all imperfect conscious beliefs washed over into it, the one pure truth could never be found by anyone ever again. The one perfect truth, can never be compromised or added to, ever, except to its imperfection. Can the source and holder of the one truth of all reality convince and conquer free-willed rebellious desires? No. The perfect truth held up beside the endless physical desires in this world, is seen by most as arbitrarily judgmental to this world's prized individuality. The true way of eternal life has a fascinating weakness, in that it holds no force over the prideful choices of free-will belief and individual desire. Yet FATHER controls the required physical half of reality, so he wins. The one truth requires that each soul *always* sees the conscious and the physical as dual aspects, separate, and warring for control; and that our soul has a responsibility to keep the body in perfect subjection. GOD'S truth of reality is not in any way an individual or democratic or humanistic principle, it is what it has been since its creation, being pure and eternal. So, GOD'S truth can only rule and dwell in his eternal Kingdom, and only over his loving inhabitants. It dwells intact nowhere else in appearing antithetical; just as the pure concepts of light and darkness cannot co-occupy or co-exist in any single place of reference. If one's desires and beliefs about life and reality are built upon a physical primacy, their soul's force polarity is antithetical to GOD. We can choose pure truth and light, or pure darkness and lies, or any unaccepted mixture of both. Most souls in their early effort to change to light, stumble back into darkness and shame on occasions, and maybe even flicker at times. But this decreases with growth in GOD'S love, truth, and faithful responsibility, being the roots of spiritual wisdom. We must be together and pure for strength. We all must hold each other up, or we all fall. Our temporary set-backs are referred to by GOD, as backsliding and falling of our countenance. Being a certain loss of brightness within our soul as it is seen by GOD. And the dreadful Scriptural term "*to fall away*" means one's irreparable loss of truth and brightness, best described in Hebrews, Chapter 6, (KJV). We see how so many beliefs among so many souls, brings so many false definitions to so many things. And we see the many reasons why GOD'S eternal truth and love must stand apart, unique, and high above all temporary things falsely called truth or love. As to our holding GOD'S truth, just one believed inference that GOD'S truth is not perfect, can cause a soul to fall. One perfect soul and one imperfect soul, with enough contact, will change towards each other, and neither will be perfect. GOD said to flee from sin, lest we be overcome. We must lose all imperfections to live in GOD'S perfect truthful ways. By GOD'S omniscience, to them there is only their one pure truth, and one pure love.

GOD'S love is relative to their truth, being eternally bound together and forged as one. The one pure truth of love, and the one pure love of truth, leading to shared eternal peace and life. Separate love and truth, and they can never stand alone. If you separate these within yourself, they both fade into the confusion of endless lesser definitions. Now, we see all *cause* as having a power of possession by having influence. And one strong rule in all physical reality, is we cannot possess or affect anything physical, without it in its own way, possessing or affecting us back. Science keeps finding new evidence that points to *design* playing a role in physical reality; so, they conceived of a biocentric view to straddle and blur the truth that GOD exists. They see our consciousness is entwined and holds influence over the physical, but they desire a soulless conscious knowing guidance to have its affect in the creation of all things. And biocentric beliefs see all consciousness through a physical perspective. They can then open the door to the influential force of universal conscious guidance slightly, holding it tight against their foot to view a sliver of this stranger at the door; who they will not allow in, as they war against GOD. Around two-thousand years ago, GOD'S mature forceful soul and mind came down to this physical world, adjoined to the flesh of a boy to grow to be a man. We saw JESUS projected desired forceful control over the physical realities around him. The world saw the creative ability in the projected forceful desires of his supreme will, as he suddenly restored sick and dead fleshly aspects; and by his words of desire, he calmed a raging stormy sea and sky in an instant. And in those days, there were many Disciples, who in growing close to GOD, held GOD'S love and truth to the extent it opened a door, allowing them to be conduits of the same forceful effect of healing many people. But over the ever-changing moment, these souls passed from this physical realm, and mankind's restored desire for this conscious force principle faded as those manifestations ceased. Quantum mechanics when seen properly, does not lead us to any model of infinite universes. It shows us there are two distinct realms of lively existence in FATHER'S vast created reality. And we see a huge number of gifted lesser *creative* spheres of life, being billions of souls. And from this firmer physical side, we see these two distinct, intertwined, overlayed ethereal and firmer realms of lively existence within one physical aether. On the firmer side, many hold a secular religious hope and faith that someday, somehow, they might unify these two realms into a single physical reality, without needing GOD or an aether. Yet these same many, can leap to entertain a purely conscious reality alone, where all the physical world is a conscious illusion. By their failing to physically account for so many things, secular mankind is pushed to accept insane notions like a Holographic Universe, and the Anthropic Principle, or Multiverse Theories, and ever-more insanity. But please stop and see that any idea of endless-universes with life, must each then have their own fine-tuning problem, and their own quantum and classical ambiguities. So, such a belief and hope, now needs an infinite number of differing theories to unify those infinite finely tuned universes. This does not solve their initial insurmountable problem, it just multiplies it by orders of magnitude beyond measure. We forcefully use and control the body we are adjoined too, to sense, affect, and alter the realities near and around us. We also need a body's sensed patterns of quantum information to create our representative perceptions, and then find our experiential awareness in them. We all consciously sense for us to see, hear, taste, smell, touch, desire, learn, understand, ponder, and to believe in faith and awareness. Yet many give the credit for all this, to a lump of wet brain meat. It is strange that those who believe in a pure physical beginning, ask how could there be the great forceful soul of FATHER, consciously projecting his patterned forces of desire, causing his physical points of reality to align and consist, forming his immense firm created perception; it is impossible. And all who know GOD, reply how can one believe a vast wondrous orchestrations of Abiogenesis, and all the finely tuned mysteries we see, and say all this vast profound amazement could occur on its own, minus purposed reasoned forceful guidance; it is impossible. From where does all force originate? I find a real forceful creative GOD is no more astounding, than the fact that you and I are here now, creating our own perceived particle-wave based experiential awareness. All we are talking about is a scale of conscious size of desire, force, reach, and understanding. A *dis*belief in GOD, is a belief that there can be no soul, that much greater than ours. What is so impossible about that, we see and accept that a scale of the lesser to greater applies to all thing in reality. Secular mankind accepts an *unreasoned* mystical, infinite increase of mass and energy called multi-verse, being based upon a math error; yet many ridicule souls who believe otherwise. Awareness is a soul's desirous ruling forceful creative aspect; it is as the old expression says, mind over matter.

Now how does the created universe and Scripture together, interpreted by HOLYSPIRIT, define the sum of GOD? Before FATHER divided any souls out from his own vast soul, he was the only self-aware knowing desiring soul, who had only his internal dreaming existence; and he still has his *internal* awareness *within*. But FATHER desired and dreamed and spoke forth out beyond himself. So now there are his internal and his gifted and created external projected qualities and quantities of perceived awareness. FATHER is self-sufficient and has no external physical needs. So, if he should desire, FATHER can exist in his eternal lonely oneness. And in seeing our own Dualism, such a lonely existence vaguely can be imagined. Prior to the very first moment of FATHER'S first gifted portion of his soul, he was alone, not seeing himself in size, reach, or power. Being big nor small, looking outward and finding no other and nothing to compare. Now consider your own soul again, being a minute gifted measure of FATHER'S forceful awareness. But what if *you* did not require GOD to provide you with a body, his aether, or his creations? What if every tiniest primal point within the body you use, was yours, and every point aligned to *your* every stronger more complex projected patterned desire. You would feel your soul's control over not just moving your legs, arms, hands, fingers. But further down to every cell, atom, and every singularity, or point, or string if you prefer. You could effortlessly desire, believe, and forcefully perceive to alter any part of your non-conscious aspects. You could look at your finger, and perceive it to not just bend, but lengthen or bend in ways it currently does not. You could know the composition of steel and perceive the quantum structures of your finger to realign as steel, in knowing, desiring, generating, and projecting that patterned force to align it just so. Whatever you desire and consciously see, must forcefully align to become physically real. Not just within your abstract awareness, your projections could extend to your physical aspects and appearance. Now imagine you are truly alone, having eternal loneliness. And being forceful enough with near countless points to work with, you could speak forth outwardly, and perceive a created physical reality in the direction of being *infinitely* smaller than yourself. You could create amazing things by layers of perceiving a vast complex work. And as long as your hands hold it, you would be consciously entangled to the sphere of it, before your face; as long as your desirous sustaining force flows through it all. *If* you were as big as

FATHER, you could eternally sacrifice, partition, and gift many adequate size unlearned free-will souls. These soul's essence of awareness, would *have been* a former part of you, so they too, are eternal. But in being eternally gifted, they would be blank unknowing souls, with a learning and believing potential. For the physical requirements of their created perceived existence, first you would forcefully speak forth vast numbers of points to perceive an aether, then build a perfect environment, and then bodies for these *gifted* souls to exist through. You could do this all in a step-by-step, controlled way. And if these souls came to love and need you in the eternal way you eternally love and need them, you would never be lonely again. But *if* you gift these free-will souls to find love, it will introduce a "Free-Will Antithetical Principle" and an "Uncertain Probability Principle" as these free-will souls might rebel against your ways and desires, and reject you. Can you just give more and more of yourself to eternally satisfy all their ungrateful individual desires. Could you remain rejected, lonely, and abused in your needs, until you give them your last measure, and cease to exist. But in sacrificing nearly half of yourself, laws of proportionality and conservation say you could not sustain such a creation, and it must dissolve back into you. So, a true real GOD can only give and allow so much. Now when we peer *up* through the quantum gateway into the higher quantum side of reality, seeing the small and smaller created particles; for most who deny GOD'S existence, they falsely assume they are looking *down* into the small dead-end foundations of a purely physical reality. But the quantum realm, like this physical realm, is an entangled continuum of all its relative aspects, being a realm of lively existence. The quantum aspects *above* mankind's comprehension, form the great quantum complexities and continuity of the higher realm of Heaven, under the distinct laws and consistencies found on that side. In looking up into the quantum realm, we are looking *up* at the foundations of all that exist down here. All things perceived to be firmer and condensed, are formed and dropped down out from that higher realm. This is the true directional perspective. In a proper eternal perspective, we look down from the quantum to view and know this physical existence, as GOD does. To believe you look *down* into the *dead-end* small physical aspects of mankind's standard model, you must see through a small slow physical perspective. The quantum realm and its laws, where our souls reside, is woven together in all six dimensions, but this physical side, minus our souls, is woven together in only three. The quantum realm offers the Dualism of all awareness, while this physical realm is the adjoined firmer side of GOD'S two created realms of reality. There is much to be learned from Scriptures that say mankind is created in GOD'S *image*. Meaning we are reflections of FATHER'S creative attributes and qualities, and not just his external semblance. Our learning potential is amazing once we see our creative conscious knowing is limited only by the aether within us, and not the neurons in the body/interface. Imagine the spiritual, intellectual, loving relationship with GOD and others, we were each in hope, created to attain. See your understanding expanding and growing in broader purity towards GOD and their Holy Angels through eternity. Back in 2008-9, hearing of a rising interest in biocentrism, for a moment I thought it might be a step that opens more souls to the omniscient truth. But having witnessed 60 years of life, I asked myself how can I have such hope. The far greater free-will powers of desire and deniability within individual hearts, are not weakened by such a theory. The omniscient truth of all Scriptural Science has always been easily denied by every simple desire to do so. I have a sadness that few will want or hear GOD'S truth I try to share in these carefully chosen words. Any perspective that denies GOD'S full truth is anathema. Yet such perspectives comfort and seem wise to many. In seeing this spectrum in mankind, I find my comfort among a very few; and those who have faith in mankind, find comfort among the many. I ask all to consider and allow the motive of not a worldly love, but *eternal* love as the purpose for creation. Eternal love is the only love that brings souls into an eternal shared focus. After mankind fell from grace, their love now starts and fades in endless lustful temporary prideful and envious ways. And all worldly love must be allowed to temporarily continue in its diminishing courses and returns, as GOD saves all found eternal love, out from this conscious this mire of hell and death. Now please consider how sharing the one Godly sacrificial responsible faithful love, is our purest evidence of free-will. As in every *individual's* way of life, we see only endless *subjective* motives applied to all their interactions and actions. A choice to become one, far outweighs a choice to be random. So, a misalignment now increases between the two great divided realms of existence. Mankind's mysteries and wonders of life are all misaligned as their endless desires rise up through their subjective false perspectives. Fallen mankind's individual lives, exist as false fragmented temporary memes, seen by them in a desired way to support their beliefs. Why do they go from one physical theory to another searching for truth, when that is the definition of insanity. We must see the evidence of GOD'S desire. What result has FATHER said he works towards, and will accomplish? And even how and why there is only this narrowly defined, painful way for us and GOD to go through this process to hopefully find each other in a shared desired eternal objective. Just as with light, and all good things coming from GOD, these loving desires must forever be felt and known as they pass through us, and allowed their eternal journey and eternal purpose. If any of the good flowing desires and creations that perpetuate love and life are held onto in selfishness of fear, pride, or envy, or restrained in any manner. By what they are, they must fade and die within whoever restrains them. We are commanded by love itself, to let the spiritual things of life we receive, flow to all who need them. We are *also* commanded to not cast these pearls before swine, such as fallen angels and reprobate souls; as they cannot desire them, and will trample our pearls into the mire, and turn to attack us. If you encounter the soul of a fallen angel or reprobate mind, do not try to preach to them or engage them. Never believe that all souls can be saved, when the Scriptures say most cannot. Most souls in confusion and fear of physical death, fight desperately to hold onto their physical life. And in fear, they gather physical sustenance and protection. And all too late, these find their imperfect physical life is like sand or water which are never held onto in any measure, as they slip through our fingers. *This life* is only our learning moment. Imagine the trust found throughout GOD'S family in sacrificial love, allowing the very force of life we cannot exist without, to continually pass through to others, never considering ourselves. It is only in these faithful responsible trusting ways, set by the eternal flowing love of GOD, that eternal life in peaceful understanding among many can eternally exist. We know this bicameral reality requires FATHER'S forceful desires for it to consist, as the rebellious forces of all the fallen, work against and destroy that desire. According to all Antithetical Principles, GOD *and* we feel antithetical forces of desire projected at us, and even by us to differing degrees. These forces we project, and sense projected at us, are perceived as brightness or darkness by all Holy souls living within the heavenly realm. If we feel love or hate projected from another, or ourself, we are sensing conscious force projections; and not just reading body/face expressions, words, inflection, posturing, or actions. As an example, think of love and hate like the Doppler shift. And just as we see and hear light and sound differently as it is coming towards us, or away; GOD and all those spiritually connected to them, do in all ways learn to feel, qualify, and even quantify, this broad spectrum of conscious forces. GOD senses love in that a soul wants to come and be one with them in all ways. And hate in their wanting to go from them. The bright forces desired and sought by GOD, are those that bind and perpetuate eternal creations and life. And dark antithetical forces are those turning from GOD, causing GOD'S reciprocal turning from them; leaving them confused and all alone in their individual divisive insufficiency of love and belief. Now only the few see that our part with GOD is to teach all children to see and avoid this world's rebellious ways; and to deny or obfuscate our created responsibility, is to help destroy all these children. GOD hates destruction, and *all* those who push, enable, or allow it! Many believe that because *they* do not desire GOD'S ways, that no one else could truly desire to exist in such an eternal loving way. They are wrong, there are always the fewer here and there who sense and greatly desire GOD'S ways of eternal life above all things. These few are *always* strongly repulsed and disappointed by the way of this world's confusion, pain, loss, destruction, hell, and death. We ask, how could any desire such an imperfect painful life of loss, and loss of life. GOD'S Children desire only eternal peaceful love. But individuals, find certain levels of conflict, deception, destruction, and selfishness, make life dramatic, interesting, challenging, and exciting; so, most want these things to some degree. Individuals see a narrow lively perfect peaceful order as boring. So, it comes down to how much our soul sees, understands, and values eternal love, peace, and life itself, above physically sensed experiences and drama. We see the fewer value life itself more, and the many valuing the physical experiences more. The earth itself is seen by most individuals as their true provision, so for them it is more valuable than GOD. Such desires are symptomatic of the devaluation of life. We value GOD and life so much, that they are our core eternal desire. The ways of eternal life are a found treasure of joy and peace to our souls. We see nothing boring in perfection; to the contrary we see eternal wonderous value and excitement for our souls. To us it is finding a most valuable other-worldly treasure, and feel far more joy than the fallen feel in winning a huge lottery. In this world, *we* see chaos, anger, individual beliefs, and each lost soul's diminishing perspective of reality. And to see all this, is to have the mind of CHRIST. We have no desire to be lost into any confused descending individuality. By accepting GOD'S omniscience, we find *his* undeniable truth. We see most souls in this world, looking in different directions for their individual truths. We see the masses have broken from the one true reality, to accept a confused illusion of mass delusion. And all accepted delusional beliefs, are levels of Insanity. Insanity is a word once commonly used, but that we now rarely hear. Is this because it sounds judgmental to an increasingly insane world? Is this word insanity, losing its firm meaning and relevance. This world has gone, and is going more and more insane. Most don't realize this in themselves or each other. It is insane for souls to think GOD does not exist, or they can deny, or overthrow GOD. Do they stop to consider how Satan and his angels, must by now be so insane, delusional, and psychotic in their rebellion. Satan was first to go insane; so, how delusional must Satan now be? Insane, also means reprobate, or one so confused they are unable to ever-again see, understand, or hold to GOD'S truth of reality. And in GOD'S merciful search, until the very last good soul is found, all rebellion is allowed for yet a moment. In our growth towards being one in the perfect belief of GOD, we strive to leave all rebellion behind. Now in Scripture, we see certain men of old, were searched out and chosen by GOD. Men who in decades of prayer and worship, sought, received, and held an adequate measure of GOD'S omniscient truth. These men saw this world clearly in its demonstrated graduating rejections of GOD'S truth. GOD instructed these men to chronical in perfect reiteration, all things he spoke and showed them. This created a *treasured* map of understanding if you will, for those coming after, who search for GOD. We are schooled by life, that the truest of agape, or sacrificial love, born of true spiritual desire, is born of pain and suffering, and forever perfected in long sacrifice. As in bearing and responsibly raising a child, such love is about our inherent inner desire to give all we know and have to the child we love, to perpetuate *their* eternal good. This love was in GOD before any other, and there is within this nature of sacrificial love a great want and need for such a love to be returned. I consider all things deeply, and in wanting to know GOD, I have tried to imagine what it was like for FATHER, to at a prior moment, be the only soul to exist in true utter loneliness. I also wondered if it is wrong to consider what it's like to be FATHER in any situation or moment, even if in wanting to know him more deeply. In my process of imagining myself all alone, and then creating an external existence filled with souls I gifted in my need to find an eternal loving family. I found *my* desire would be to *intimately* give, share, and teach, each and every new child *all* I know and have done; and *why* and even *how* I have done all these things. For this is intimacy, and loves greatest attribute is that it is most intimate and spiritually shared. And we see this loving need is in GOD. But this thought process scared me, as I found I did not like the thought of being GOD. I wanted for GOD to be there above, teaching and amazing *me* with the how and why of all they know and have done. This has been my life-long experience, and it pleases me immensely. By my desire to be and follow in the lineage of the *taught*, I saw my deep core free-will desire is to not be the GOD above, but for GOD to be GOD. I wondered how others might not feel this way, as they may want to be a GOD above as Satan did. So, I saw how much these two contrary desires contribute to the final conscious divide of all souls. I hope you all desire and find your learning *following* purpose. I know and love that GOD is there. And if not for GOD being there *above* me, in all my profound utter horror and devastation, my soul's purpose itself will break and die. I would then be less than nothing, coming from nowhere, going nowhere. I could have no eternal meaning or consequence. FATHER long ago, found no one above beside or below him, being alone. I see how that greatly saddened FATHER, and my loving need is to spiritually ease his loneliness and please him. As JESUS said, this *is* our love *for* FATHER. And when the hate and violence in the earth, beat, tore, and pierced the very heart and image of GOD'S greatest sacrificial love beyond all recognition; and lifted his impassioned torn broken bloody image up high, and nailed GOD'S great responsible faithful love in a shameful way upon a cross, to publish before the world their firm evil rejection and ridicule towards him. Yet *then*, his great redeeming loving merciful blood flowed down upon those fewer wanton souls in the earth. The power of GOD'S love, calls us in a hopeful provisional desire; but it can never force or coerce, that is not the nature of love. We find this truest most painful Spiritual sacrificial love given us by GOD, to be the most rare and valuable of all realities. This alone in all that exists, gives life its eternal meaning. But great woe and fear to any soul, when FATHER sees no hope of their returned sacrificial love for his SON within them. Those who without care, shame, or fear, trample upon and just discard CHRIST'S act of love and sacrifice. For any soul to be found in that conscious state on our LORD'S Day, I tremble to think about it. It requires our fervent study and contemplation to see the eternal strength of GOD'S love for what it is, and to know its full reciprocal requirements and ways. We know that in searching for GOD'S love, *we* find in each step along the way, our own pain and sacrifice. In going to GOD, we find a deeper understanding, that sharing love brings our desire to feel and share others pain. Bringing expressions like, I wish I could ease or bear your burden, or take your place. And sacrificial love in this world, is rarely equally returned and is most often used and abused, rendering only more pain to the giver. We see that in our knowing and sharing the pain of GOD, they feel, share, and ease *our* pain also. In this we share and ease the burden of loves pain which brought all things to exist for a great hopeful purpose. By our knowing and feeling the ache in our heart for GOD'S unimaginable pain, we find oneness in them, as when another shares our pain, we are helped. An easing of GOD'S immense pain and burden, comes by our loving obedience. We see Scripture is all about GOD and those who desire their great eternal love, finding each other. It is an eternal love story of the highest order, and nothing less. Materialists say our choices and our love, are illusions of chemical-electrical complexity, born of chanceful events. But both choice and chance were actually created out of great planning and effort, and for a great purpose. We see that before all that now is, when FATHER was all alone, that random choice and chance and all the plural beliefs we see around us, did not exist; and still do not exist in GOD. And the responsibility for another did not exist. We all know what responsibility is, but most have no idea what is eternal responsibility. Do you see how FATHER, for love, has promised to take on the eternal work and responsibility for our provision. Eternity is not just a long, long, time, it will never end. This is an amazing promise to make, even for eternal love. But FATHER has made and *defined* his promise to his SON and SPIRIT first, and now to others. FATHER knew what he could promise to eternally do for us, long before we existed. We have zero say, in defining or altering that eternal promise towards us. FATHER is a vast soul, with an eternal perspective far higher than all gifted souls. We all were gifted in the hope that we would desire to share in GOD'S pure eternal promise. So, how and why is GOD'S created truth, being replaced by billions of individual lies. We have all heard it asked, what is truth, and also, what is love. We have heard it said, the truth is what you make of it, and love is what you make of it. And, beauty and all things are defined in the eye of the beholder. By this individual temporary slow small godless perspective of belief, reigning and increasing in this world, we see all these evil ways preclude GOD from associating with any soul projecting these contrary thoughts and acts. So for the moment, FATHER averts his eyes from most evil, as he blindly sustains his creation until that certain point of his promise to end all evil. Even sustaining in pre-ordained ways, the physical life of those who hate him, in hope a few may return. What is it about free-will choice, that makes most souls choose to be singular dying individuals? Can one ever be joined to GOD in rebellion? No. All the rebellious will be utterly alone. The great eternal free-will choice we must make, is do we desire and accept how and why FATHER has created all things to be, and do we want

GOD to reign over us? Herein lies all envy and pride to rebel. But this still does not answer why, or what rebellion ultimately provides. Does one feel superiority by rebelling against a common omniscient belief of life? Can wanting and choosing to be unique unto ourself, bring or mean anything more to life than in being one in GOD'S shared belief of understanding? Can a desired belief that the one truth does not exist, be held without accepting the ultimate destruction of all you desire? Souls watch hell and death swallow up this world and all they love, all while refusing GOD'S common good. These souls place their *temporary* physical desires above an *eternal* love for their children, siblings, parents, friends, and self. I wish one soul could explain to me in a reasoned less than insane way, why they rebel against GOD'S ways and beliefs. Can any rebellious souls, having blinded themselves to the truth, even ask or know this about themselves? In choosing and putting the hope for the eternal good of others above ourself, GOD says this will bring us an eternal peace we cannot yet comprehend. So, we see the full created nature, essence, and purposed summation of GOD'S given choice between eternal love and worldly lust. Now the physical aspects of our eternal biased moment of awareness, are relative in all ways to our created perceptions being memorialized in the aether book of *our* existence; while the current moment of FATHER'S physical creation, inflates, spins, rotates, and orbits, to envelopes all that ever occurs within it. But we can't think of our *conscious* existence in these physically relative terms. Our initial moment of conscious awareness, was a small vague unusual event. But our learned growing awareness is a *potential* equation, as our true or false understandings of belief form and grow. If you see life as a static or temporary equation, you don't know life at all. Now we all in our process early on, come to question about what and where we are, and why are we here. Most souls quickly grasp for the illusion of the firmest things their physical hands and eyes can sense, and accept they have found the truth of their reality. But these do not know that they grasp a temporary reality. In materialist beliefs, every aspect of this classical reality is subject to all the laws of conservation and thermodynamics. Yet in those same rules they exclude any aether medium, being the substantive essence of all space itself for conveying, allowing, and regulating all creation within it. Yet they then accept that something unchanging and basic must underly all sensed firm complexities (hence string theory). They do not desire a foundational aether, being subject to conscious force, because a uniform aether leads to a need for GOD. Some believe Special Relativity dispels the need for an aether; and to date, no experiment detects what they *expect* and *predict* could substantiate an aether to them. Yet they can't imagine or test for an aether that exists beyond their standard model predictions. Again, they don't know what to look for, or where and how to look. They look for what they want and need to see; so, they look for the wrong thing in a wrong way. All patterns are there for souls to interpret through their desire and belief. And the basis for the dying remnants of FATHER'S once perfect created physical reality, lies just there beyond mankind's limited physical senses and beliefs. This vast dissolving universe, in all eternal sense, exists between disappearing, and having never been. It has become a lesser diminishing beauty of a temporary aspect. Can anything known to be temporary, have any part in defining any eternal certainty? What is eternal certainty, and can anything temporary or phasing, be truly certain? No. Yet most see the temporary as their only beautiful certain consideration. The Creation is only meant to be beheld and experienced in wonder as our platform for finding GOD'S truth. It is to be looked at and through, to know and see the great creator of it all; and never to be our final understanding of reality. Souls can believe that the physical is the bases and sum of reality, but their belief means nothing eternal. Now certain secular concepts pose ideas of traveling through time to a past physical reality. Well, the created substantive patterns of every created perception formed by everyone, arc in the aether. And if all the original created perceptions of many souls, from many different angles of any single long-ago event, could all be brought up and congealed into a filled sphere to be shared and re-perceived by us. In this we could in moving around in it, be a witness to a prior event. We may then have a strong sense of traveling back in time, but in truth, this patterned information of these re-created things of the past, would actually *come to us* in our ever-present moment, as with all recollection. The created past can always come to us, but we can never go back to the past. And GOD'S Prophesied future can come to us in visions, but we cannot go to it. The Laws of this expanded new term for all current "Presentism" says that not even FATHER can escape his present moment of reality and awareness. So, this is the only form of what might be considered time travel there can ever be. There is zero evidence that says space, time, and reality, form a linear physical continuum that allows the present to curve around to the past-or-future. To believe such bizarre notions, one must believe the entire physical reality creates copies of its current self, frame by physical frame. Now just as we internally plan to do external things, FATHER being at the front of all creation has planned a Tribulation period he will bring upon the earth and mankind. Yet GOD cannot know mine or your future free-will responses or choices when confronted with that Tribulation. Prophecy is when FATHER knows and tells us what he will in a coming moment, physically do within his creation. Millennia ago, FATHER shared these things he will soon do with help from trusted souls, to advance his Kingdom. But written Prophecy, can only be truly interpreted by those indwelled by HOLYSPIRIT. This is the only way that the souls of mankind can ever know any future events; and this too, *necessarily* requires the existence of GOD. Now when we perceive a re-created re-experienced memory, we are not experiencing the pure past, as we re-run a *former* patterned information, through who we creatively have grown to be. In this, we add new unavoidable touches of current desired biased belief upon our old information. Otherwise, in our recalling something bad, exactly as we did before, which we maybe did not regret in the moment, we could not sense current regret. It would be a strange uncharacteristic visual emotional reiteration. All the lesser point particle waves filling in between the original greater defining point pattern of our old and new created remembrance, are our added touches of current belief and desire. These reveal the changes to our belief in our new moment, so inevitably our memories change with us. A stroll down memory lane, or planning future events, are the only form of time travel that exists, making time travel an everyday occurrence for everyone. The abstract essence of time, lies in sensing changes in one's created moment, and in entropy, fission, divisiveness, and hell; these all being a part of a slowing shrinking dying temporary perspective, seeing a beginning and an end to all things. Satan and his followers want all souls to see an impending final conscious end; but GOD wants you to see your eternal soul. Scripture says angels and mankind's bodies can be translated between the quantum ethereal and classical firmer consistencies. But we always exist through our soul's sphere in the quantum spiritual realm, and can be aligned to any-body FATHER creates. The bodies in this physical realm, have far more condensed mass that provide for far stronger condensed physical sensations. Because of this, most souls on earth live for the physical. But in the upper quantum realm, the angel's bodies are a vast ethereal expanded form, having far less physical sensation, and far greater awareness of their soul. Falling from spiritual oneness, makes a souls an individual. And experiencing our rich physical feelings, allows souls to more easily forget GOD and their looming judgment. So, the far more condensed intense sensations from firmer human bodies, are what the fallen Angels also desire. Scripture says that *among* all prior translations of messenger angels to Earth, who felt these strong physical attractions, a few were drawn to women, and they fathered children. Scripture says these children became mighty men, who gained power, privilege, and many possessions. Scripture says this will happen again in the last days, which have begun. In case you wonder, after this world of procreation is dissolved, there will be no more physical congress. This act of intense physical sensation, will never exist again. Now, we read that when *fallen* Angel are not translated here bodily, they will seek the strong physical sensations of being here another way. Through continuous efforts in the quantum realm, these more powerful souls can acquire a shared to complete access and control over a human body already here. Do not think a more powerful fallen soul cannot work its way in between a human body, and a weak unbelieving quantum soul. And then share or intercept all experiential sensations and control. This is called demon possession in Scripture; but the Psychiatric community calls it Multiple Personality Syndrome, and other such dissociative States. There are now many true physical neural diseases, injuries, imbalances, and drugs, that alter and impair our patterned information; yet demon possession is still a real phenomenon. Short of a firm controlling possession, fallen angels can struggle against a soul to gain degrees of shared feelings, control, and influence. Souls hear interjected voices or thoughts, not theirs, some see two perceived images (hallucinations) projected through one interface. Sadly, those who struggle against demons today, are being drugged into submission. So, having free-will, we must keep a firm solitary control over the body. Now more often than not, if a soul wants two different things, this is an unclear choice of warring desires within one's self. As when one wants to smoke out of habit, yet in self-preservation wants to quit. But if you strangely can't focus on something you desire and need to do, and are being drawn to something *you* desire far less in that moment, or feel distracted by an unusual force. Do not ignore this, you must immediately pray and turn your life over in great obedience to GOD'S protection. Demons invade and alter this world by influence and possession, just as they did in Sodom and Gomorrah. Soon a multitude of fallen angels will be cast into this world in their own translated bodies. Demon possession makes no sense, until you see two souls within the quantum realm, waring over the use of one physical body. If you ever feel external influence, pray and rebuke that spirit in CHRIST'S name who has absolute power and judgment over them. But as with mankind, there are degrees of confused rebellion that angels have fallen into. The rebellious angels that have not fallen to deeply, still sense that if they do not respect and heed the invoked name of JESUS, there is a greater fearful price to pay. These usually flee and go elsewhere at any earnest rebuke. But as with mankind, some angels are lost so deep in the insanity of individual denial, that they have lost their understanding and fear of even these greatest realities, and have no sense to flee. These require a true intervention by GOD, requiring the fervent prayer and fasting of a true Child of GOD. When the Disciples said to

JESUS, that of many, only one evil spirit would not obey them and leave a person. JESUS said that some require GOD to actually physically move them by force. JESUS said in Matt 17:21, "*HOWBEIT THIS KIND GOETH NOT OUT BUT BY PRAYER AND FASTING.*" Prayer and fasting are our *ordained* offering of faith, presented before any request for GOD'S help. As science peers into the quantum realm, having all its mysterious different lawful consistencies. They see a surface glimpse of the creative possibilities in GOD'S great conscious domain. Scientist see excitations in GOD'S aether, as forceful created wave and particle certainties. Now the knowledge that time and the speed of light limitations, are not real restrictions or considerations in the quantum realm, helps us hear the following verses. Acts 8:39, "*AND WHEN THEY WERE COME UP OUT OF THE WATER, THE SPIRIT OF THE LORD CAUGHT AWAY PHILIP, THAT THE EUNUCH SAW HIM NO MORE: AND HE WENT ON HIS WAY REJOICING. BUT PHILIP WAS FOUND AT AZOTUS: AND PASSING THROUGH HE PREACHED IN ALL THE CITIES, TILL HE CAME TO CAESAREA.*" John 6:20-21, "*BUT HE SAITH UNTO THEM, IT IS I; BE NOT AFRAID. THEN THEY WILLINGLY RECEIVED HIM*

*INTO THE SHIP: AND IMMEDIATELY THE SHIP WAS AT THE LAND WITHER THEY WENT.*" Just two of many Scriptures, that speak of FATHER translating physical consistencies into quantum consistencies, including living bodies and ships, and then back into their firmer physical consistency at a distant physical location faster than their experiential awareness could know. These Scriptures *must* get credit, as the earliest inspiration for all literary science-fiction portrayals of time travel, and all mass to energy transporters. In body translations by FATHER, the soul, which remains an aspect of the quantum realm, never feels altered or disturbed in any fashion or manner. In our soul, receiving a new body, or translation experience, it is like changing a TV channel to see receive, sense, and see new external information. In Scripture, intra-realm movement and inter-realm translations of mass, occurred many times. JESUS said, if you walk *filled* with HOLYSPIRIT, and say to a mountain be thou cast into the sea, and do not doubt. It will literally be cast into the sea by FATHER. This is meant to show us GOD'S conscious force controls reality. Now anyone who *could* say this in a pure faith in GOD, would also be in complete obedience. And would never say this, unless GOD told them to say it for a Holy reason.

GOD would never tell a disciple to cast mountains into the sea. But he did command them to love, heal, and restore his truth. The projected healing forces of GOD are now rarely understood as most see and believe only in a natural physical life force and process, and science and doctors. Consider again the conscious correlations that exist within your created reality. See how it was not your hand, but your eternal soul that picked up this book in control of that hand. It was, and is your soul which opened this book to contemplate its meaning. There exists nothing more to you or I than our soul. And we are gifted to learn, know, and experience far more than a body's physical instincts to live and die as flesh. Our conscious perspective is changed from the timely to the eternal, at our loving connection to GOD. But new connections to GOD, are at first cautious and minimal on GOD'S part, as expected. And as our love and belief are nurtured and grown, the strength of our tested and trusted connection grows. Most souls sense and know only their connection to this physical world, and most do love the things and sensations of this world of the flesh. Souls in love and worship of the flesh give credit for their love and belief to a wet fatty electro-chemical relay interface formed from the dust of the ground. The errors of worldly understanding and belief are so insane, that mankind's individual divisions of hell's confusion have become countless. If one dropped a crystal vase from the top of the heights of the world, down to its bottom. It could never shatter into as much confused broken division, as we see with GOD'S one pure truth in this world. And how can such awe and beauty of a once perfect creation, now lost, be restored. Once the spiritual and physical perfect order is broken, and take on a timely appearance, can it ever be restored? No. That once created beauty of fallen perfection must pass away, and all things must be created anew. In this day, the world is filling with a false belief that the physical aspect gave rise and birth to our miraculous self-awareness. This world, in endless errors, tries to explain our awareness as originating through physical means and processes. And every effort to date to do so, has failed to find or show one aspect or reason for their belief. Their ongoing theories continually arise and fail. If asked, what is it that *I* brightly visualize with eyes and a brain, I answer truthfully. I *brightly* see nothing with eyes or a brain. If asked what *I* hear with ears and a brain; I *hear* nothing with ears or a brain. I do not taste with a mouth or brain, or feel touch, or pain, or sense hot or cold, with a body or brain. I remain honest to GOD in that all my experiential qualia, rest in my soul's interpreted awareness of realities dark cold created moving patterns of dark silent unaware physical information. The utterly dark waves and photons called light, stimulate an utterly dark eye, to then send reformatted patterned information, describing all unaware things in GOD'S creation. After that, GOD'S physical light has not one thing further to do in correlation to my bright awareness. The wave-particle representations I create in the aether filling my flexing experiential soul, followed by my awareness of them, are a phenomenon of a far higher other-worldly process. A photon has no bright quality, it is an utterly dark particle. And as we ponder these truths, one might say the physically blind see the true dark reality surrounding us, while most sighted souls are blinded to this fact. Materialist try hard to imply otherwise, but if we understand and remain within GOD'S truth, Scriptural science imparts a perfect consonance of belief and understanding. Scripture and GOD'S science, hold enlightenment that only HOLYSPIRIT can reveal. We see at GOD'S will, the highest temperature that has ever existed, could kiss the absolute frozen zero without affecting each other. Everything physical is now the evidence of a long-fallen creation. HOLYSPIRIT first showed me the science of FATHER'S created external manifestations, then the truth of my own created inner manifestations. From that moment on, I have been shown the endless correlations causes and effects between the two halves of reality which have not one time in any way, failed to perfectly align in all relative ways. But if you believe your soul projects no creative force in your every newly generated desire to exist; being our teleological determination and intrinsic purpose, then Scripture *and* Science from your very wrong perspective, become filled with problems and paradoxes, requiring you accept many suppositions of physical mysticism. Now by the far greater perspective of EUD's far greater Observer Effect, and Observer's Frame of Reference, along with the standard model's Measurement Problem, the Quantum Zeno effect, Quantum Non-Locality. And adding the fact that mankind can never observe a wave function collapse. These points all together reveal our souls have their forceful creative influence, and do create our perceptions. Yet scientist turn from these certainties, not wanting that one final answer, waiting down this conscious force projection path. They sense only great repercussions to their beliefs down that path. I heard a second-hand story of a soul renowned among their peers. Who professed that when they were a child and had lost someone close to them, that they were told by an elder to not fear death, because they would then live again forever. This person declared, that as a child they were afraid of the thought of living forever, and cried. For me this sounds like eternal life appeared to them, as an eternal trap they hoped to escape. I still wonder at this, for in all my life I have witnessed the fear of death many times in many souls. But this was the first fear of eternal life I had ever heard of. So, I then wondered, are most souls trapped in between? Do most never want to die in every current moment, yet also not want to live forever? Can the consensus of souls on this earth be walking a tightrope of two desires, wanting the continuity of life in each moment, yet needing a distant looming permanent end. Does this comfort them with a belief their mounting shame and pain will end minus eternal accountability. Many materialists and atheists in their last days, coming near to their death, find they hope to find and experience something more and good they don't know or understand. But these find a faithless blind wish, based solely upon the physical aspects they sense. The body's death is deceitfully cruel to lost souls upon their being resurrected to an eternal lonely anguish. Now mankind's knowledge suffers from endless uncertain changes, while GOD'S knowledge is eternally defined. The true spiritual realities are hidden by all fallen temporary perspectives that sense and know only the physical side of reality. Yet in any moment, some may find a new eternal love for life, and a new revelation of clarity that redefines their perspective. I warn all, that walking on a tightrope of imperfection, not believing in the conscious force projections of GOD and all lesser eternal souls is treacherous. There is no safety net, you can fall forever. There is a Godly-walk of understanding and belief on one side, and an endless dark abyss of questions as to reality and life's purpose on the other. And in not fully wanting *either* side, many choose that physical tightrope they walk upon in temporary hope and denial. The eternal soul is now to each of us a firm reality, an ambiguous possibility, or non-existent. And we are the sole desirous arbiter of choosing and building our own belief. In all of this, our physical existence is still in all true ways, just our created dream, and life and love are only about our desires and choices to find *eternal* meaning and purpose. Will we each desire and choose that *one* good provided beautiful eternal *shared* dream with GOD; or choose our own individual dream that will not be fulfilled. The innate inherent creative forceful qualities of a soul, reside beyond physical understanding. I wonder if even FATHER can teach this understanding, as explaining anything requires knowing all its formal forming aspects from the very beginning of anything's existence. So, seeing FATHER has eternally existed, how can he define the qualities of his perfect and eternal spherical soul and body, that had no discernable beginning. So, again, we see linear time is a false concept of belief that tries to describe in a false linear way, the beginning, the end, and the spectrum of changes and continuity of experiential awareness in between, that is our growing spherical reality. Linear time can only exist for those who believe that the seat of all awareness had a true beginning, and believe it will have a true end. But all basic physicality and every soul has always been and are eternal. Our souls were once an eternal part of FATHER, before they were gifted to us, our souls are nothing new or created. We know that time does not exist for GOD or his Children, so, eternity *must* exist. Please see that something can never come from nothing, and nothing can never be truly defined, and now by individualized conscious souls of differing size and belief, all things can now be perceived in so many subjective false ways. We can't just define nothing by what it is not, and where it is not. Abstractly, *nothing* does appear completely antithetical to something. So, how can the *contrary* abstract aspects of nothing, ever be a basis for anything real. Nothingness is impossible to define, so there was never any true beginning to reality. But there are beginnings for all newly created complexity. Physical reality is consciously deterministic, so it can be eternally fixed or be transitional. We can never escape the fact, that the farthest edges of all *mankind* can currently see, never reveals what lies beyond, raising every great doubt in all materialist believe. Our current moment of awareness, exists upon our crest of desired certainty. In our creative perception process of rising and descending fluidity, each new forming moment rises up out of the aether, till our indwelling awareness is realized; this as our prior created moment of awareness lowers back into the aether to be stored. Do you wonder what level and reach of creative projected forceful desirous ability you truly have to project, build, indwell, and become one with your created perceptions. You create, sense, and know the 3D multi-sensory particle wave properties of the streaming perceptions you effortlessly create, covering and defining the surface of your soul's spherical moment of reality. Do you see how effortless, tireless, and Godlike this is within you. Can you see the required conscious projected structural elements of force, math, logic, desire, and purpose, that must come together in an instant for a streaming rising particle-wave perception to be created. Do you see that by each quantum of omniscient understanding we each acquire, the size, force, and beliefs of our soul increases; revealing the eternal learning creative potential of our soul with GOD, a potential we cannot, and may never fully comprehend. Can you begin to comprehend the levels of understanding we each might hold after ten thousand, or a million years of expanding awareness. Imagine a strong loving personal relationship with GOD and all three hundred million souls we will spiritually know in Heaven in the greatest and most intimate ways. Every streaming *multi*-sensory particle-wave perception, *must* be formed by one to five or *more* entangled mathematical arrays, layered together. To experience our awareness within our soul, is to desire, choose, create, and define our faith and belief. To hear, we create and indwell a sound in our awareness. To smell or taste or touch, we create and indwell our perceptions that bring our soul's awareness of those experiential qualia. Once we fully comprehend the complexities occurring at the highest and deepest quantum levels, and realize these miraculous phenomena are beyond physical orchestration or explanation. Then in GOD, we can see the many innate inherent qualities, quantities, and elements required for all of this to occur. And then we see and find the true answers to all of reality. Materialist say proof is in seeing, hearing, touching, and measuring, yet they can't prove they have ever literally *physically* seen, heard, touched, or measured anything outside their consciously created, second-hand, here-say, perception of doing so. They can never jump this inherent hurdle, so, they deny it exists. The problem most get into with physical reality, is the more they focus on the firm and changing, the more they believe it is reality's foundation, no longer seeing the wonder of it all being consciously created, or perceived. Most souls forget how miraculous and mysterious it all is, even as the enlightened declare it all to be most wondrous. Most still choose to believe in time when there is no past or future. And like all things real, we see love and hate, good and evil, can never exist in a past or future that does not now exist. Science mixes linear and spherical concepts together within their theoretical linear foundations of physical reality, when all its true aspects are purely spherical. These qualia, and all the physical evidence we have of our own existence, live in our created perceptions. Science tries hard to convince us all to accept godless beliefs, but in many logical free-will agents, they fail. So, they work to control a financial consensus, while among themselves, they debate their own variations. We see even GOD, cannot teach one true single thing to every soul once gifted free-will souls rebel. If you think this is not true, why can't GOD teach all souls that loving and obeying them above all, will bring us all eternal happiness. GOD places love, truth, and our eternal need for them, right in front of us all. But this cannot overcome most free-will rejection. Such evidence should preoccupy all souls; but in holding contrary desires, and like children running in abandon, most souls forget about their created perceptions as they run through one into the next, and the next. We see overwhelming conscious/physical evidence of two lively realms of existence, yet in Scripture, we see a third true separation in reality. There exists a form of division within the quantum Heavens, putting a guarded separation *between* GOD, the seven most powerful Arch

Angels, the four Seraphim, and all the lower lively orders of Angels. This third highest Heaven, is called the Holy of Holies, being separate for the reason that a lesser but epic rebellion among the lower angels in Heaven occurred first. Such insane fallen angels cannot have unguarded approach and access to GOD. Now consider FATHER'S semblance, formed out of his needs and desires. Imagine that you are the only soul, and considering dividing your soul to gift another soul as a son. What would you choose as a perfect visual form, to show your SON your loving wisdom and benevolence in being his FATHER. The word FATHER means Creator; and FATHER created his own semblance and ours, entirely for the reason of loving each other. FATHER chose an ancient and wise semblance; he chose forward facing eyes, to convey looks of love, and see love returned, a mouth to speak eternal words of endearment and loving warnings, ears to hear the echoes of his loving common understanding, hands and arms to hold, help, and comfort, in gestures of love, legs and feet to walk with in love, or turn and leave in the hurt of being unloved. Within FATHER, rested the greatest breaking heart of all souls, wanting only to share *all* he is and has. FATHER has an immense bright white-haired ancient appearance, so we can know he *was*, before we *are*. His ancient appearance projects great love wisdom and responsibility. FATHER'S ancient bright white-haired visage gives the understanding that he is in every way, only spiritual towards us, being beyond all considerations of physical need or desire. FATHER'S ancient appearance, full of light and fire with every white hair, sitting on his Great Throne before all angels and mankind, is meant to say many things to us in a visual way. It is the perfect semblance for us to desire his loving spiritual wisdom. FATHER'S semblance cannot be improved for his eternal purpose. And for all good reasons, FATHER chose his appearance, and sits upon the Greatest Throne. By these, FATHER is seen as Supreme above and before all souls and all things in force, understanding, and love. Yet all who believe in their personalized gods, and unbelievers alike, ask how can a loving GOD hate and bring eternal separation to so many. But these same souls bring only hell and death, as it is they who continually draw first blood and inflict first pain. And it is they who desire and choose to separate themselves in rebellion as individuals. These believe GOD should provide all their desired eternal physical provision, as they live in their spiritually rebellion. These refuse to see how they work and live in sinful ways that separate themselves, and steal all GOD'S Children to themselves. These reject all truth of eternal life, and they envy and hate all who truly love GOD, and whom GOD loves. So, how can GOD not hate and separate these, if he is to protect his Children, being his loving purpose. Why has GOD said, I love those who love me, and if you love me, obey my commandments. Can sinners bring their destructive ways into Heaven? No. Satan immediately shared his evil destructive ways in the current Heaven and earth. Eve immediately shared her chosen evil with Adam, as does everyone holding selfish individual sinful desires and beliefs. The most common liberal attribute of evil souls, is to spread, share, and demand acceptance of their divisive evil desires, as being good. They use every means to subvert all who disagree and separate themselves away from their sinful ways. Self-preservation and dissemination of individual beliefs are a core aspect of individualism and liberalism. Liberals are confused envious angry activist who hold existential ideals they don't understand and cannot define or defend. They live to be contrary, fierce, and to silence opposition. Angels and Mankind for the reason and purpose of love, were gifted and created at *our* beginning in FATHER'S visual, emotional, and creative, free-will image. This gives us insight into FATHER'S good desires for our eternal created purpose. In GOD'S work and plan, we see a great initial founding piece of reality is the reason for it all was FATHER'S greatest loneliness and unrequited love. This entire created reality, is the greatest of eternal love stories, being FATHER designed all creation as a mechanism to allow eternal love to arise in as many as would love him. Love can never be forced, it can only be allowed and then nurtured. All lesser souls combined, have not sacrificed or endured as much pain and loss in their search for love, as FATHER. *And* his SON and HOLYSPIRIT have suffered more than all others combined also, but not eternally like FATHER. JESUS loves FATHER so much, because he knows FATHER'S love intimately, and what FATHER has suffered and endured to bring us all here. Real eternal love cannot exist but by free-will desire, and all its own choices. So, GOD allows all physical chance and free-will rebellion to be temporarily woven throughout reality. This creation of conscious free-will choice and chance, must be entered as evidence of FATHER'S love. As these allow all beliefs and ways in which all things have now come to be known. Including a spiritual spectrum, ranging from pure sacrificial radiant love for GOD, to a pure evil, being those most narcissistic insane individuals. FATHER needed the existence of choice, chance, and many souls having free-will. After losing a third of his angels, he needs to replace them with around one-hundred million *eternal* holy souls from among mankind. And GOD needs to do this out of between 12 and 70 billion souls, that will have been gifted since Adam to the end of this world. So, *on average*, we see 1 of every 120 to 700 souls that have ever lived, may be saved. These averaged odds, were far less at first, and are far greater in these last days. Now please understand that free-will is not the absolute control that many wish or think it to be. Free-will is our forceful controlling desires, deep in our soul's core, that we have formed and hold towards other souls and reality itself. One can try to lure and assuage another soul; but none can *will* or force another soul, or their self, to instantly desire something they don't want. We each are subject to our core strong free-will desires; so, the great faithful responsible work of eternal love haunts us all. GOD knew when he gifted so many free-will eternal souls, having sovereignty over their own desires, that an unknown probability and uncertainty for rebellion must exist in allowing and providing them an actual true choice in their existence between two desirable realities. Meaning and requiring FATHER must give all souls, not just the *concept* of a choice, but a second very desirable path, other than GOD'S ways. True choice does not exist until a true choice is seen. It must be between two wondrous pleasing things, that self-aware, capable souls might greatly desire. If someone asked you to choose between a grain of sand and eternal blissful life, you would say that is not a true choice. A true choice must be a choice between GOD'S ways, and the allowance to play a god of one's own ways and beliefs in a powerful, temporary, yet pleasing way. That is a true choice. And this choice *was* created by FATHER, and in each heart a choice is made between desiring to be under GOD, or desiring to be a god over what they attain, believe, and how they live in *this* world. Scripture and Creation are our map to GOD, with HOLYSPIRIT as our guide. All corrupted and corrupting aspects are now evidence of the ever-growing rebellion to GOD'S created ways. We must see how truly great and complex the creation and our senses and desires for awe and grandeur must be, to constitute for each and every soul, a true choice between loving GOD most, or loving the creation the most. In our choice of love between GOD, or playing a limited god in this physical world, our choice will be the most powerful pull of desire deep within our soul. And at a certain point, it will become set and eternal. A choice for a certain love, must be precious in the eye of the beholder; and precious also means costly. This choice between two loves, is offered by FATHER in his hope of his love always being chosen. But he mostly gets the great pain of billions of free-willed rejections. These choices and this great cost, just must be or love is not real, let alone the greatest valued treasures to be found. Creating and giving to all souls a true choice, requires GOD be hidden to a degree. No other process, could yield as much eternal love for less cost, than that which FATHER has instituted. Many do hold a love for GOD; yet it is not in this moment their greatest love. Their greater love is still for family, self, friends, and the physical world. The many cannot see or know this, or understand why their greatest love has to be for GOD above family, self, and any created reality. They cannot see that rejecting this order is the initial root of rebellion. FATHER'S gift of a shared physical reality, can never receive more love than the giver of that gift; or the giver has been set below the gift. We must love every conscious giver, far above every physical gift. This created choice for defining our eternal love, must offer *only* two opposite directions or polarity of desire. Once our choice has solidified and set as stone, it is eternal in being sealed by GOD. If you are still interested and still reading at this point, you are most probably *not* a spiritual reprobate. And you may through a fervent nurtured eternal desire, love GOD and be loved by GOD. Both conscious directions of choice have their horizon events, or points of no return. At our horizon event, we are eternally found, or forever lost; and forever lost are the saddest words in all creation. The ongoing work of GOD, extends beyond our imagination. And by rebellion, everything physical is no longer in its perfect order and balance; it has all succumbed to laws of growing disorder under GOD'S fading desire to sustain it all, as he pulls it apart. All things physical that reside in perfection, according to these laws, must endure eternal, being minus any and all disorder. This "Conscious over Physical Principle of Perfection" that equals eternal life, is more than a high probability under GOD, it is a loving spiritual certainty. Yet it is not desired by those who love imperfect temporary things. When GOD was first betrayed in their Garden by Eve and then Adam; FATHER'S hands which touch and hold the edges of the universe, pulled back at being rejected. Stretching the spherical creation out from its created stable dynamic perfection, and its farthest reaches fell into disorder. The dynamic motions of the former *perfect* fine-tuning and care, were no longer found. The laws of shared perfection, are vitally important for a shared conscious existence of many souls, tied to one shared created physical platform for eternal life. Beyond FATHER'S ordained created ways of life, his laws of perfection lose all reason, purpose, and hope. We see the eternal reason GOD dwells in all perfection, and pleads for and requires perfection in us, is to sustain the eternal perpetuation of life. We must come out of all individual imperfect ways. I ask all to understand and hold tight, that GOD can only desire and seek out love that is of an eternal perfect nature. Perfect souls by their nature, are void of hurting and causing loss to GOD. Yet GOD and all souls must bear and share the hurt and cost in first *finding* this eternal love. The rebellious think the things they love, are good. But they only think those things are good, because those are the things they love. They have denied and lost the eternal lively Godly perspective. We ask you to accept that we are seeing a great conscious division of desire. And at the last we will see two opposite realms of mind formed, as the oneness of all eternal loving souls are sealed away from all individuals. In individuals, we see all conflicting descending beliefs; but in the oneness of GOD'S belief, we see the ways of perfect enduring loving life. Between these two perspectives, we see a simplistic law of their being complete opposites, antithetical, and anathema in every way. We see in this world the desire for receiving and experiencing physical pleasures, rules the hearts of most, bringing all fading illusions of joy. These divisions among souls, bring the lost their prized individuality, as their godless liberal standards bring allowances for more and more imperfection and insanity. Today's unacceptable fringes of societal deviance, become tomorrows defended norms as insanity grows. Yet in this world of progressive moral decline, certain perceived injustices will always be punishable. The moral standards and ways of GOD, are becoming the prosecutable acts of today and tomorrow. We see every injustice and violation to the perfect truth, projects hate on all that is eternal, as each evil injustice of individual belief tears at the fabric of life. FATHER at his core, hopes to love every soul, but his being rejected by the many is set beyond his control. Every requirement now exists to allow every soul to fulfill any conceivable desire, and seek the fulfillment of every possible choice of belief within each soul. There are endless paths for the many that blind themselves; but one bright straight narrow path is traveled by the few. The possibility for all things of choice and chance that can be imagined and acted upon, must be temporarily allowed so the core desires of all souls may be revealed. Only a chosen desire for GOD, can open our eyes to see these things. James 2:9-10, "*BUT IF YE HAVE RESPECT TO PERSONS, YE COMMIT SIN, AND ARE CONVINCED OF THE LAW AS TRANSGRESSORS. FOR WHOSOEVER SHALL KEEP THE WHOLE LAW, AND YET OFFEND IN ONE POINT, HE IS GUILTY OF ALL.*" Our only *loving* choice, is to desire eternal perfection. So, if we disbelieve or disobey GOD'S truth, for even one moment, even to spare an uneasy thought or offence of another person, it is a technical sin; it is in its way, our abandonment of GOD'S ways. The world is flailing in imagination and self-rule, unshackled from GOD'S truth to run through and get lost in its abyss of physical pleasure and unbridled fantasy. But they find only hell, death, and destruction. Lost souls believes in error that the body and the world to which they are adjoined, belongs to mankind to control, rule, feel, play with, and be a god over in what is truth and love in their life. But this evil is allowed for only a moment. All evil desires, raise this disagreement of ownership of the adjoined body GOD created and sustains. If a soul thinks they own their body, they have lost touch with reality. GOD sadly knew that creating a true choice of love, meant allowing a rising probability of delusion; even as GOD cries out in every Holy Righteous way for them repent and be restored to life. Now, our strength and weakness, in being gifted *and* created, is knowing we had a beginning to our awareness, and then having our desire to continue. GOD knows the limit of helpful inspiration he can give us without skewing our true choice, and that only some will want his help in love. GOD cannot exceed a certain point of true righteous help without diminishing our free-will choices required for our love to exist. GOD loves every soul's gifted potential and created purpose. But GOD loves no soul in their destructive selfish choices. At the last, GOD must not love those who will never love the eternal ways. Many ask, so what can a highest love be like in Heaven. Forgive us, but to try to convey the heights of a pure higher eternal love, we must invoke the remembrance of your life's greatest pain and loss. Remember please, the moment you realized you lost a loved one, remember the breaking pain in your heart, and endless tears. This was the sum of *all* your love for them, deep within, rising to your soul's surface at your loss of them. We usually sense only a small portion of our true measure of love for other souls in our daily routine, as it gets crowded out on our soul's surface by other competing momentary experiential considerations and emotions. The spherical surface of our expandable soul, holds the sum of our current moment of experience, yet it is finite. So, in our limitation, we can only experience what fits upon our soul's surface. In this fallen world, we now have many competing emotions and considerations swirling on our soul's surface, crowding each other out. But in moments of tragic loss of love, we completely forget this world, and only the fullness of our love for the one lost, covers our soul's surface, bringing a full sense of love and loss. Now imagine if your most dear someone died, but weeks later, GOD came and told you he would bring them back, and they would come walking down the street in one hour. Imagine only your joyous love crowding out all other emotions and considerations as you look out the window waiting. Then you see them coming and you run to them with all your tearful love at your surface. Imagine your waves of love running all through your soul, stronger and purer than you have ever experienced it before. In that moment, you have no other cares upon your mind but your love. Now go even further and imagine the joy all would have and feel, in sharing this same strength of love for everyone always. Most have never felt this fullness of their love in daily giving or receiving. We can have strong daily emotions of love while in this world, but again, they are diluted by other concerns. Imagine intimately knowing and loving around 300,000,000 souls at this highest intimate level, bringing tears every time you see them, and every time they leave your presence moving among the many. You keep meeting souls you haven't seen for a long moment, and when they move on, you know you will see them again, but it may be a while in eternity. By being eternally unburdened of every other daily physical need and emotional competition, imagine feeling only this highest pure love, projected from everyone towards you in this way. Imagine being adjoined to a new eternal body in between the consistency of the ethereal body of the angels and a firm physical body. A body that physically senses, but in no strong enough way as to distract, tempt, or lessen our spiritual considerations of life. But in this current worldly reality, most loving spiritual desires and needs struggle to reach our surface, as other desires control and dominate most souls. Most assume their current greatly diminished fallen experience is the level of life and love, that GOD created for us. They forget, mankind has fallen down so very far from where we were created to be, to the shameful depths of what they have become. Some do hear Scripture say mankind fell from their heights; but because they love this world, they can't see the true meaning and distance of that fall. Sit at a table and see high above your head, your clenched fist holding and obeying every word of GOD. Now see it opening, letting go of GOD'S loving ways, and falling into an open flat palm, becoming one with the surface of the physical table. This is the literal fall of which GOD speaks. From this visual, what do you think GOD'S redemptive plan is for each fallen soul? Is it to leave us there, and keep forgiving our endless destructive divisive rebellion, or is it to restore us up on high, to again be one, perfect, and eternal. Imagine pure love ruling us, and being in GOD'S presence, living in a new eternal perfect environment of peace, beauty, inspiration, and wonder forever. The new City of Heaven, New Jerusalem, is described as a city of mansions, with all city floors and outer walls made of gem stones, about 216 feet thick, sparkling and radiating ever-changing brilliant beauty and color. Streets are laid in fine pure gold, lined with beautiful trees, fruits, and plants that never fade, all filled with a warm bright light. These sights and our heights of responsible loving emotion and senses can come to be; but only under the eternal perfect provisions and ways of GOD. Heaven is a far more loving spiritual experience, and a far, far less physical experience. *Fallen* mankind can never grow or hold a love that could rule and never fade. We see in FATHER'S promise, that only he has the size and force to project and perceive this flow of eternal loving provision down through all who choose GOD above this world. This great love must flow down in great force and sacrifice from above and through us. And then only as a whole, can all the souls in Heaven together, return GOD'S great measure of love. We see two-hundred million Holy Angels, and around a hundred million from Mankind, being resurrected as angels, will fill the City of Heaven. We will sense our physical aspects of life become almost irrelevant under such strong love. No one will ever wonder or care where they will dwell, or what they will eat, or wear. Our sustenance from the tree of life will be received in thanks and spiritually enjoyed in a far less physical way. Let's dissect further, the current choices made by us all concerning our desires for GOD, or for this world and physical body. If we consider our every thought and act in each day. What percent of our thoughts and acts are about how we and our family and friends physically feel and live, and our health, home, and food, etc. When we think of our family or friends, do we focus on their physical needs and attributes. Where are, and what should be our constant concerns for their souls in all our daily considerations. A conscious spiritual way of life, cannot and will not be suddenly gifted and instilled within our soul to suddenly become who we need to be in heaven. This we must build, here, now, and forever. Love is the highest emotion that consciously rules within GOD, and it always will. And their love can be eternally known, felt, and shared by us. FATHER in omniscience, knows the conscious and physical boundaries, souls living together, need to observe without fail to live in his eternal peace, love, and provision. This is a great trying purpose of the coming Millennial Reign, prior to the New Heaven, Earth, and New Jerusalem. I just touched against Holy Doctrine, so, I *must* impart the importance of Israel and Jerusalem being representative of JESUS and FATHER'S Kingdom. I have a familial command from GOD to warn *all*, or I risk your blood upon my hands. So, while here in this fallen world, whenever GOD'S Holy Land is raised in any issue; I warn *all*, do not be found among those who do not hold for Israel's true borders to now and forever be, as they are set by FATHER in Scripture. FATHER and Israel are symbiotically joined as one in the earth; Jerusalem and JESUS are one in the same way. Israel and Jerusalem are never to be divided, diminished, or compromised, not even for a false hope of worldly peace. *Any* compromise in *any* heart or mind, concerning Isarael's Scriptural borders, *is* in part with the abominable spirit of antichrist. I have now fulfilled my responsibility to all who read this, and your responsibility in this, now rests with you. GOD tells the world to do, and not do many things; yet most say, do, and believe differently. Do these have a better way than GOD? No. They just refuse to believe and follow GOD. Do not say, but these are different times, there are no changes in GOD'S eternal omniscient plan. In life, if one chooses to be an individual and look physically and temporarily at life, they can say whom does it hurt if I do something. Or, why is it wrong if I acquire something that answers mine or my family's physical desires, or increases our physical pleasure, health, or safety, seeing all have the option to do the same. Individuals define the term equality to mean their *right* to enhance and fulfill their own physical life and beliefs. When individuals seek equality, they always work to outlaw GOD'S Holy ways to allow their own pleasures, comfort, excitement, and control. Liberals see equality to mean all their individual ways are accepted as good in the eyes of GOD, but this can never be. A personal redefining of life, creates those who want this, and those who want the opposite, bringing divisions of confused struggle and strife. Most all who accuse others of having too much, would embrace the same if it were theirs; so, it is pure unrighteous envy. Those who don't have, feel a moral superiority in themselves by not having. And if they succeed in taking it, others who don't have, will see them immoral, and try to take from them. The world operates under this "False Moral Entitlement Principle" that rules within individuals and groups, and all larger groups called countries. There is no morality in having or not having physical wealth and security. And there is surely zero morality in trying to take or demand it from another. There is no righteousness in giving any soul any unneeded measure of abundance. True acts of righteous morality in this fallen world, rarely exist. There is FATHER, who painted a perfect picture that could endure in loving peace for eternity. GOD desires this perfect picture with every soul being perfect in this picture. But as most souls come in their free-will desire to see themselves in GOD'S picture. They are either thankful to be in such a beautiful eternal picture, or they pick up their conscious free-will paint brush and change and add to themselves, and others, and the world around them, and their beliefs. They repaint their environment in the ways of life to fit their desires. Many try to make the body more attractive; some try to appear smarter, or otherwise superior. Most want to appear as having wealth, health, and all false temporary securities. But they deny the initial perfection FATHER created, in wanting to greatly widen his created narrow path to allow their evil ways in which they want to live. These souls seek power, privilege, and possessions; many seek drugs and unnatural vile physical relations, being contrary to GOD'S perfect design. And more and more love to feel power in being violent by hurting and controlling others. There is no end to the evil unbridled hearts can accept in changing those boundaries of GOD'S once created perfect picture, once they no longer see or desire GOD. And if they cross their final fallen *horizon* event, they become what GOD calls reprobate; being souls so confused in belief, they can never find their way back to see, hear, or believe GOD'S truth. It is that zombie blindness we now see in so many. If GOD revealed himself to these reprobate souls in every way, it could not convince them beyond all their desired doubts. Can GOD, against a soul's desires, remove every doubt within them, as to their being GOD. We all had our gifted, created beginning, and did not witness the manner of our creation. So, if the strongest desire in anyone is to deny GOD, how does GOD, without removing free-will, remove a desire they firmly chose and hold. CHRIST has already loved us all in his own suffering death, so GOD'S love is already sufficient. If FATHER took each reprobate, and through miraculous acts, scared them to the point of death. That reprobate could still think, if he is a god he evolved, or maybe there are more preferable gods to serve, or GOD may be vulnerable to overthrow as Satan thought. It is *impossible* for any gifted created creature, to know any single thing beyond all and every possible doubt. And so, by GOD'S great design, it is our *desired faith of belief,* born through our core true love, that forever enters into the equation to establish our eternal steps. We can never truly believe in GOD and who they truly are, if we do not love GOD for all they truly are. And how can one truly love GOD, unless they truly know GOD. We must truly know the souls we love, or we love a desired imagined image. It is hard for many to see the many free-will dilemmas GOD has to try to reason with, to allow and find a true eternal love among free-willed entities. A true free-will love for GOD cannot be created, coerced, or forced, by GOD; and as it turns out, it is allusive, intimate, and rare. Love cannot be created, but a vast complex environment of desired choice can be. A created choice and chance filled reality, in which a spiritual love, *minus* all contrary lusts, can and must arise in some, by free-will desire. Surely GOD'S conscious forceful efforts are required to create such vast complexity, and all these vast complex requirements are exactly what we see. We see every single possible belief, can now be held, desired, and lived. The evidence for or against GOD, is the one same evidence, perceived from many desired perspectives, being either GOD'S eternal conscious primacy, or fallen temporary physical primacy. One's eternal perspective and desires can fade, and a temporary perspective, and desires, arise when one enters the confusion of individual beliefs. If one does not believe in eternal love and life, one will see things only in the near and present; saying things of belief like life is short, so I should do what makes me happy. Or, I don't see how what I like to do, hurts you. If you deny or forget that the one truth of love exists only in the ways of eternal created life, you have no true perspective, and you are trapped in the physical aspects of your created moment. All things are seen subjectively through individual desire and truth; and if this rules over you, an individual is who you now are. An individual can sense this, but they can never know the extent of their own confusion about reality. So, individuals just accept certain levels of confusion as a part of life. To understand reality, visualize the universe as a greenhouse, where free-willed spiritual seeds of hope for pure eternal love are sown into the grounds of hope, choice and chance. Yes, this is the perfect moment to bring up the Scriptural parable of the "Sower". We see in that parable, that many by self-desire, quickly reject the omniscient truth of love, and are defined as tares, weeds, and a virus against the eternal truth. A parasite puts first it's needs and wants, above those of its host. And if all these agents of rebellious disorder, were allowed to mingle with GOD'S Children for eternity, there could never be peace or love; but only the same aggressive battle of confused individual beliefs that permeate this entire world. Mankind's reductionism says the end of every physical state, brings a new physical state within a colder inflating reality; in this, there is no eternal lively perspective. Consider if one sets out upon a conscious journey to find their eternal purpose, where does their true journey begin. Did it begin with a desire to find, or with their search, or at finding their purpose. Again, we see the eternal Principle of Conscious Cause and Effect. All Materialist theories, exist between the two terms of cause and effect; but what are their arguments against GOD'S Conscious Cause Principle. One may *try* to reverse every chaotic turbulent physical thing that has occurred, back to a supposed physical beginning, yet how can they ever be correct. Seeing that competing free-will forceful projected interference is unpredictable. Can any predict or reverse the wave interference and convergence patterns of the ocean? No. And predicting the forceful desires and choices of billions of souls is impossible. If one believes in the determinism of a physical big bang; one holds a *faith* that this bang was an initial cause, and what followed was a physical chain of events. One then accepts there is no cause of anything, other than a physical cause. But seeing mankind only sees 6% of all physical cause and effect, this is less than useless in a predictive sense, and it offers no true description of reality. Physical cause and effect are useful in *trying* to organize many aspects of fallen physical life, yet *we* still see a free-will conscious determinism variable exists everywhere, again bringing uncertainty. So, in conscious determinism, it is not so cut and dried. We see conscious determinism can be reasoned with, it being malleable and fluid among hopeful desires, understandings, and choices. Our point being, there was at the creation of the Universe, a founding eternal conscious reason, purpose, and desire; and since it fell into endless evil choices, its complexity is abandoned more and more to chance, entropy, and fission, as all these conscious effects of contrary choices increase. Only the soon end to contrary conscious effects of evil choices will allow GOD'S eternal purpose to start. All the now firmly physical vanishes, and profits nothing. Then the conscious reasoned end for this lost physical reality, will have traveled full circle to be one with the conscious reasoned cause of a next new physical reality. And the LORD of all, will bring the final division of good and evil conscious determinism. CHRIST said, *"I AM THE BEGINNING AND THE END"*. Do you hear and love his firm purposed existential statement. For now, we see a continuum of evil conscious effects unfolding, as most can't see anything but this physical reality. At the beginning of all creation, there was either our conscious creative FATHER, or what? A physical *miracle* of condensed matter, energy, and force, that for no reason or purpose, just existed there in the middle of nothing. By all that you see, are, and know, how could that ever be? And of these two scenarios, which do you believe could create the universe, and all physical life, and gift us our souls of awareness. Can conscious souls design and build things? Yes. Can non-conscious soul-less things build themselves? No. Complexity by its semblance and structure, is designed, and never a chance occurrence. GOD'S true loving wisdom is here in our current moment, but for what purpose does it come to us. GOD will not forever allow the free physical exercise of souls to commit evil. Once GOD'S purpose for finding eternal love, ceases to yield any more new love; any continued allowance would then be a reward to evil, and be an evil in allowing evil its continued free reign for no eternal loving purpose. All evil souls are allowed to keep dividing into their own desired purpose; yet there exists no way for anyone to add any true good measure to their soul, outside of GOD'S perfect design. And FATHER'S perfect eternal purpose remains unstoppable, as most souls and a once perfect physical creation devolve. By all true reason, I know FATHER sees the Kingdom he desires and needs; and he knows it will be fulfilled and filled no matter what amount of evil arises during this process. We know by our loving obedience, that JESUS, HOLYSPIRIT, and FATHER, do lovingly, responsibly, and faithfully, see and hear all who seek them in earnest pain. GOD knows in great patience their love will be fulfilled without over-riding any soul's desired free-will choice. GOD'S eternal Kingdom is a matter of when, not if. While the vain religions, politics, and science, of this world abound, GOD remains the faith and love of the fewer. But all lost souls hope for an eternal physical paradise, or their final non-existence. Now some believe GOD sees and controls *all* things as they happen, and some claim GOD has compromised to accept their sin. Such souls deny their accountability in desiring the things of this world; and so, in affect, blame GOD for *their* evil. Oddly they see righteousness and justice in celebrating their unrestrained temporary worldly lust. But GOD seeks only his defined eternal spiritual form and standard of love. The mind and love of GOD are completely incapable of unprovoked thoughts of bringing arbitrary free-willed separations or divisions. FATHER knew by proclaiming and publishing his ways of love and truth that perpetuate eternal life with him, he had then defined everything that arises beyond those same created and sustained bounds, to be contrary and destructive. Yet in defining evil, as being outside of their good, GOD committed no evil. Evil is defined as all things beyond and contrary to the sustaining loving purpose of GOD. So, FATHER did not create any act of evil, he only created what evil is. Isaiah 45:7, "*I FORM THE LIGHT, AND CREATE DARKNESS: I MAKE PEACE, AND CREATE EVIL: I THE* ***LORD*** *DO ALL THESE THINGS.*" In forming light, all antithetical darkness was secondarily created by exclusion. In creating his peaceful ways of life, free-willed contrary chaos is created by all souls having contrary desires. All things antithetical, arise by an ever-increasing probability as each *fallen* soul rebels in their moment of choice. GOD knew the horizon of all things antithetical, created by free-will choice among many souls, would be limited *only* by the imaginations of all those rebellious souls. And so, GOD unavoidably must witness some evil events, and sense the over-all levels of evil. But GOD does not desire to witness each evil act or thought as they occur. FATHER'S Omniscience (meaning all true science) is the sum of all perfect knowledge and understanding of reality, and that is all. A false religious view, that omniscience knows or guides *every* physical action prior to its occurrence is a twisted blend of physical mysticism, that is incompatible in every way with free-will love ever being real. Some ask how does GOD see his Children differently from the lost. Again, like the Doppler shift, GOD sees this fallen dark corrupted physical world and all *lost* souls as drifting and fading towards pure darkness; and those who are growing in love for GOD, as increasing points of brightness. How could GOD ever choose to watch every heartless murder of over 400 million defenseless babies being torn apart as he grows them in the wombs of women; wombs he also grew and created. Or witness each lie about his creation, being indoctrinated into the hearts of naïve children. Or every deceit, pain, rape, murder, adultery, death, and war; and all the poor, weak, old, and sick being robbed or left to starve in a cold dirty refuge. Who could bear to see the entire sum of the continuous horrors committed by all evil men women and children in every expanding moment. Many ask, if GOD is there, how could GOD allow anyone to *physically* hurt a child. I say to all who ask this, should FATHER stop *all* who *endorse* ripping out tiny unready children, he grows in the safety of a womb. And stop all who endorse or allow a child's spiritual destruction in teaching them lies against GOD'S six days of creation, placing them on a path of eternal loss, pain, and suffering. Seeing how life under GOD is meant to be sacred, perfect, nourished, protected, and eternal; do souls who ask such questions see their own culpability and responsibility for the pain and confused spiritual destruction that *they* have brought upon children. If any think GOD should kill all who have ever hurt a child, GOD should kill everyone now. Such souls see themselves as judge over what is hurt and evil, and over what is life and death, and even over what GOD should do according to their false individual understanding. I have not my own judgment, belief, or understanding, as I work to accept and reiterate GOD'S perfect Scriptural ways of created life. Many souls say *only* certain destructions are evil, being selective and bowing not to GOD, but to their own desired beliefs. If we each could feel the sum of all the pain we have brought upon GOD and others. It would bring us a guilt and shame of such a magnitude, we would fall to the ground with a broken heart we could not bear. We could never draw another breath for the welled-up shame and spasms of pain in our throat and chest. We would cry unending until our last drop of moisture was forced from our last dry tear duct, leaving us dust in the winds. FATHER, throughout Scripture, states his pure reasons why he has gifted to us our free-will desired choice; and he also set responsibility and accountability upon our soul for our every unrepented, unforgiven evil choice, made within our heart. How can so many believe GOD accepts *them,* just because they are not as bad as others? Do many believe GOD is just looking for the lesser of two evils to be his children? Who can believe that GOD takes all the evil, divides it, and punishes the greater evil souls, and greatly rewards the lesser evil. GOD hates all the lesser to greater evil attacking these children; and are unavoidably aware of the ever-growing extent of this world's evil. Now GOD is focused on rescuing all true eternal spiritual love for life, and they are not *as* concerned with all temporary physical suffering. We are taught by GOD'S own words that he is surprised and greatly grieved at the evil acts most free-will souls are capable of. Gen.

6:7, "*AND THE LORD SAID, I WILL DESTROY MAN WHOM I HAVE CREATED FROM THE FACE OF THE EARTH; BOTH MAN, AND BEAST, AND THE CREEPING THING, AND THE FOWLS OF THE AIR; FOR IT REPENTETH ME THAT I HAVE MADE THEM*." The rapid increasing evil, which free-willed men, women, and now little children, are capable of, is hard to believe. In his perfect creation, FATHER created all things and life in his good orders. And gave to all souls, their choice to introduce free-willed disorder into all levels of life below the Trinity and the 11 Spirits within the Holy of Holies. As an example of GOD'S Holy order, visualize an immense long perfect funnel, far larger than the universe. Having a vast opening at the top, which rapidly narrows at first, and then more and more gradually until it narrows down to a tiny opening. And this funnel is made of about 300 million rings. The first vast immense ring, represents the size and responsibility of FATHER, and the next two smaller rings are JESUS, and HOLYSPIRIT. We then see seven gradual much smaller rings being the Archangels, then the four Seraphim in descending order, all being smaller than the prior. We then see the greatly smaller and smaller rings of the Holy Cherubim where Satan once resided (Ezek. 28).

Then any lesser order of angels being smaller still. Lastly, mankind appears as rings of Godly men and then women, and then children. But mankind can grow to be among the lowest Holy Angels; or be as the fallen angels, and fall out of this order completely. This is just a visual to see the gifted ascending and descending size and responsibility of souls under FATHER. We see under GOD, *no* orders of power or privilege or possession, just a greater to lesser created responsibility and ability to eternally sustain love and life, running up and down within it. All Angels and Mankind are formed in their lesser-to-lesser semblance of GOD, yet all in an adequate measure to allow us to grow to know all things of GOD if we desire. But if any soul, undeservedly desires to see their self-rise up, as equal in quality and quantity to GOD or another soul set above them, as Satan did, it is the Spirit of Antichrist. And if any soul rejects FATHER'S created order, please take your objection to the creator of it, and not to me; I have zero authority in this matter that we see declared throughout Scripture. I do faithfully accept, abide, and obey GOD, in my place, far down this chain of responsibility and purpose. I am thankful for my soul and life, no matter my size or ability. I seek my full responsibility and ability, and

I have no desire as Satan and others do, to rise above my ability and responsibility. Each soul must receive and bear, not individual power privilege or possession, but their great eternal faithful responsibility over all below them. Our eternal purpose is to fulfill our faithful responsibility towards eternal life for all. I know in my gifted current conscious size, force, and reach, I can't serve any higher purpose or responsibility above my ability. We can only be true and happy where we serve our eternal purpose, or we can rebel in sorrow and anguish. There is no third option of desiring and being equal to, or more than GOD. I accept my created purpose, and I know that all GOD'S loving truth will fill every soul in Heaven. This is GOD'S work alone. I have not placed anyone in any order, and neither has anyone else in this world. It is up to us to strive learn and accept where we belong, *or* rebel against GOD. Now in their objection to FATHER'S order, we see a fast-growing number of godless male and female feminists, who feel women are more stable and compassionate, and think women should lead the world. Now to see if such a superior female compassion exists, I compare the far less than 400,000,000 combined murders, and deaths in wars throughout *all history* by men; to the far over 400,000,000 million women who kill their own defenseless babies each year. These most innocent souls striving within, as GOD knows and grows them in strength, which these women were born to love and protect. Of course, we know women hurt and kill adults and older children too. For feminists, compassion is a subjective term; and a less compassionate insane soul can never know they are less compassionate than another, as compassion is experiential. When women claim a right to kill the most innocent of life within them, I see the greatest narcissistic lack of compassion that ever existed. Jer. 2:33-35, "*WHY TRIMMEST THOU THY WAY TO SEEK LOVE? THEREFORE HAST THOU ALSO TAUGHT THE WICKED ONE'S THY WAYS. ALSO <u>IN THY SKIRTS IS FOUND THE BLOOD OF THE SOULS OF THE POOR INNOCENTS</u>: I HAVE NOT FOUND IT BY SECRET SEARCH, <u>BUT UPON **ALL** THESE</u>. YET THOU SAYEST, <u>BECAUSE I AM INNOCENT</u>, SURELY HIS ANGER SHALL TURN FROM ME.*" Murdering babies in the womb is nothing new. From seeing the first miscarriages and still births, evil souls have sought to artificially cause these deaths. The closer you get to the very moment of physical conception and the soul's union, the closer to innocent that life is. I am surely not saying today's rebellious secular, or vane religious male leaders are doing far better. But if a soul wants to do good and follow GOD, both men and women must accept that men were, not according to *me*, but GOD'S words, created first and closer to GOD in their orders of responsibility than women. And as with GOD, Godly *men* can only lead the free-willed who follow. JESUS came in pure loving responsibility to lead and correct, but was rejected and not heard or followed by divisive individuals. By this created chain of responsibility by FATHER, to pass down and share his instruction and care, men are meant to follow GOD and the Holy Angels, and to lead women and children. And if GOD'S created chain of responsibility offends any soul, and I am sure it offends *so* many; your hate for, and your battle is not with *me*, or my pure iteration and obeisance to GOD'S WORD, but with your Creator and Scripture. I gave up personal understandings long ago. Gen. 2:18, "*AND THE **LORD GOD** SAID, IT IS NOT GOOD THAT THE MAN SHOULD BE ALONE; I WILL MAKE HIM AN HELP MEET FOR HIM.*" Gen. 2:23, "*AND ADAM SAID, THIS IS NOW BONE OF MY BONE'S, AND FLESH OF MY FLESH: SHE SHALL BE CALLED WOMAN, BECAUSE SHE WAS TAKEN OUT*

*OF MAN.*" Right after the fall of mankind we read. Gen. 3:16, "*UNTO THE WOMAN HE SAID, I WILL GREATLY MULTIPLY THY SORROW AND THY CONCEPTION; IN SORROW THOU SHALT BRING FORTH CHILDREN; AND THY DESIRE SHALL BE TO THY HUSBAND, AND HE SHALL RULE OVER THEE.*" 1st Cor. 11:8-9, "*FOR THE MAN IS NOT OF THE WOMAN; BUT THE WOMAN OF THE MAN. NEITHER WAS THE MAN CREATED FOR THE WOMAN; BUT THE WOMAN FOR THE MAN.*" I know many women, and men, say these words were written by men for men. I imagine godless feminists closing, dropping, or throwing this book at reading this; and if so, they throw the Bible also. These are GOD'S true iterations about FATHER'S omniscient design. If one sees life in this world as a struggle between men, women, cultures, and races, for power, privilege, and possessions, they are blind to the ways of GOD. Their slander means nothing to me, because I know what they desire and cheer for, and it is evil. If I should think or say I am equal to a higher created Angel, or JESUS above me; am I righteous, or rebellious? Can I responsibly accomplish all of the requirements of a higher place; or is that falsely seeking power, privilege, and possession, and destroying GOD'S created order? If a man, woman, or child, assumes they have a right to murder a defenseless growing baby they find undesired, or inconvenient, or any who condones or allows such evil; I truly fear such an uncompassionate cold evil soul having control over my life. They probably find me inconvenient and undesired, being another soul to be gotten rid of and killed. Each and every life is FATHER'S gifted creation of undefinable value. Yet, free-will means exactly that, and many take it upon themselves to be the arbiters of life and death. All souls are free from GOD'S will to choose and do what they want. Right up until their body dies, or the moment that earth yields no more new eternal love. Then FATHER intervenes to stop all sin and horror. If FATHER were to focus upon our souls *as* we create our choices, his supreme forceful *will* of creative perception would over-power our will to create our choice. But we see this does not happen, because FATHER does not do this. FATHER only sees us after we make our choices. We cannot blame GOD for anything we do if GOD did not see it before we did it. GOD can see everything we have done, after we choose to do it. And from what he *has* seen, FATHER repented his decision to make mankind. He could never repent, or be surprised, or regretful, if he sees controls and knows everything before and as it happens. Minus a true complete free-will and chance among many, all choice, love, and responsibility, cease to exist. I refer you to the record in Genesis. JESUS came *daily* to visit and teach Adam and Eve FATHER'S truth in the Garden of GOD. JESUS came one day, *not yet* knowing they had sinned. He came in hope as always, for the joys of the free-will love he had found each visit. But calling out to Adam, where are you, and hearing certain evil things; JESUS asks of Eve, in sorrow and pain. *WHAT IS THIS YOU HAVE DONE!* In that moment, Adam and Eve did not know the destruction they brought, only GOD knew what they had started. We now see clearly, the long results of what they started. We see love and life destroyed by that first tiny sin. We see for love to exist, FATHER has to trust, hope, and eternally let go of our souls, and at most times hide his eyes and presence from us. So, we must fervently pursue the belief, understanding, and ways of GOD, by going to where they reside in eternal truth and love. If we run to the loving ways of FATHER, he can in spirit come near to commune with us on a personal loving level, as he does with our LORD. Many think they see, feel, and know the love GOD holds; but the love most feel is mingled with their desires for this world. With every worldly desire, our love for GOD weakens. No mixture of Godly and worldly love can survive forever, it can barely grow and then only diminish, as everything else in this world does. Only GOD'S love can eternally erase the suffering and loneliness in each soul. Now, it was at *least* 11 orders of magnitude below the physical/conscious size and perspective of the Trinity, where we see Lucifer was gifted and created. He being the first to turn inward in a desire to be his own god, who shared his liberal desires to tempt and lure away one third of the perfect number of the Angels in Heaven. A loss of one third of the Holy Angels, caused by Satan, destroyed a third of the eternal love for GOD in Heaven, and introduced entropy and fission to its created complexity. Yet entropy and fission do not control the desires we hold, or the choices we make today. In the earth, Satan succeeded in first luring away the weaker Eve from following GOD. Then using Eve's newly confused belief, and Adam's love and trust in her; Satan lured Adam out of his shared perfection with GOD. Then the earth and all physical life in it, being under Adams dominion. Fell from perfection as FATHER recoiled and pulled his hands away from Adam's sin, stretching the physical creation out from its dynamic perfect courses. Adams dominion over the earth, given by GOD, was there and then surrendered to Satan. Do you see why and how, that for *eternal* loving peaceful life among three hundred million souls there must be only *one* shared lived existential truth. All spiritual and physical choices to be made in a New Heaven and Earth, must and will be made to support all shared eternal loving needs under GOD. We all exist forever, and we will either be in, or outside of GOD'S loving provision. We see all dynamic created physical perfection endures, only if the spiritual perfection of GOD'S love is lived within it. Now we see the Trinity is the Perfect balance of GOD'S love. See the three sides of a perfect triangle, being the strongest of shapes, with one side always seeing the other two sides equally. In a four-sided square, one side is always seen more than the other two, and the more sides you add, the greater this problem becomes. Also, two sides become one line seeing only a reflection. Three, for many spiritual reasons, is the perfect number of the GODHEAD. Yet, they are purely aware of each being one of three distinct souls, bound as the Trinity of GOD by their free-will eternal *oneness* of love, truth, and believed understanding. This Triune Spiritual Symmetry and oneness of immutable truth and love, has total control over all that is physical. This is why these three alone, are the eternal GODHEAD; and why all they gift, create, and know below, can never rise up to or above them. Of course, GOD is not a triangle or a funnel shape. The Trinity is the true perfect numbered order of GOD, existing in eternal forceful mathematical logic, *and* pure indivisible eternal love. GOD has not one weakness or external physical need, and all loving souls are gifted and called to become one in Spirit (belief and understanding) with GOD and GOD'S Family. If we look at the TRINITY in their created size and force fashion, being the top of that ringed funnel, we see FATHER being far above, and all supreme omniscience, force, truth, and love, remain with FATHER. And only by GOD existing as three subsequent souls, can they see and reach down into the imperfections and impurities of mankind, to grab our repentant hand reaching up out of hell and death. We see in the Trinity, three great facets that allow all angels and mankind to aspire, and become one with GOD in spirit. Without CHRIST'S sacrifice making a miraculous loving bridge. And HOLYSPIRIT to lead us across that bridge. And FATHER planning, designing, gifting, and creating it all from the other side, *all* fallen souls would remain beyond the reach of FATHER alone. Within this Trinity, rests all the expressions and mechanisms of the strongest possible forgiving love. FATHER alone, could not put off his eternal body to come in the flesh of mankind to suffer and die for our sin. But CHRIST could physically suffer and die under FATHER'S supreme assurance. And only HOLYSPIRIT can do other certain things for GOD'S children that FATHER and JESUS cannot accomplish alone. We see a qualified quantified mechanism of loving restoration through the three entangled souls of GOD, seeking out and supporting our return to all eternal perfect love. Sin *and* its forgiveness, both have great cost, or they are meaningless. Now is there a perfect number of souls required to return GOD'S great love, and fill the New City of Heaven; or is Heaven adjustable to house almost any number of souls? We know Scripture defines the New City of Heaven's size, as 1500 miles wide, deep, and high; so, it is not infinite. It does sound perfect for the three hundred million angels implied in Scripture. There is much to this loving logical equation we can't see or comprehend at this moment. Please see that all confusions of hell, which bring physical death, at the last, these must be separated and imprisoned. Satan, in visiting GOD'S Garden in the earth as the serpent, offered Eve subtle lies, being his seeds of envious evil individualism. And by Eve falling to his temptation, all possible false beliefs were loosed to fill the earth. All endless individual beliefs, were described thereafter by GOD, as the growing endless conscious confused abyss of hell. Which FATHER says, in Isaiah 5:14, "*THEREFOR HELL HATH ENLARGED HERSELF, AND OPENED HER MOUTH WITHOUT MEASURE*". And in Proverbs 15:24, "*THE WAY OF LIFE IS ABOVE TO THE WISE, THAT HE MAY* <u>*DEPART*</u> *FROM HELL BENEATH.*" I plead with you to *depart* from all confused lies of hell while you still can. As to these two conscious-states, or conscious estates, described as heaven and hell, we must study the spiritual and physical evidence around us to see how to depart our own confused state of *hell*. For a visual, we first ask what is the one shared conscious state of Heaven. Of course, Heaven and Hell, are not primarily physical places, they are the only two eternal conscious-states and paths of conscious outcome, laid out before us. The conscious state of Heaven, is found in our desire and need to perpetuate eternal love and life. The desire for eternal life is not born or held by thinking to one's self that a perfect physical place for eternal physical life sounds wonderful. This true eternal desire, is found in our deep search to understand and accept our own responsibility to sustain eternal life. It is choosing here and now, GOD'S perfect eternal created ways; and rejecting all individual ways and beliefs. There is GOD'S promised ever-increasing spiritual peace and love to be found, but most of mankind, including most who believe in a god, just cannot want or trust in this enough to turn from their individual desires. All their lesser dying physical pleasures of lust and life, they still cling tight to. Heaven is and will forever be many spiritual things we cannot now imagine. But the truth is, that our loving desire for these spiritual things is not just searched for by GOD in our desires to receive these, but in our free-will desire to faithfully and responsibly protect, share, give, live, and perpetuate these, *now* and forever. The perfect conscious estate of Heaven *is not* the impossibility of many physical places, made physically perfect for soul's different physical desires. Heaven can only be where we are all made spiritually perfect together in one way of life. Heaven is the conscious state of finding the shared eternal love and joy that GOD offers. But we see those souls ruled by individual physical desires, can't accept that there exists *only* one omniscient belief and way of eternal peace and love. This *one* desired conscious state and estate of Heaven, could never exist among individuals. Now, having only barely scratched the surface of what is the conscious state of Heaven, we turn back to define the conscious state and estate of hell. Simply put, it is contrary and opposite, in rejecting the understanding and subsequent desires to live in GOD'S one omniscient truth. It is being set adrift, unanchored and lost in one's endless unattainable individual desires. It's playing a god to define one's own truth of existing, even defining GOD for one's self. It is choosing to be central to one's desires, and to live different from GOD in any way one wants, and causing the timeliness of one's endless fading physical life. It is a soul denying all responsibility to sustain eternal life and love. These are souls that feel entitled to eternal unreciprocated care and love, having no desire for the created ways of GOD. Being lost in their confusion of enjoying all the physical pleasures they sense between all their prior and approaching pains and losses. Consider the entire *physical* reality, and how its smaller aspects are drawn to join that which is massive, to be part of one larger system. And like-wise, in the spiritual realm, GOD is the massive forceful polarity of eternal love that draws our souls together in one eternal loving system of belief. The rejection of GOD'S spiritual ways, belief, and love, leaves each soul as a free radical agent, whose only eternal attraction is to their own mind; a black hole of a soul becoming more and more contrary to others forever. We see the physically aged, separate from the physically young, seeing very different disfigured physical pictures; and both holding to theirs in confused fierceness, as if either were of eternal value. Mankind sees the ghostly as an illusion, and the physical as the real. But at the last, the ghostly and spiritual are the real and eternal, and the physical world is the fleeting illusion. We see that each individual conscious state of hell is best described as a state of confused conscious denial, as to what is true, good, and set by GOD. Being deeply lost in a conscious state of hell, is like wandering into the center of a million square miles of tall dense forest. And no matter which way one looks for the far distant existential truth of all reality to escape, they see only their dense physical surroundings. They know only that which is physically near, having no understanding or guidance to escape. But if they bow and repentantly look straight up spiritually, they may yet see the truth and light of their rescue. Now the thing about being an individual, most never consider, is that being an individual means to separate, and put space and difference of desire and belief between you, GOD, and everyone else. But most don't see how in doing this to a lesser pleasing degree now, they will have severed all the unseen eternal conscious moorings, necessary for our soul's good and shared eternal spiritual life. This leaves them set spiritually adrift and changing forever. And in rejecting all further efforts from GOD to rejoin their family, these have chosen their final judgment to be forever alone. They are set adrift in the conscious hell of their own making, as their beliefs and understanding forever change. Hell is not only a state of ever-growing confusion as to the one truth of eternal life, it is a falling descent into a never-ending insane ever-increasing loneliness, to be imprisoned and trapped in an individual sealed conscious confused reality of self. The fullness of Scripture, does not describe hell as a *physical destination* of the lost; but as their confused malleable conscious state. Rev. 20:14, "*AND DEATH AND HELL WERE CAST INTO THE LAKE OF FIRE. THIS IS THE SECOND DEATH.*" Death and hell are adjoined. Death, being our conscious separation from GOD; and hell, being our conscious confused separation from GOD'S truth, and everyone else, are neither one a physical thing. At the final divide of all souls, we see all conscious states of death and hell are cast into a *physical place* of endless confinement called the lake of fire. This is a final physical prison for all unredeemable conscious evil, to be secured away from all who chose GOD'S ways. Please turn to emulate the loving perfection of GOD, and leave the conscious state of hell. Proverbs 15:22-24, "*WITHOUT COUNSEL PURPOSES ARE DISAPPOINTED: BUT IN THE MULTITUDE OF COUNSELLORS THEY ARE ESTABLISHED. A MAN HATH JOY BY THE ANSWER OF HIS MOUTH: AND A WORD SPOKEN IN DUE SEASON, HOW GOOD IS IT! THE WAY OF LIFE IS ABOVE TO THE WISE, THAT HE MAY DEPART FROM HELL BENEATH.*" Meaning; the council and wisdom of all eternal oneness, comes only from joining to those responsible souls above us. And we can then escape the conscious hell of individualism that reigns below. While young and rebelling in the flesh, most souls love the freedom and power they sense in their early separated individual state; but as their confusion grows and grows forever, they will hate and fear their hell more and more. Yet most don't know to fear this conscious hell; they only fear that physical lake of fire. And even this they do for false physical reasons. Though there will be eternal torment, being conscious misery. There will be no searing physical pain of flesh being burned by flames. This lake is not a place of eternal cruelty coming by the wrathful hand of a rejected GOD. It is a just required wrathful separation of all divisiveness, to protect GOD'S found love. All individual eternal punishment is self-inflicted, just like all the painful things of hate and death they experience here in their time of choosing. All souls at our judgment, are resurrected to an eternal body, having our known outer semblance, yet having eternal properties like the Angels. Such bodies are not susceptible to fire or other elements in the ways of this current flesh. Isaiah 43:2, "*WHEN THOU WALKEST THROUGH THE FIRE, THOU SHALT NOT BE BURNED; NEITHER SHALL THE FLAME KINDLE UPON THEE.*" Luke 20:36, "*NEITHER CAN THEY DIE ANY MORE: FOR THEY ARE EQUAL UNTO THE ANGELS;*" All pain of being burned, is felt not by fire or heat, but in it destroying and consuming the fleshly nerves. Every resurrected body is eternal for both those in Heaven, *and* for the lost. It is impervious to destruction, or searing pain, or any strong feeling. Being eternally lost, means being trapped in an unfeeling prison of a body. The great individual torment for the lost, comes in not having their former freedom of movement, or strong physical senses they currently have, ever again. Mobility and these strong sensations are required to satisfy their physical desires. All physical beauty will forever fade by their tiring disfigured perceptions of it. Their never-ending anguish will come by their own growing confused conscious perceptions of hate for GOD, all others, and their self. Their punishment is in wanting and searching, yet never having or finding. Heaven is in having found, and eternally having all good spiritual and physical loving things. To visualize these repelling and attracting forces and effects that come from holding to either the one pure truth of life, or an individual belief of life. Visualize souls as bi-pole magnets, that when aligned, their polarity joins together as souls become one in a chain of shared truth and love. Now if you reverse one magnet, or soul, we see their opposing force of direction pushes them apart. A shared forceful flow, represents FATHER'S flowing omniscient love and truth, sustaining his creation. And its great source, sits rightly upon the Throne of care and responsibility above all created eternal life. FATHER holds and defines reality, life, truth, and he alone fulfills all requirements to create and rule over it all, so he rightfully belongs there. Now all souls below the Trinity, do *not* own or sustain their physical body of eternal life. So, they do *not* belong on a throne, not even a throne over their own life. This term "force polarity" is a visual to help see the design of forceful oneness that perpetuates eternal life in a shared direction of attraction and bonding. And if we do *not* belong on any throne, having no power of life; yet by inward selfish prideful envy and free-will choice, we sit upon the throne of *our* life. We have reversed our polarity to be a conflicting, repelling, conscious force, pushing away the source of eternal life, and falling into conscious confusions of hell. If an individual prior to becoming a true reprobate, repents and reverses their love to GOD, and are forgiven, they then become one in the direction towards being saved. This example is simplistic, yet analogous to what GOD is trying to repair in each of us by giving us eternal understanding, reason, purpose, and being our example. Wanting us to choose to accept his required responsibility and guidance to lead us back to his pure created model of eternal life. We see that if we are true and loving, we cannot help but be forcefully pulled directly to GOD. We are all here forever, so why not desire eternal love and life. We ask you all, is not your self-awareness of true understanding of reality and life itself, your most desired gift. Or, are your true joys of life found in temporary physical sensations and experiences. Why are the true spiritual fulfillments of love alone, not enough for the many who must have their physical experience of qualia. If we do not resist the evil fading physical temptations, we take on the semblance of a purely physical organism, even as a group. Life is then more and more, a pure physical exercise. I see only self-centered motivation in *individual* ways of life; and I see only responsible faithful loving motivations in GOD'S *one* way. All should clearly see the simplicity of this matter. Our trust, born of true sacrificial love for GOD *and* his ways, is our strength and guide. If this sacrificial love and trust is not growing in us, by what other force can we be led and drawn to GOD. In knowing we are eternal conscious souls, we fear losing all love and shared sanity, far above losing this temporary physical world and body. If we don't love and observe GOD'S ways, it is because we just don't believe, love, or fear GOD, more than this physical world. We all do the things we desire out of belief, love, and fear; what other reasons can there be. All souls, families, and even nations are defined and guided by their shared beliefs, loves, and fears. If our ways of life differ from those of GOD, how can we say we love the ways of GOD. And if one cannot say we love GOD'S ways, how can we say we love GOD, except for what we hope GOD can physically do for, and give us. Matt. 15:8, "*THIS PEOPLE DRAWETH NIGH UNTO ME WITH THEIR MOUTH, AND HONOURETH ME WITH THEIR LIPS; BUT THEIR HEART IS FAR FROM ME.*" At the last, despite chance and all free-will choice; just as water seeks its own level, and as water and oil separate, the polarities of the perfect plan and the loving needs of GOD and his family are being fulfilled. And yes, eternal love comes at great cost to all through our remembered, or never-ending pain. The formula before us is this; to come out of our imperfection, requires the exact same desires ways and means required for us to remain in perfection. Being that if GOD tells us to commonly believe and do this and that for the good of us all; and commonly don't believe and do that or this for the good of us all; we obey GOD in our eternal love and trust. Why, because we see FATHER is loving, in that in his search he does not try to force anyone to love anything. GOD is long suffering and omniscient, and trying to preserve us with them in their promised loving eternal peace and joy. The only true question is not will we exist forever; but will we spend our eternity in loving unity, or as singular minds in an ever-growing confused lonely hell. Every whisper from GOD is in nurturing love, sowing seeds of truth into all his elements. He weeds, waters, digs, and covers at the sight of any young loving plant breaking through, reaching up to the good warm light of Heaven. Now around 4,000 yrs. ago, all mankind was on the verge of being eternally lost as reprobates; yet FATHER'S plan was not fulfilled and he desired more loving souls. So, to save the last eight souls who held and looked up to the truth, he cleansed all corrupt life from the earth. Every other soul, had by their depths of rebellion and lust, grew to be reprobates beyond GOD'S reach. GOD truly needs to love us; but by their great love, they need us to truly love them back in the same eternal way. With HOLYSPIRIT, we are strong, reasoned, and purposed to live eternal. *And* we do relieve GOD'S added suffering he would feel, watching us also fall into the self-destruction of the wicked. 2nd Peter, 3:9, "*IT IS WRITTEN, IT WAS AND IS NOT THE WILL OR WANT OF GOD THAT A SINGLE SOUL SHOULD PERISH, BUT HAVING THE HOPE THAT ALL WILL COME AND*

*FIND EVERLASTING LIFE*." Have no doubt that all the evil in this world, comes by lesser prideful envious souls and their desires for the non-spiritual things of this world. Every soul that truly turns back to seek the one pure truth, brings to GOD two-fold, being more love on one side, and less pain and loss on the other. If you feel great pain within your soul in the loss of those dear to you, and you feel pain and compassion for others who hurt. How much more does the one true loving GOD feel the pain and cost for the sum of all lost souls. We are all gifted in the basic likeness of FATHER'S emotions and outer semblance, yet most souls never consider the hurt they cause GOD. Eternal love and life is the only concept having true value. And by evil's prevailing law of *diminishing* returns, eternal love appears more painful and costly to GOD and all for each soul lost, and for each soul saved. The cost and truth of what evil is, and that for a while it is allowed, is an indisputable harsh fact, as evil again fills the world to over-flowing. We know beyond doubt, that if eternal love could have been found any other way, our loving suffering FATHER would have embraced it so dearly above what we see. Now going back before the very beginning of all known external creation. We see the eternal lonely self-awareness in FATHER all alone at the moment of his great choice of going forward into uncertain cost, to dived his soul to find a loving eternal family and to not be alone anymore, to risk unknown pain and cost to all in finding his eternal love. We must see how costly eternal love can grow to be for all, and for GOD, the cost being beyond comprehension. We see weak inferior, asymmetric, divisive, dying loves, that GOD can't sustain, abide, or desire. And though these weaker desires do exist and are called love in this world. They are not the higher form or standard of eternal faithful responsible sacrificial love that GOD holds, desires, and needs. Impure desires, are partial, weak, and confused in their ways, and will all some-day succumb to adversity. All lesser loves will turn to tolerance, and to hate in the end. The unstable individual fractions of affection and emotion this world calls love, are all divisive in their ways. Souls disagree and argue, as most blind themselves to their endless strife, and can't see how they destroy all eternal peace. The battle between the eternal good and all *im*perfection has only grown since that first seed of individual envy was born in Lucifer that now consumes him in pure insanity. It was a fuse lit to all dying destruction. All evil is the dying cause that brings a final dissolution to this current reality. FATHER'S search for eternal love was set to unfold after his creation of *choice* being within his first gifted free-will soul. And by the glorious sacrifice of MESSIAH, the end of that search is now set and approaching fast. From the first soul to fall from GOD'S truth, what test has this great battle endured? Is it the test of time? No. It is the test of eternal love. We see here and there, that weak worldly love starts and stops, it comes and goes. Weak love is born, and just as often it dies. If love is not eternal, it is not perfect. We see how the first three, then fourteen, then eighteen perfect loves aligned eternally; and at the last we will see around three hundred million eternal loves align perfectly. Now let's discuss how as our soul's mature, we sense physical life is speeding up. This is said to be a conscious anomaly, due to aging; but none can test that theory in any physical measured way. So, every generation says the same thing, being that as they grow in experience and understanding, the feeling of their reality occurs in faster way, having less conscious space between all thoughtful experience. Each generation is told and believes, this is only a trick of the light that comes by age. But this is not a scientific answer. Our soul's sum and surface grow in size as we experience the elements and chance, and all choices of reality. All as the far reaches of the universe expand away from us faster than the *slowest* speed of light. So, why wouldn't we sense a minute time dilation and acceleration effect upon our own expanding relative experience. Each new soul enters at a *current* point of the expanding state of reality; and without proof or reason, are told time is a real and constant fourth dimension. Yet through their life experience, each generation perceives their moment expanding faster. And again, are told it is a personal misgiving of age. Yet each generation enters at a faster pace of reality, having more anxiety urgency and insanity than their predecessors. After Eve and Adam fell from their far faster eternal perfection, they only lived for 900+ years in a new slower smaller perspective of reality, being aged by the now imbalance of FATHER stretching the universe. What did those 900 plus years feel like? They endured a first certain stretching of a perfect dynamic creation into an altered imperfect dynamic creation. But not being a continual stretching like we now experience. There was a second greater stretching at the flood, and a third continuous stretching, followed CHRIST'S Crucifixion. But through all of this, our spiritual and physical choices, and chance, and all the sensed information to make our choices, remains relative to the evidence. All entangled reality now inflates faster, even down to all entangled particles, which introduced fission and increases all the destructive effects called time. Once you get beyond mankind's slow small perspective, you see why they can't properly define the true constants. The entire universe, from its least aspect to its entirety, is a conscious variable in flux. But most of these entangled variables cannot be seen from within mankind's limited understanding. Everything quantum and classical, from the outside in, expands all at once in a stretching relative way. Many variable aspects appear to the fallen as stable and constant. Mankind's limited search for understanding relies greatly upon contrasting information to comprehend anything of difference or change. So, as every near aspect inflates in a relative way, mankind cannot see all these entangled classical/quantum changes. If the oscillations of cesium atoms in atomic clocks change, their measurement of time must change; yet, it goes unseen. But if a cesium atom could be measured using the aether, or any contrast that touches GOD and never changes, that contrast would allow materialist to see the atom's changing oscillation. Mankind could then measure our soul's increasing relative perspective and perception of reality's passing, we only now notice after 2-3 decades. After mankind fell into their confused conscious abyss of hell, and GOD initially stretched this universe, that initial loss of perfection, introduced disorder first, *then* fission, predation, disease, cancer, and all newer physical confusions, appearing in their predictable order of infecting all aspects of physical complexity. FATHER'S orders of perfect dynamic stability and physical life that once permeated all physical reality, is contrary to those who do not desire GOD. Yet they look inward and backward to find any secrets that might extend their physical life. These do not believe a consciously created perfect physical life was ever behind them. They hope a consciously created physical perfection is in front of them. They do not see how they oddly try to overrule physical death, through the very same rebellion that brings death right to them. In all their godless efforts, they only destroy themselves. They fight for every illusion of adding a few months or years to their physical side, while throwing away eternal life on the spiritual side; a wisdom to which they are blinded. Is reality rising to find mindful omniscience, or falling into confusion? Reality is consciously splitting apart, so both. As Scripture says, the cosmos and all physical life are moving faster and faster towards final dissolution. Each generation becomes less and worse than the last; yet believe with all they are, they are more and better than the last. Such an odd dichotomy of growing insane delusions. We must see that eternal joy cannot be found for all souls, by FATHER'S never-ending work to fix all endless destruction. Eternal joy, for us *and* for GOD, *rests* in zero corruption projected into GOD'S created lively perfection. The Dualism of physical and spiritual perfection of life existed before, and will exist again, once all has been accomplished. The *im*perfect temporary perceptions of time are so cruel, as each grain of sand drops through that narrow now of the hour glass, into a dark abyss of hell below. With each grain, leaving perfection further behind, bringing souls the physical death they dread. Slowly here, and suddenly there, ripping from their heart and hands all the things they desire in this world. Through its power of death and destruction, time promises at the last, to take it all. It is written throughout Scripture that in the end of times, which are at our door. The world will see a sudden qualitative and quantitative creative conscious eruption. Scriptures say the ancient exercising of the greater powers of creative forceful souls, interacting upon reality, are about to once again be let loose upon the Earth. Rev. 12:7-12, "*AND THERE WAS WAR IN HEAVEN: MICHAEL AND HIS ANGELS FOUGHT AGAINST THE DRAGON; AND THE DRAGON FOUGHT AND HIS ANGELS, AND PREVAILED NOT; NEITHER WAS THEIR PLACE FOUND ANY MORE IN HEAVEN. AND THE GREAT DRAGON WAS CAST OUT, THAT OLD SERPENT, CALLED THE DEVIL, AND SATAN, WHICH DECEIVETH THE WHOLE WORLD: HE WAS CAST OUT INTO THE EARTH, AND HIS ANGELS WERE CAST OUT WITH HIM. AND I HEARD A LOUD VOICE SAYING IN HEAVEN, NOW IS COME SALVATION, AND STRENGTH, AND THE KINGDOM OF OUR* ***GOD***, *AND THE POWER OF HIS* ***CHRIST***: *FOR THE ACCUSER OF OUR BRETHREN IS CAST DOWN, WHICH ACCUSED THEM BEFORE OUR* ***GOD*** *DAY AND NIGHT. AND THEY OVERCAME HIM BY THE BLOOD OF THE* ***LAMB***, *AND BY THE WORD OF THEIR TESTIMONY; AND THEY LOVED NOT THEIR LIVES UNTO THE DEATH. THEREFORE REJOICE, YE HEAVENS, AND YE THAT DWELL IN THEM. WOE TO THE INHABITERS OF THE EARTH AND OF THE SEA! FOR THE DEVIL IS COME DOWN UNTO YOU,*

*HAVING GREAT WRATH, BECAUSE HE KNOWETH THAT HE HATH BUT A SHORT TIME."* Rev. 14:13-14 *"AND HE DOETH GREAT WONDERS, SO THAT HE MAKETH FIRE COME DOWN FROM HEAVEN ON THE EARTH IN THE SIGHT OF MEN, AND DECEIVETH THEM THAT DWELL ON THE EARTH BY THE MEANS OF THOSE MIRACLES WHICH HE HAD POWER TO DO IN THE SIGHT OF THE BEAST; SAYING TO THEM THAT DWELL ON THE EARTH, THAT THEY SHOULD MAKE AN IMAGE TO THE BEAST, WHICH HAD THE WOUND BY A SWORD, AND DID LIVE*." The Holy *and* fallen Angels, all have greater exercised forceful creative reach than mankind, and many are coming here. Is their greater creative reach, entirely in the gifting of their souls? No. Like ours their control has been learned and exercised in faith. The aether is provided by GOD for creating, and it is one shared aether. So, how large firm or real, can an Angel's perceptions be and *affect* us. All but FATHER, exist immersed in his external aether together. JESUS, HOLYSPIRIT, and every other soul are perceptually bound together under FATHER within his created theatre of reality. JESUS, HOLYSPIRIT, or an Angel of great size reach and desire, can by command, or the allowance of FATHER, project great patterns of force that can alter our local physical reality. Or, they can create light patterns of information traveling through the aether, giving to many, a true illusion. Great realities and illusions will both soon return. We must be one with GOD to know which works are of GOD. CHRIST came here to see, hear, walk, and suffer, among mankind's endless evil. Through JESUS, FATHER showed his total conscious control over, not just patterns of light to cause illusions, but over the perceived physical reality itself. By the projected forceful desires of CHRIST, many saw restoration and reanimation in bodies that had been severely damaged, or lost to the sleep of death. Also instantly seeing a raging wind and sea, suddenly go calm at his command. At his projected desire, a ship at sea moved through the aether to its destination instantly. At FATHER'S desire, Angels, Apostles, and Prophets, are translated between the two realms without time. CHRIST explained FATHER'S control over reality to his disciples who accepted and lived by that omniscient truth. JESUS taught them why they being of GOD in mind and belief could pray to FATHER for certain *external* things to occur, but not certain other things. Explaining why some were harder than those they had experienced.

Matt. 17:20, "*AND* ***JESUS*** *SAID UNTO THEM, BECAUSE OF YOUR UNBELIEF: FOR VERILY I SAY UNTO YOU, IF YE HAVE FAITH AS A GRAIN OF MUSTARD SEED, YE SHALL SAY UNTO THIS MOUNTAIN, REMOVE HENCE TO YONDER PLACE; AND IT SHALL REMOVE; AND NOTHING SHALL BE IMPOSSIBLE UNTO YOU.*" This substantive faith as a mustard seed is a rarer pure angelic faith. Do not think because you say you believe, you have this ardent solidity of faith. And if one without doubt, loves, believes, walks, and lives as CHRIST did, such a faith calls them to obey every word from FATHER in pure love and fear of losing him. It is our Godly way of life that declares the strength of our loving faith. If one believes, yet loves GOD a lesser amount, they cannot obey GOD'S every word faithfully, and are in each moment of rebellion, a devil. Matthew 16:23, "*BUT HE TURNED, AND SAID UNTO PETER, GET THEE BEHIND ME, SATAN: THOU ART AN OFFENCE UNTO ME: FOR THOU SAVOUREST NOT THE THINGS THAT BE OF GOD, BUT THOSE THAT BE OF MEN.*" We see Peter's strong desires temporarily over-rode his loving obedience to GOD, and in a moment Peter's love was not strong enough. Many times, in

Scripture, we see FATHER'S projected guidance create, restore, sustain, and reorder physical reality. We see fallen mankind's hopes and dreams in movies, based on blurred science, where people gain super powers in this world. But many think the Bible saying these powers always existed, is an ignorant fairytale. They reject our conscious forceful projection as part of their equation. The angels do not have the forceful reach and control over the vast faster heaven they have way down in this tiny slower physical earth. If we are one with GOD, we can ask FATHER to alter the physical existence in the ways he has said he desires to do, just as CHRIST, HOLYSPIRIT, and the Holy Angels do. Science detects and measures GOD'S original projected desired force *affects*, but never that invisible desire itself. Science is unable to qualify or quantify the strength of our soul's projected creative force, and they are only aware of the effects of our created particle/wave-based perceptions. They sense the created affects as evidence, yet deny the soul's creative force of perception. FATHER said for a moment, he allows mankind all the desired delusions of their heart. Meaning he allows all individuals to form their own biased beliefs. Allowing each soul to interpret, use, even abuse and destroy any physical evidence in almost any way they desire. So, yes, people do murder babies and each other, and they rape and are abominable, unnatural, and sexually vile and insane. We see the world going insane, faster and faster. When FATHER told Satan that Job's physical reality was in Satan's hand to try Job's love for GOD, but to spare his life. Satan forcefully projected tornadoes and fire to come down from the sky, destroying all Job physically had. Even covering his body from head to toe with boils. Holy Angels at FATHER'S command, forcefully perceived fire and brimstone to rain down and destroy Sodom and Gomorrah. We see far larger firm manifestations being perceived by far larger faster and far more forceful souls. Yet reaching out to reorder physical reality like this, only occurs at FATHER'S command, or allowance. These long-ago events are said in this day to be lies, or mass hysteria. Materialist say we must accept that the physical half of reality is all there is. As this leaves all materialist and their followers with no ability to know the truth. For Pure Dualist, *our* horizon to sense and know reality forever expands as FATHER perceives his and our eternity before us. We can only reach one horizon created by FATHER, after he provides us another more wondrous; this is a part of eternal love. Trying to know FATHER'S truth using the physical aspect alone is impossible, and eternal love, shatters all godless theories. Through eternal love under GOD, all things reasons and purposes congeal as crystal clear. What are the definers of love? They are a responsible faithful desire to intimately communicate and spiritually know, respect, honor, enhance, and sustain another. And in mankind, to then bring children to GOD. And then in fulfilling all of these, only then, love needs to feel and receive all these same things in return. To want or live only the second parts, or the first part in a lesser way, is selfish narcissism. But now love exists in a spectrum from fallen godless temporary perverted forms, up to that strongest spiritual eternal Godly love being loves highest purest and strongest form. In final judgment, all souls lost in rebellion against GOD'S purpose will feel an injustice is passed upon them. But they can't see that this is more of their own denying thoughts, putting their selves outside of and holding themselves up above the whole and Crown of Creation. FATHER would have to eternally serve their every individual desire, for these to feel justice and peace. And each individual would need to be served in a different way, requiring a personal creation to match their desire; making GOD their eternal servant. And making all who disagree with FATHER, to rule over him. FATHER would forever bow to and serve the lesser, never receiving any love he searches for and needs. FATHER desires nothing that forever-changes from one imperfection to another. Multiple beliefs, and chance occurrence did not, could not, and had never existed before multiple free-will conscious-entities were gifted. To reverse our separation from GOD, requires we denounce all such division and come off our throne. Our rebirth does not come by knowing FATHER'S physical ordinances and spiritual oracles, or by reverent Church attendance, or meeting the physical desires of the poor, or singing words of praise every day of our life, or claiming the name of JESUS while living in disobedience. Our spiritual rebirth comes in trading all *our* individual ways and beliefs, in love for GOD'S ways and beliefs. Seeing that *if* we each have our own ways of life, we will be divisive in GOD'S Kingdom. Many think they have changed for GOD, not fully, but enough. These wish to share a throne with GOD. To half-hearted religious souls, a desire and need to surrender *all* their individual beliefs and ways of life, does not seem right. Have you ever thought that if everyone believed just like you, there would be no wars, fights, murders, or acts of violence? But you're not omniscient, omnipotent, or omnipresent, or a provider of life and all reality, so, that could not work. Yet there is one omniscient omnipotent omnipresent provider of life and all reality. So, if everyone would let GOD be GOD, to lead, define, and show us how we can all believe and live together in his ways of eternal love, that will work eternal. Souls holding organized factions of belief in GOD, feel an aversion to GOD'S total rule over their ways of life, just like the materialist do. Even if they say with their mouth, there is only one omniscient truth of GOD, individuals will rationalize a way to form a personalized religion, in which they believe their way of life aligns with GOD'S way of life. These draw GOD'S line to allow their ways, and do not see they form an individual personal belief, contrary to FATHER'S. These say with their lips they are under GOD, but all their works and ways of living are not. Titus 1:16, "*THEY PROFESS THAT THEY KNOW GOD; BUT IN WORKS THEY DENY HIM, BEING ABOMINABLE, AND DISOBEDIENT, AND UNTO EVERY GOOD WORK REPROBATE.*" All hypocritical beliefs come by a desire to be a physical creature, trying to add just a comfortable amount of GOD into their life. We only see the true reality, when we learn to see it from the perspective of its Creator, and establish ourselves in his eternal nature and perfect picture. We must never see ourselves compared to other lesser souls; we must only see ourselves as compared to the love and ways of CHRIST. One soul says I believe this way of life is good, another says I believe differently, and most think this is no big deal as long as one does not encroach to much upon the other; being impossible for eternal peace. So, individuality is the source of all spiritual division. After hearing GOD'S pure truth from me, most come back saying things un-Scriptural, and defiantly go on about their own ways and perspective. Great pain and suffering are unavoidable parts of our existence, being grounded in our purpose of finding love, surrounded by souls having free-will choice over their desired ways to live. The good *and* the evil feel the painful loss of love, and both always will; but from differing perspectives and to different extents. Once we see how and why all physical reality and physical life required the force projections of FATHER to form them, we also see his sustaining work is required for them to continue. No materialist can show any physical controls beyond chaos, random chance, and disorder, as a cause to force, order, and align the elemental structures of life's complex needs in so many different complex cells, forming many different complex systems. There are no pure *Physical* Principles or Laws that can orchestrate the endless complexities of life we see. Science offers vague partial theories of what may occur in cell replication. Yet *we* see more *than* physical complexity; we see intelligent motivated directed forces are truly required to orchestrate all we see. And the only two candidates are: an intelligent design by patterned projected force over and upon the physical aspects; or unknowing chaotic random chance physical occurrences, battling against all constant harsh corruption and disorder. All under the four ordained forces, mankind sees as controlling all they can't explain. We all know there exists no mystical physical complexity principle, or law of physical knowing guidance. Conscious forceful intelligent design, and pure physical singularities are woven together. Our forced creations of particle/wave-based perceptions are a known certainty in each moment. No one has begun to show how a purely physical reality might create our visuals, sounds, and more, or raise our awareness of them, as they can never define for us how or even what a physical knowing awareness might be. Show me any pure physical complexity that has any observed way to become aware in their theories. Show me their source of newly generated force. No physical quality has any internal guidance to achieve complexity. Yet all our conscious projections of desired patterned force that create our perceptions, shows a >99.999% scientific probability, that FATHER can create and gift all the forced physical, and forceful conscious complexities we observe. Physical complexity out-racing harsh disorder by physical chance, delivers at best a <0.000% chance for its trillions upon trillions of ongoing required steps. And after the first step, the odds diminish with each next step towards complexity. Random complexity has no purpose reason or meaning, until it becomes a true forceful informational design. So, we see all building rising complexity has to *know* where to stop building, or it builds past its purpose to be of no use. What orchestrates each design and knows where and when to stop building. The odds of one single protein forming by chance, is beyond rational belief. So, the odds to create a functioning cell by chance, needing 150 to 300 different proteins, can't even be calculated. Yet materialist, by pure evil desired faith, accept chance orchestrated Abiogenesis and biogenesis, and a discredited evolution of species. We do see their great godless hopeful desires and need for it to be this way. But to forever continue their facade, they need to concoct a theory of a greater mystical force than chance occurrence. They say this minute physical thing *knows* to do that physical thing, describing a realm of physical *mysticism* that conjures up complexity by some unseen omnipresent physical omniscience. Their believed process for complex cells of life being created by chance *are* all miraculous, unprovable, and do not meet any firm long-standing scientific criteria. To this day, Science can't define what any founding guiding force of life might even be, or how it may have been formed in all this mystery. Yet they can say with a straight face, that for them, GOD is *not* scientifically necessary. Materialist theories do not require consciously guided miracles, yet they do require enigmatic physical miracles and mysticisms of chance. If unwound, human DNA is around two meters long, so it is tightly folded a great multitude of times into a microscopic bundle within each cell. Yet materialist say each bundle is miraculously by physical means, loosened and severed down its entire length at its center, and separated without unfolding. Can anyone show a geometric design that could allow this, as I really need to see a visual of how this can be accomplished. They say this miraculous act of cutting it down its center is done by a miraculous enzyme or something that sounds just like a micro highly-intelligent surgeon with decades of practice, finely tuned senses, and miraculous maneuverability, and also having scalpels and sutures as hands. A micro-surgeon with an understanding to perfectly divide something having billions of parts in half. Then more miraculously, we see two new identical folded halves appear, being copied and knowingly made out of what, and by what? These four tightly folded halves are then seamlessly knitted together into two new identical sister bundles. Can you see all of the impossible mystical physical conjuring's of design and guidance required at every step of that process. Such a constant miracle work this is for GOD, let alone chance. Materialist use constant physical miracles throughout their theories; yet they only admit to using physical miracles as the initial foundations for *all* their theories. They describe a chain of physical miracle after miracle in Cell Mitosis, from beginning to end. Materialist say things like, this enzyme *knows* to do this, or a certain gene *knows* to do that. A certain *knowing* event, only occurs in conscious guidance. Materialist say our knowing awareness, mystically emerges through vast-numbers of neural interactions. But within cell meiosis and mitosis, we see no vast neural network, or a single neuron.

How can any believe that genes, or enzymes, or proteins within cells have a knowing ability to go into a genome bundle, perfectly center cut it, create two new identical halves, and precisely splice two new bundles together. The term "to know how" means to be educated in an awareness of certain specific knowledge and purpose. A cell, a gene, or an enzyme know nothing. Yes, chemical compounds have forceful properties that locally exchange, repel, or bind charge. These alter properties to form compounds; but these chemical actions and reactions are all unaware. They know nothing of what physically occurs through, or to them. So, they too, are ruled over by ongoing conscious force, *or* physical chance. The sciences must use an honest word, other than "know", but they can't. There is no word to replace the intelligent purposed guidance they observe in every knowing process. We surely see much knowing going on in each process. Science greatly abuses the word "know" in every guided physical process. The truth is firm and defining, and never to be compromised. We all see consciously designed gears, levers, conduits, and switches everywhere. These are consciously designed to cause a *desired* chain of events. We consciously design and build planes, bridges, satellites, and cities; yet mankind cannot build life. And we know physical chance could not build any of these examples; especially life. Are planes more complex than physical life? No. Every aspect that occurs in the miraculous chain of events for physical life, requires design and guided orchestration from beginning to end. I do not like the religious proverb, asking could all the parts of a 747 come together by chance in a tornado. Though accurate, it's too absurd. What about a chance occurrence of a *perfect* polished sphere or square made out of a single pure element. Such shiny elemental shapes would be far more likely to occur by chance, countless times, before physical reproductive life will ever occur by pure physical chance. All physical complexity only comes by conscious projected forceful patterned guidance. Science will admit the odds of a perfect polished sphere or square of 100% pure gold silver or diamond, occurring by chance among the agents of entropy is impossible. As the chance process of forming anything perfect is over-run by the agents of disorder called entropy and fission, being about their work of dismantling everything. If an ancient one in a quintillion chance event, did bring the required elements together in a perfect order and proportion to form a unique compound in a pond of water, which temporarily stored an unnatural abundance of static energy from a nearby lightning strike, or sunlight, or any energy source. That stored energy would just dissipate and fade away. That would not be any slight example of any definition of any simplest buildable first step towards life. It is only the simplest useless one in a quintillion chance occurrence of a natural battery, to be quickly destroyed by entropy. Which would not be seen again for ages, and could never know or guide or add more complexity to itself. Theories of Physical Abiogenesis, are tales of rising complexity that require a mystical knowing patterned principle must reside within all quantum aspects in reality, forcing complexity to out run entropy, fission, and chance. And *again*, it must *know* exactly where to go, and where to stop. How could a purely physical principle create life, minus a pre-existing design. These far too high, endless hurdles, stand in opposition to all pure physical complexity beliefs. And paradoxically, the unravelling of complexity is now seen to always win. So, any mystical *physical* complexity principle or mechanism that might have once been, must now be dead. And all cells now replicate under a dying principle of diminishing returns. I proffer the following most seriously. Imagine a long-lived tree, once planted to grow in a valley right upon a very shallow fresh immense mass human grave; so, the tree adapted to a harsh decaying flesh diet. This tree then grew and spread to become a competing entangled forest, where local people went to die according to their religion. So, as these trees evolve, some grew a physical advantage of a hollow center area for luring dying humans in to become food when they die. And after eons, as people went into these woods to die, they saw some trees had an advantaged shape of having a smaller opening for privacy, and soft bed-like leafy areas inside. Yes, this is so foolish, yet I see *no* difference between this and anything else *said* to evolve as advantaged through natural selection. Show me one true difference. Now for arguments sake, let's remove forceful intelligent projected purpose and guidance, and allow only pure chance, fission, and entropy, by flipping a coin. Now every time your coin lands on heads, you can add one chanceful step towards physical complexity; but every-time it comes up tails, a step is destroyed. So, to construct one simple complex single cell of a virus by chance, having an unknowing animated processes to take in and use energy in a purposed way, and having no awareness or ability to replicate. To create this simple cell, your coin has to come up on heads hundreds of thousands of times in a row. And again, you would require a design to know how and where to go, and where to stop in making each component of its cell structure. Yes, DNA maps so many physical aspects; but DNA has no knowing awareness, or a forceful desire to implement its instructions. With no forced knowing design, destruction rules over all physical construction. Entropy comes from every direction within, without, and around every step, tearing it down as fast as chance can build. Life must be designed and created suddenly and completely. Countless orders and a perfect environment are required, and this was all created around 12,000 yrs ago. There are real chances and truths found throughout this fallen reality. And they all say life could never form on a harsh fallen earth. Science has been forced to turn from life originating on this most rare planet, to its being seeded from another planet. But can you believe another far more-rare planet, light years away, had perfect conditions, and lacked the problems that say life did not originate here. To believe this, you must imagine a perfect planet of life was hit by a monster asteroid or moon. Busting that perfect planet into massive pieces, having carbon based complex life deep within a huge piece containing lots of water. Now this life formed in perfect conditions, was formed to survive *only* in those same conditions. Yet catastrophically, its perfect environment is hit by a massive force and heat that must *incinerate* and *crush* all life. Yet, a vast piece of planet flies out at a cell crushing velocity of between 25,000 and 160,000 miles an hour, burning through its atmosphere, crystalizing its outer layers to form a seal. This remnant quickly cools and freezes in the -450+° of space, leaving behind its perfect conditions and source of heat, light, and water. And miraculously travels through the cold darkness of space for millions of years. Mankind has tested micro-organisms being fired from a gun at just 1600 mph. And every cell was totally crushed by the acceleration. For an organism to survive an instant 25,000+ mph acceleration, it must be metallic based and machine like. And even then, its survival would be a miracle. But let's just fantasize by unimaginable chance, it found its way to impact again at between 25,000 and 160,000 mph upon another rarest of planet in the whole expanse of the universe. And say at the very center of this asteroid a few frozen forms of life are not crushed in a red-hot reentry and impact into the earth, right in a spot conducive to its former needs, destroying those conditions. We see the odds *against* life evolving elsewhere and coming here, are many orders of orders of magnitudes higher, than the impossible odds of evolving here first. Each requirement to bring life from anywhere else to here, infinitely multiplies the already incalculable odds. How can any sane soul hold such belief, just to deny the certainty of FATHER'S consciously created reality. Consider this one certainty, that before the beginning of all known physical realities, there was only FATHER'S desire of conscious forceful projected pattered certainty, being his conscious forceful designs upon his body of physical singularities of uncertainty. From this came his every desired imagined complexity of substantive reverie within his person, being his created dreaming perceived reality. The odds of FATHER having always existed, are better than the odds of us existing now. And we *are* here, creating *our* forceful desired perceptions. GOD'S forceful pattered projections of desire, is the only possible formula, reason, and purpose, for all the created complexity we see. And rebellion against GOD is the only possible reason for all individual hate, hell, and destruction. There are only two mechanisms at work, being creation and destruction. We see every soul's consciously desired guiding intent of choice; and by GOD letting go to a certain great extent, the allowance of all physical random chance occurrence. And no matter how hard one wishes, there is no third option called knowing physical guidance. Every false physical guidance belief is hated by GOD, and formed in hatred for GOD; and spread by Satan and all conspirators to deceive naïve children around the world. GOD'S truth says our every conscious thought and subsequent physical action, are good and creative, or bad and divisive towards *why* we are all here. We Physically and Philosophically call this a *Forced* Eternal Choice, as reality does not allow any *true* benign third grey area to exist, except in one's confused error of belief. Every understanding under the sun goes awry, when the line between the powerless unknowing physical, and our knowing conscious force projections are blurred or switched. There is no boundary to the errors that can be believed in one's deep desire for lies to be true. Materialists say how can forceful knowing souls exist, and forcefully create all purposed reasoned things. Then they believe these same miraculous things occur within the inanimate, without reason or any known purpose or way to do so. Yet we see great reasoned ways and purpose, rule over all that exists. I see no tiny measure of any common or rarer logic or sense in their fractured beliefs. Consider that at the very edge of current micro-biology, how in so many times in sterile laboratories using their protocols of *conscious* guidance. Mankind has taken and manipulated cells, and nourished their growth outside the body. They find they can get life from life, and alter life; but they cannot in their same perfect environment, under their same conscious knowing guidance, get life from non-life. *But* if they ever do, seeing we *are* minute and godlike; I need to see them do it by pure chance in a harsh environment, minus their consciously learned knowledge, guidance, tools, and perfect sterile controlled conditions. Only then can I discuss a godless Abiogenesis. For life to emerge by chance, means life must be an easy thing to accomplish. Secular science implies that *if* they *learn* to *consciously create* life in perfect sterile controlled conditions, that means chance and disorder could do it in extremely harsh chaos. They accept an absurdity, as all *conscious* questions remain unanswered, and are avoided in their efforts. Now GOD'S true children know our everyday physical healing comes by GOD'S design and ongoing controlling will. And as each body lives grows and heals, not one new cell exists, but by FATHER'S predetermined patterned force over all life's mechanisms. All of life's forceful complex guidance flows from GOD, down through all living bodies. Every time we physically grow or heal, including our body's first cell division; it was by a pre-determined complex divine love upon creation. Science consciously designs and creates complex molecules in labs that have desired purposed affects upon pre-existing life. These are learned by *conscious* trial, error, and contemplation; and in sterile environments. Allow me to hold up a mirror, so they might see their *continuous conscious guidance* in all their experiments at each single step. Put those same basic elements of life in a harsh environment without conscious guidance, and watch for trillions of years, and a living organism is never going to appear. Tell me, is it possible to split a person down the middle of every cell, organ, and system, and create two new halves from nearby energy, and rejoin them as two different identical people. If this is impossible, tell me why? Do you respond with the only logic, that a human body is alive and too complex to be split in two; but each single cell is complex and alive. All we are talking about is *scale* of size and function. Smaller complex functions vs larger complex functions. Can a molecule or atom spontaneously split and duplicate on its own? No. Physical self-duplication of anything is just impossible. Many physical things can be split in two, but two new identical halves can't spontaneously appear and be adjoined. Spontaneous replication is an absurd term in science, created solely to replace the term *miraculously created.* Spontaneous occurrence can never describe the miraculous creative work seen in mitosis, it just jumps over and ignores it all. The word "spontaneous" means a sudden unexplainable phenomenon, arising strictly within all the qualities and quantities of a subject itself, and not caused by any external source. Using the term spontaneous occurrence instead of "a creation event" explains nothing. A forceful guiding intelligent design is required for the appearance of all things complex. We see GOD'S work of cell duplication, maintains even the bodies of the souls that work against his purpose. And we see this maintenance becomes insufficient as we age. Seeing a cell duplicate, is seeing GOD create. Hebrews 3:4, "*FOR EVERY HOUSE IS BUILDED BY SOME MAN; BUT HE THAT BUILT ALL THINGS IS GOD.*" For physical life to start *and* continue, every aspect of each cell's design, requires an *ongoing* complex force guidance. FATHER is our creator, designer, and sustainer of life, but most are blind to his work, even as they see it all the time. Most won't attribute life's everyday functions to GOD. For most the more GOD does, the less they see him. Created bodies are adjoined to us for our good use, by a certain reason. If FATHER loses his forceful desire, reason, and purpose for a body, or for this entire creation to continue, they will cease to consist. Even weak believers remain blind to the constant work, required to keep a body alive. Most believers think their body was created and is now theirs, living separated from GOD by a self-perpetuating physical life essence. These believe the body is their property of life, it being who and what they are. For these to see FATHER as their healer, he must before their eyes, speed up his set rate of healing, to make it instantaneous. Then they say this is of GOD. The vainly religious, prefer the secular belief that life and the body's healing are set under a physical guidance from the interface/brain. For healing, a religious soul seeks out doctors and chemicals they trust above GOD. And upon healing, to save face, they give GOD the credit and thanks. This is self-deception as to what we trust and believe. If a prayer for healing is not answered in a fast way, beyond any other explanation than GOD, we must never doubt in GOD. But in fear, look into our soul to examine the state of our conscious closeness to GOD. If one desires the ways and things of this world more than GOD. Doubt arises in their trust and reliance upon GOD. In all our unanswered prayers, we must doubt ourself, not GOD. Isaiah 59:2, "*BUT YOUR INIQUITIES HAVE SEPARATED BETWEEN YOU AND YOUR GOD, AND YOUR SINS HAVE HID HIS FACE FROM YOU, THAT HE WILL NOT HEAR.*" Jer. 11:14, "*THEREFORE PRAY NOT THOU FOR THIS PEOPLE, NEITHER LIFT UP A CRY OR PRAYER FOR THEM: FOR I WILL NOT HEAR THEM IN THE TIME THAT THEY CRY UNTO ME FOR THEIR TROUBLE.*" Job 35:13, "*SURELY GOD WILL NOT HEAR VANITY, NEITHER WILL THE ALMIGHTY REGARD IT.*" It is evil to expect GOD to consider or do a specific physical thing for any soul in a rebellious state. And it is so humbling and fearful for an obedient soul to receive such healing. After healing a man, JESUS said in John 5:14, "*BEHOLD, THOU ART MADE WHOLE: SIN NO MORE, LEST A WORSE THING COME UNTO THEE.*" For GOD, physical healing is an forceful thing. If we have a disease, injury, or tumor, and FATHER finds his desire to gaze upon us in mercy. As he scans and sees corruption to his original design, faster than the speed of photons, that wrong is forcefully perceived back into its correct way. The body to which we are adjoined is a tiny most-complex aspect of GOD'S vast particle-wave-based perception. At his glance of projected forceful desire, it aligns to his forceful projected conscious pattern, just like when we will through the neural system to move our fingers before our eyes. Imagine a cut on your hand, that past experience says GOD'S current rate of cell creation, will heal over the next two weeks. And compare that to an 80's computer processor and service provider, taking forever to download a file. Now a modern processor and service, does the exact same thing in a fraction of a second; it is nothing new or different, just faster. Do you see how FATHER'S preset level of cell replacement, is insufficient against the accumulating destruction brought by all sin, up until the body no longer functions. A sudden healing by GOD, as with modern computing, is the same thing, only faster. The cell replacement in a cut hand at GOD'S preset rate, is our everyday experience. Yet most do not see this slower set rate to be the work of GOD. I have experienced sudden healing, and other frightening miracles. And if you ask why I say frightening, you have never had a personal undeniable intervention of GOD in your life. It shocks, scares, and weakens you down to your core to experience such a thing. And you know beyond all doubt, that GOD is looking right at you in all his knowing in that moment; so, in fear you fall to your knees. It also inspires, edifies, and strengthens at the same time. Of all my experienced miracles, only two have I asked for. GOD answers all my prayers, but most often the answer is not what I expect, or it is no. Do you think GOD is supposed to give us whatever we ask? GOD gives us what we need, whether we know and ask, or not. GOD *never* gives what we selfishly ask for in the flesh. These finer lines in our prayers, we can't always see. Please consider exactly what Scripture means when it says GOD *can* do all things. Learn, pray, respect, *and* expect the wondrous personal touch of GOD, *only* in his Holy ways and Righteous abilities, and *only* in his time and perfect fashion and his proper context according to his truth and loving purpose. GOD leads *us*, they are not at our childish beckon. GOD never answers unholy unrighteous prayers. FATHER'S Children know the body is fashioned not from the substance of stars; and like the stars, they are made from the forceful point filled breath of FATHER'S spoken words. We must retain the truth, that by this fact, these bodies are in every respect the property of FATHER. Accept this now in love as your choice, or eventually he must severely restrict your sensations and motions forever. So please don't tell me, my body, my choice; it is not your body, yet it is your vile sinful choice. Scripture says GOD composed our flesh from the dust of this earth, and the now fallen defiled flesh must return to the same. Matt. 3:9, "*AND THINK NOT TO SAY WITHIN YOURSELVES, WE HAVE ABRAHAM TO OUR FATHER: FOR I SAY UNTO YOU, THAT GOD IS ABLE OF THESE STONES TO RAISE UP CHILDREN UNTO ABRAHAM.*" GOD made all flesh from the simple dust of the stones in the Earth; and science has found those exact elements in the dry dust of all flesh. GOD put within all DNA, system signals to cause hunger, thirst, pain, preservation, propagation, and a desire towards common life. Our senses of physical life are described in Scripture as desires of the flesh. And since the fall, these have by sinful-exercise, grown stronger. These physical desires, acts, and sensations, are increased in the flesh beyond their perfect reason and purpose, to now be called lust. Instead of all soul's spirits being holy and righteous and the flesh being in total submission, we see desires and sensations within the flesh, keeping most souls in submission. All physical life on earth was built by FATHER, starting from the least to greatest, using a rising design of ever-more complex DNA systems in each higher species for its lively attributes in propagation. Why did FATHER create DNA? It is a required aspect of physical chance and choice. GOD creates each new body according to our choices, as each new blended physical body is not GOD'S choice. FATHER created mankind's physical complexity last; and even that old question of which came first, the chicken or the egg, is answered in Scripture. Gen. 2:4-5, "*THESE ARE THE GENERATIONS OF THE HEAVENS AND OF THE EARTH WHEN THEY WERE CREATED, IN THE DAY THAT THE LORD GOD MADE THE EARTH AND THE HEAVENS, AND EVERY PLANT OF THE FIELD BEFORE IT WAS IN THE EARTH, AND EVERY HERB OF THE FIELD BEFORE IT GREW*:" Empirically stating all things were first made mature and prepared to give seed *before anything grew*. Scripture also says, Adam and Eve were formed a man and a woman, not as single cells or infants. All living creations that require two parents, were meant to produce combinations of traits and strengths, yet all life bears only similar fruit after its kind. Scripture says in the perfection of GOD'S Garden, the fruit and leaves on trees did not fade, indicating initial zero entropy. But after falling from their spiritual height, mankind began over-exercising and abusing the physical information sensed and preferred by their soul. This over-exercising of senses, increases the input of information, raising a soul's sense of gratification levels. GOD says physical gratification, above the basic requirement to sustain and propagate physical life, enforce and increase our soul's desire towards a physical purpose. GOD warned Adam and Eve of the corruption to their desired purpose that would come, if they went beyond his perfect design of their provision. Choosing to do so, was, and always is a choice of desire to no longer live in GOD'S eternal way of life. Each act of physical rebellion, will for a while, give a soul a stronger-sensation, and raise their desires to sense their next rebellion. By this, souls choose a sin nature, and have strong addictions. Scripture says if our soul desires something evil, it brings a forceful inner physical pull of temptation. Please know, even the strongest of physical pleasures most adults now seek, was initially a minimal sensation, done for reproduction only, and not recreational pleasure. In the beginning it was easy to control, but now it is so over-exercised, misused, abused, and desired, it has become mankind's strongest sensation and addiction. And fallen angels also desire the over-powering sensations of pleasure, by gaining conscious possession and control of a human body, even animal bodies. GOD warns, that unless we turn from physical slavery, these desires can become encoded in our genes as strong proclivities. By this, these evil desires are passed down to our children, to the third and fourth generation. Repent and turn for your children above yourself. The strong signal responses of physical to soul gratification that have spread and grow everywhere in the world, sound unfair to souls realizing they are captive to desires of their flesh. But we see more than just choice and accountability on one side; we see great understanding and strength of love on the other side, once we overcome these with GOD. We gain great understanding strength to resist, and in the mercy of our escape, we know the forces we must resist. As we become firmly settled in GOD'S ways and understanding, we sense sin differently. And as we sin less and less, *if* we do sin again, we feel a powerful crying shame for betraying GOD'S love. With growth, at the first hint of a former temptation arising in us again, we recall our pain and shame, and flee at any desire to sin. In our growing faithfulness we are repulsed by sin, till one by one we overcome our sins in love for GOD. GOD has promised, and we attest that in our enduring loving efforts,

GOD brings us healing from all physical desires that imprison us in addictions to lust, security, comfort, beauty, drugs, sex, food etc. In my youth, I, to differing degrees, was addicted to many sensations. I loved fast cars, many women, loud rock music, food, money, the beach, movies, and yes alcohol and other substances. And at the age of 20, GOD'S love strengthened me to fight these adversaries; and now most are strangers I know no more, or very little. My battle is ongoing, but I am in this moment winning more than losing. We don't just get addicted to certain or many things, we get addicted to an over-all level of excitement. When we overcome one sin, we feel a stronger need for others. So, our last battle will usually be the hardest as it contains all our inherent desires for the flesh crammed into one. If you overcome 9 of 10 physical desires, the 10th is stronger than before. We recommend you work on food last, it being among the least offensive. Years ago, I lost my only son, and that tore through my very being, and brought this body within one heart-beat and one-breath of death, as my soul was savagely pierced clean through. GOD held me through that horror, that no matter how I try, I can never describe. But months later, I recalled that through those long darkest moments, I never *once* thought of drugs or alcohol. So, I knew my love for GOD had healed my soul and this body of those desires completely. To this day I have a hatred for smoking, drugs, alcohol, and all sin. And my great love in this world, my son, GOD has promised me another chance with him in the Millennial Reign. GOD does all they can to honor and sustain all their children's heights of love we each find and hold. Again, FATHER'S designed pre-set rate of cell correction and mitosis was perfect for creations initial state, but is now *in*sufficient against all the ever-increasing disorder brought by sin. This adversarial predatory world comes against us in all its physical confusions, with cancer being the highest order of confused genetic corruption in DNA. Science knows our repetitious choices, can alter our genetic disposition towards things. But they deceitfully deny, that with our righteous change, said dispositions can be reversed. Why do they say addictions and perversions are set for life? Souls no longer lust for things they have been denied for a long enough time. If we can become more-evil through repetition, why deny GOD can change us back through our denials and avoidance of the same. Such hypocrisy is an unreasoned unscientific position to take. GOD warns these last days will bring unimaginable anger, aggression, disease, war, hell, and death. And by most rejecting GOD, we see the world-wide confusions of DNA in all life, turning it all towards ever-more confused mingling and predation. We also see by this world's same rejection of GOD, his fading desire to sustain or repair it all. GOD warns us of coming famine, disease, weather phenomenon, and great death by a third world war. Luke 21:25-26, "*AND THERE SHALL BE SIGNS IN THE SUN, AND IN THE MOON, AND IN THE STARS; AND UPON THE EARTH DISTRESS OF NATIONS, WITH PERPLEXITY; THE SEA AND THE WAVES ROARING; MEN'S HEARTS FAILING THEM FOR FEAR, AND FOR LOOKING AFTER THOSE THINGS WHICH ARE COMING ON THE EARTH: FOR THE POWERS OF HEAVEN SHALL BE SHAKEN.*" We were warned thousands of years ago of a soon unstoppable great distress to all life, born solely by spiritual pollution, not physical pollution. Souls through their individual perspectives, will surmise their own reasons for all they now and soon will see. Each individual will see the causes they desire to see, through the perceptions they choose to create, all being experienced through their own slow, small, temporary, perspective of reality. We know GOD does not arbitrarily bring *all* plagues. And *most* destructions before the Great Tribulation are the reasoned effects of mankind's sinful ways, while FATHER turns away in being rejected and allows all free-will choices to occur. I ask all souls in scholarly respect to scientifically see and not discount our next chronological account of the forceful events of GOD perceiving all reality. We start at the *very* beginning, being *before* GOD spoke forth the aether of all external physical reality. Now in *our* souls being adjoined to the physical reality, we see these bodies we use need air to live, and to speak as we force air across the vocal cords while contorting them, the tongue, and the mouth to speak. But we see GOD consisting of all that is, does not breathe air. We read how not for his *internal* created perception, but for all his *externally* perceived gifts and creations, that FATHER verbally spoke forth out from his personage. For instance, saying the words *I AM*" and "*I AM, THAT I AM*" and at other times "*LET THERE BE, etc*. Prior to all gifts and creations external to FATHER'S person, a need to speak and project outwardly beyond FATHER'S known existence, held no purpose. Scripture shows FATHER'S voice to be a release of, *not* air, but patterned force and the pure physical point properties of his body that form the *breath* of GOD. Psalms 33:6, "*BY THE WORD OF THE LORD WERE THE HEAVENS MADE; AND ALL THE HOST OF THEM BY THE BREATH OF HIS MOUTH.*" Science and Scripture, both describe a required primordial aether of pure singular points having a forced minimal resonance, and being the foundation of all things physical. FATHER'S patterns of force frequencies and physical points spoken forth, are the dual essence in the words of the voice of GOD. Now the *first* act of FATHER speaking out, was clearly the *first* act of FATHER'S design to find eternal love. Scripture says FATHER'S very first spoken word of eternal love was not for the aether. FATHER'S greatest love was spoken forth in a solemn eternal promise of sacrificing a *substantial portion* of *not* just his soul, but of his soul and body spoken forth together. In this gift of sacrifice, FATHER spoke out, and let go the beginning of all creation, being the second person of GOD. We see in Scripture, a gifted form of complete partitioning, out from the soul *and* body of FATHER. Bringing the very first new eternal free-will soul *and* body. Being far smaller and less powerful, yet eternal and known by FATHER. Now FATHER knew before he did this, that no matter the free-will outcome, this will be an eternal separation. FATHER did not know what would happen in this first new distinct unlearned eternal conscious force and body of desiring free-will. And in adding a second free-will soul, *then* and *there* was the first existence of an Uncertainty Principle. And we see how uncertainty towards FATHER increased with each newly gifted free-will soul. In FATHER gifting the first new soul to be beyond his desired control, we see the uncertainty of a choice *for love* between these only two souls to exist, was far higher than a 50/50 probability. FATHER knew that for eternal love to be returned to him, his only begotten SON must make every choice to hear and accept FATHER'S truth of omniscience and eternal perfection out of his free-will desire. And for his SON to learn and know FATHER'S omniscience, the SON must go through his own process of learning and finding eternal love. A process all who *follow*, experience. Omniscience and faithful responsible love are both learned and desired. These cannot be imparted by FATHER in his gift of a soul and body. If he could, and did do that, the beliefs and desires imparted by FATHER, would be a reflection of self without choice. By this Principle that Choice defines Love, we see why eternal loving understanding can never be irresistibly imparted into another. Like the knowledge of truth, love must be received in desire and nurtured from a seed. Any *eternal* love must come by two souls meeting and finding one shared desirous belief and purpose, and not by FATHER causing it to be. This second most special of all souls, came into existence, having only a conscious certainty of hearing the echo of those first loudly spoken words "I AM GOD". Being FATHERS spoken release of his first conscious/physical gift, being JESUS'S start of his new eternal awareness. This eternal conscious and body of the only begotten SON was at first, a hovering uncertainty and possibility, knowing only those first heard and felt words. All far lesser souls, including ours, were spoken and gifted out of FATHER in this way, but most can't remember their very first created perception of minute understanding, in hearing that whisper "*I AM GOD*". These words are spoken in different strength, to gift each new soul. The soul of CHRIST was brought forth with FATHER'S first loudest crying out for love, releasing and imparting much soul and body. Those first great words were spoken in a more desired force of love, than they would ever be spoken again. Comparatively our tiny souls were brought in a barely audible whisper. We all started our existence as FATHER'S imparted gift of a soul, adjoined to a single fertilized cell in the womb of our mother. After Adam and Eve, this was every human's first minute conscious physical experience. Now, after JESUS was spoken forth, his perceived understanding and love grew and grew in being taught and loved by FATHER, until JESUS found all true loving wisdom. So, by this first *free-will soul*, being beyond FATHER'S control, came the first existence of true love, choice, *and* a possible desire for an individual rebellious belief, being contrary to the omniscient belief of FATHER. CHRIST was the exact *point* and *place* of the very first Uncertainty Principle of conscious creation and desire, that appeared beyond FATHER'S control. The only begotten SON is the true beginning of all creation. The SON was consciously self-aware, and in all essence and aspects, just as, yet less than FATHER; having conscious control over every point of uncertainty within his *person*. Initially the SON was an undistinguished unlearned six-dimensional form, set free before the face of his FATHER who whispered an eternity of truths of patient love and understanding of his own existence into his SON'S soul and body. FATHER taught YESHUA the understanding of perfect eternal life and sacrificial love, and that FATHER is the only source of the omniscient truth. And FATHER loved his SON dearly. The only begotten SON, grew in his desired sacrificial love for the eternal life he received, and both were strengthened and pleased. JESUS trusted the loving life coming from FATHER. And through this experience, JESUS grew to intimately know the desire of love from FATHER. This great eternal desire within FATHER, was once the greatest ever of all *unrequited* Love. Now again, in all soundness of adding force, motive, and opportunity; love goes directly to a lawful motive for all FATHER'S unfolding events of creation. The SON grew and formed as love consumed him from FATHER until it could not be contained and must be returned in all its fullness. The SON, then perceived himself in the image and likeness of FATHER, having a face, eyes, mouth, arms, hands, etc. all for expressions of love. Scripture clearly says, that CHRIST is the first born of all creatures, and he is the beginning of the creation of GOD. Colossians 1:14-17, "*IN WHOM WE HAVE REDEMPTION THROUGH HIS BLOOD, EVEN THE FORGIVENESS OF SINS: WHO IS THE IMAGE OF THE INVISIBLE GOD, THE FIRSTBORN OF EVERY CREATURE: FOR BY HIM WERE ALL THINGS CREATED, THAT ARE IN*

*HEAVEN, AND THAT ARE IN THE EARTH, VISIBLE AND INVISIBLE, WETHER THEY BE THRONE'S, OR PRINCIPLEITIES, OR POWERS: ALL WERE CREATED BY HIM, AND FOR HIM: FOR HE IS BEFORE ALL THINGS, AND BY HIM ALL THINGS CONSIST.*" And in Revelation 3:14, CHRIST says, "*THESE THINGS SAITH THE AMEN, THE FAITHFUL AND TRUE WITNESS, THE BEGINNING OF THE CREATION OF GOD*;" At this strong point of pure eternal love. The SON needed to return love in more than eye contact and facial expression or touch. JESUS then sacrificed out of *himself*, and spoke forth expending his own force and points of eternal life's essence back to FATHER. Eternal life's resonating force, coming from one in love and trust, received, and spoken back into the other in love and trust. So, this sacrificial form of vocal communion became their highest form of sharing love. This continuous shared oneness of their flowing eternal life between them, became the next act of creation. This back-and-forth communion of their life's eternal essence, grew to continually exist between them. And it was determined by FATHER, that their oneness of spirit shall eternally be between them in abundance as the third person of GOD; existing in all his own eternal communicating measure. The HOLYSPIRIT of love and truth enables all eternal loving teaching communications between FATHER, the SON, and *all* those yet to be gifted and created. And FATHER called this third Eternal Person of GOD, his moving HOLYSPIRIT of love and truth, who guides and holds souls together as one in eternal life. We see every spoken WORD of GOD is an external gifted creation of GOD. John 1:1-5, "*IN THE BEGINNING WAS THE WORD, AND THE WORD WAS WITH GOD, AND THE WORD WAS GOD. THE SAME WAS IN THE BEGINNING WITH GOD. ALL THINGS WERE MADE BY HIM; AND WITHOUT HIM WAS NOT ANY THING MADE THAT WAS MADE. IN HIM WAS LIFE; AND THE LIFE WAS THE LIGHT OF MEN. AND THE LIGHT SHINETH IN DARKNESS; AND THE DARKNESS COMPREHENDED IT NOT.*" Our soul's awareness is our bright flashing experience; but the creations dark physical aspects, comprehend nothing. GOD'S path of life is bright and true, but the dark path of hell and death are confused. Now after their *moving* eternal forceful substantive loving belief became the third person of GOD, being then self-aware. HOLYSPIRIT felt and knew the love of the FATHER and the SON. HOLYSPIRIT was dearly loved by them above all that could come after. HOLYSPIRIT was not the SON, but someone else wondrous and unique, being the third unifying Soul of GOD. FATHER *then* spoke his great vast purely physical spherical aether into existence. So, now HOLYSPIRIT is our spiritual bridge of unifying love and truth, moving throughout the aether as his body, being the connection between FATHER, JESUS, and all who follow them. HOLYSPIRIT moves through the creation, yet does not take on an external image of a person, or receive a throne. The clean undefiled aspects of the aether itself, including our body if maintained as a Holy Temple, are all part of the body of one unified GOD, being a conveyer of true love and light. This is the gifting of the TRINITY. These highest most powerful conscious entities, each have their eternal soul, and their own eternal physical aspects. These three alone, are in this way one and above all that follow. They are indivisible, being of one mind and purpose, in all pure eternal love, truth, and reason. Yet not equal in every manner of conscious and physical size, reach of force, and omniscience. The fullest omniscient understanding, comes only from being literally alone, in being before all others. So, FATHER is the most omniscient of the three; being the first source and above all that exists in the aether. The SON and HOLYSPIRIT, Heaven, the universe, and all lesser souls, dwell within the aether, held between FATHER'S hands. After this, the TRINITY, led by FATHER, speak forth in unified desire, to form all subsequent creations within the aether. The aether being divided by laws, into the firmaments of the quantum more ethereal upper realm of Heaven, and the firmer more physical realm below. GOD, being they, then created the first firmer much larger physical elemental particles, and with these they perceived the initial watery physical sphere of the earth and light into existence. The aether and *all* created certainties, rest between, and touch the hands of FATHER. In this, FATHER'S continuity and ultimate control runs through all created physical reality as being a part of his extended body. As the SON and HOLYSPIRIT, dwell within the quantum realm Heaven. Now in our understanding the original tone spoken from GOD to form the aether; we see how all his words were spoken in perfect positive fusing loving force, being (fusion). Having no contrary thoughts, words, or acts. So, no fission disorder or decay existed. To *see* the original completed Earth, we reverse every soul's contrary force projection and all destructive chance affects, until every appearance of introduced disorder disappears, all the way back before Lucifer and other lower angels, and Eve and Adam fell from perfection into a small slow perspective. Only then and there, can we see and learn to qualify and quantify a single aspect of the true beginning. So, after Heaven, and the Earth, being at first a watery sphere, were perceived. FATHER spoke and perceived seven and four heavenly bodies as smaller acts of GOD. These ethereal quantum realm bodies remain a part of, and under the authority of GOD. Then GOD gifted the first seven and four most forceful protecting angelic souls. These are the Seven Great Arch Angels, and Four Seraphim in the Holy of Holies before the Throne of GOD. These were taught and learned the omniscience of GOD'S love. These stand between GOD and all that occurs. Rev. 1:4, "*JOHN TO THE SEVEN CHURCHES WHICH ARE IN ASIA: GRACE BE UNTO YOU, AND PEACE, FROM HIM WHICH IS, AND WHICH WAS, AND WHICH IS TO COME; AND* <u>*FROM*</u> *THE* <u>*SEVEN SPIRITS WHICH ARE BEFORE HIS THRONE;*</u>" Rev. 4:5, "*AND OUT OF THE THRONE PROCEEDED LIGHTNINGS AND THUNDERINGS AND VOICES: AND THERE WERE SEVEN LAMPS OF FIRE BURNING* <u>*BEFORE THE THRONE,*</u> *WHICH ARE*

*THE SEVEN SPIRITS OF GOD*." Then the TRINITY, in one belief and desire, returned to form a firmer universe, the lower orders of angels, and all physical life, in the direction of the infinitely smaller from FATHER'S perspective. The fallen universe, at its current point of stretching, still exists between the outstretched hands, and before FATHER'S face. For approximal understanding, FATHER'S remaining body in fullness, is at *least* seven times greater in size, mass, and density, than the entire universe. And the desired projected patterns of GOD, form a massive cooperation of quantized force perception builders; being an immutable consciously guided, puzzle-builders union under one understanding. Jeremiah 23:24, "*CAN ANY HIDE HIMSELF IN SECRET PLACES THAT I SHALL NOT SEE HIM? SAITH THE LORD, DO NOT I FILL HEAVEN AND EARTH? SAITH THE LORD.*" The forceful *conscious* laws of projected cause and effect state that every cause must be larger and more forceful than any seen affect. This law determines the size limitations and direction of all creations, by all creators. Scripture details each step of the universe's creation, which literally occur in their required scientific order of rising complexity. Each step being a required basis and platform to support the next more complex step. This required scientific order of the creation, was chronicled in Scripture thousands of years before this order of required complexity was even known to mankind. The odds of this correctly documented order of creative event after event, being recorded by chance without GOD, are staggering. Each occurrence is found in its proper order to be built upon the prior. A more complex creation started with perceiving the aethers ethereal heavenly firmament. Then we see a far firmer elemental spherical form of water and atmosphere, being the earth, and at least three speeds of light particles. If broken down into all its elements, the first day was simply creating the elements required to define a day. The firmer elemental waters, forming a watery sphere of the Earth, stood turning in its first rotation within the newly created surrounding light, standing alone in a lower firmer realm. As FATHER desired and spoke, these realms were filled with the fast patterned information in light's waves of particles. Being created only by FATHER through the aether on that first day. So, during this process, GOD saw the first rotation of the unfinished earth before his face. GOD then and there saw the requirements to define the first evening and morning, so creating the first day, on the first day. And this first day was created in FATHER'S eternal perspective, not any slow small fallen 24 hr. perspective. This layered creation process, after six earth rotations before the face of GOD, with all its rising layers of complexity became the earth and the universe. Finishing it all, and filling Heaven and Earth with all lively things and creatures. All noted living things were also formed in their proper orders of rising DNA complexity with mankind being last and most complex. Now remember, fallen mankind's false dying, small slow perspective called time, did not exist until they fell from their higher, larger, faster, eternal perspective. During the second day or rotation of the earth, FATHER in his purpose and reasons, saw the conscious conceptual distance between the greater quantum heavenly realm or firmament (ethereal consistencies) above; and the (firmer) physical firmament below. Each having its own lawful forces and aspects. And on the second day of the Earth, FATHER perceived the separation of earths waters in perceiving the many denser firmer elements of a hot core, and dry various detailed surface lands. On earth's third rotation, GOD spoke, perceived, and created all the simplest forms of life in every tree, plant, herb, fruit, and grass. On the earth's fourth rotation in eternity, FATHER created up in Heaven, all the lower Angels ethereal bodies, being as big as stars as to our comparison, and gifting to them, their greater differing size souls. These were then taught GOD'S omniscient love, as they watched them finish the heavens, with all planets, moons, stars, wonders, and signs. Job 38:4-7, "*WHERE WAST THOU WHEN I LAID THE FOUNDATIONS OF THE EARTH? DECLARE, IF THOU HAST UNDERSTANDING. WHO HATH LAID THE MEASURES THEREOF, IF THOU KNOWEST? OR WHO HATH STRETCHED THE LINE UPON IT? WHEREUPON ARE THE FOUNDATIONS THEREOF FASTENED? OR WHO LAID THE CORNER STONE THEREOF; WHEN THE MORNING STARS SANG TOGETHER, AND ALL THE SONS OF GOD SHOUTED FOR JOY*?" These lower 300,000,000 then Holy Angels, each have an ethereal connection to the first or only 300,000,000 stars GOD created. GOD then ceased to perceive physical light from the aether alone, as this responsibility passed from FATHER to all his SON, HOLYSPIRIT, and the Angels, to be the bearers of his light and signs from Heaven. Scripture says GOD'S Holy Angels countenance are seen in the heavens. On the fifth day in eternity, GOD perceived according to their orders of rising complexity, all physical life in the sea and the air. On the sixth day, GOD perceived all lower subservient animal life on land, starting from the least organism and bugs to the most complex in their order of requirements, up to a great brontosaurus, or behemoth. And lastly at the end of the sixth day, GOD perceived the very reason and purpose for all his prior preparations and work in the earth. Perceiving the form of Adam, meaning man, and dividing out for him his smaller eternal soul. All this occurred as FATHER saw the earth turn in *timeless* eternity; and not mankind's slow small fallen six-day 24-Hour frame of reference. Mankind's fallen conception of time, has no correlations or effects on GOD'S perception. What is thought to be billions of years of a false physical past to mankind, is around 12,000 X 365 rotations of the earth before FATHER'S face. After all these creations, the same teaching and loving process continued between GOD, the Angels, and Adam. We are *not* told how long Adam lived alone in eternity before Eve was gifted and created. Or how long in *eternity* before Eve, and then Adam rebelled. But we do see all free-will choices towards good or bad, were always allowed for revealing love and hate, and for bringing the true separation of each at the very end. Now all good and evil wanting souls, by their desires and choices, collectively choose, define, and divide GOD'S future Kingdom from a prison called the Lake of Fire. Souls hunted and haunted by a physical death, see *physical life* and its sensations as their supreme concern. And all who hold false doctrines, believe our physical life should be GOD'S main concern. But GOD exists in all opposing spiritual ways, offering us his preferred spiritual things of eternal life. And here on earth, we see very few strong spiritual people, and so, so many physical people. To be strong and Godly, we must first be highly spiritual and minimally physical; yet most people's spirituality can also be false. We need to be strong *true* spiritual souls. Many deny FATHER gifted and created us and all reality to live and exist in his ways of eternal love and peace. They think he should or wants to work to give them physical choice and health, and leave them to *their* chosen ways that all contradict his love and purpose. All who are honest, see who is right in this disagreement, *and* who at the last will find all their desires. Scriptures ask mankind why they swim against the current, and kick against the sharp thorns. It is evident that as rebellion spreads, the more physical these soul's desires and belief become. But FATHER made all things for his eternal reason, being to serve his eternal loving familial purpose. If you stand in opposition, you will not win, but suffer utter devastating loss. Most do not know what is actually lost, in being lost. Being part of the second death, means you can never sin again. Why? Because you are trapped in a barely responsive barely feeling eternal body, and never-ending time, denied most physical mobility and feeling as you consciously sink inward forever in ever-increasing insanity and loneliness. Losing all you ever loved and know, because you did not treasure and preserve life and love. So, you remain forever aware and trapped in a shell of confused loneliness, as that responsive sensitive body you loved above GOD, no longer works towards your desires. We see two conscious disciplines: one seeks the eternal understanding of how our conscious forces create our perceptions, and how our contrary force projections hurt GOD'S purposed creation. The other discipline is in denying this, being grounded in individual belief and physical pleasures, and lacking any eternal responsibility for life. Do we say all greater creative powers can be found within every soul? No. But we have our endless learning potential, once adjoined to GOD'S loving truth. GOD can never be fooled or tricked about our true deep desires. So, if we convince ourselves that we love GOD, yet GOD sees we do not love his ways. It is GOD who is right, and we have fooled only ourselves. GOD created the ways of eternal love, and all temporary forms of love are individually defined. If one seeks to define what is love for their self, their love is then of their own unsustainable design. Within this vast conscious battle field, we see the touches of GOD'S purposed influence here and there wherever GOD sees a pure need and justice to all that is good. Meaning GOD cannot *justly* influence my existence, if it *unjustly* influences the truth and love in his other children; and again, *we* do not define what is just. If a soul feels injustice coming from GOD, this is always a rebellious need to fulfill their own physical purposed desires. Justice, only and always flows down from above, and it was purely defined long before we came to exist. None can from the bottom up, redefine it to suite their individual desires. Now two things underly each *answered* prayer in Scripture. The first being, that within the soul praying for a forceful act of GOD, there is a firm belief that all external realities are the created perceptions of GOD, and GOD can alter his perception. The second is knowing, that before any act of GOD occurs, it must in all its ways, serve GOD'S Kingdom and not any individual desire. If we pray to GOD in all our hope and desire, and our prayer is *never* answered. GOD says they did not hear it, because it was rebellious, as our request were for wants in this fallen world, and not for GOD'S eternal Kingdom. So, we "ask amiss". James 4:1-10, "*FROM WHENCE COME WARS AND FIGHTINGS AMONG YOU? COME THEY NOT HENCE, EVEN OF YOUR LUSTS THAT WAR IN YOUR MEMBERS? YE LUST, AND HAVE NOT: YE KILL, AND DESIRE TO HAVE, AND CANNOT OBTAIN: YE FIGHT AND WAR, YET YE HAVE NOT, BECAUSE YE ASK NOT. YE ASK, AND RECEIVE NOT, BECAUSE YE ASK AMISS, THAT YE MAY CONSUME IT UPON YOUR LUSTS. YE ADULTERERS AND ADULTERESSES, KNOW YE NOT THAT THE FRIENDSHIP OF THE WORLD IS ENMITY WITH GOD? WHOSOEVER THEREFORE WILL BE A FRIEND OF THE WORLD IS THE ENEMY OF GOD. DO YE THINK THAT THE SCRIPTURE SAITH IN VAIN, THE SPIRIT THAT DWELLETH IN US LUSTETH TO ENVY? BUT HE GIVETH MORE GRACE. WHEREFORE HE SAITH, GOD RESISTETH THE PROUD, BUT GIVETH GRACE UNTO THE HUMBLE. SUBMIT YOURSELVES*

*THEREFORE TO GOD. RESIST THE DEVIL, AND HE WILL FLEE FROM YOU. DRAW NIGH TO GOD, AND HE WILL DRAW NIGH TO YOU. CLEANSE YOUR HANDS, YE SINNERS; AND PURIFY YOUR HEARTS, YE DOUBLE MINDED. BE AFFLICTED, AND MOURN, AND WEEP: LET YOUR LAUGHTER BE TURNED TO MOURNING, AND YOUR JOY TO HEAVINESS. HUMBLE YOURSELVES IN THE SIGHT OF THE LORD, AND HE SHALL LIFT YOU UP.*" GOD can lift us up out of a fallen slow small physical perspective. Now when GOD calls mankind adulterers and adulteresses, as in that last verse. Unless it is otherwise explicit, GOD is *not* referring to the physical adultery of a husband or wife; but mankind's unfaithfulness in their spiritual marriage bonds between them and CHRIST. There has been this ongoing conscious war since Lucifer became enamored with his own creative gifts of force and beautiful physical appearance. Lucifer has taken to himself a great confused insanity that he will rise up to be equal with FATHER. Lucifer formed this insane belief when FATHER had been generous and loving to him. FATHER never wanted to express his true vast powers over Lucifer or any. So, somehow, Lucifer grew to imagine himself rising above

GOD, and no longer saw his dependence on the physical provision of GOD. Lucifer came to his belief the way everyone finds false belief; by his deep envious prideful desires and evil fantasies. Free-will is an amazing, yet dangerous power; yet it has only two possible applications for all souls under GOD above. One is to love and align to all the good GOD creates; the other is to work for self and destroy all that good that GOD creates. When GOD gifted and created all that is below them, they restricted all lesser souls by their not owning an eternal body. GOD provides all souls the use of a body for sensing and movement to interact with them, each other, and within GOD'S created environment. And GOD is good in retaining final absolute control. We see lines drawn by GOD that pertain to our lesser forceful souls, and affirm GOD'S ultimate rule. We see GOD'S Children's eternal provision is assured, *only* by GOD'S ultimate control over all physical reality, and their loving compassion. Now the combatants of this battle between good and evil, are all eternal souls; but what does this battle field look like? Let's visualize the current fallen physical reality. We see patterns of GOD'S conscious force, projected into GOD'S aether to create all higher complex reality. But FATHER'S personage exists outside it all. So only the aether and the physical creations it holds, define this battle field, where all good and evil perceptions existence and are now at war. We find within the aether, GOD'S two vast realms, and billions of physical and ethereal bodies of life. These bodies are the vehicles and platforms as each soul's means of existing, sensing, expressing, and exercising their free-will desires on this battle field. We see gifted from and out of FATHER, billions of eternal forceful souls. Each being a gifted size and projector of desired newly generated patterned force, and all having their creative awareness. These then are all the qualified *quantified* combatants. In every *asymmetric* conscious force confrontation between two souls, there are conflicting aspects. And in *every* conflict, both sides are evil, or one side is good, and one evil. Yet both sides can claim they are good by their own perceived logic, reason, and desired purpose. Now two conflicting good souls are not possible, as only our perfect alignment with GOD can be good. Two conflicting sides are most often both bad; meaning both are imperfect. We see the ways of our loving Creator and rightful owner of all that he has perceived, on the good side; and on the evil side, are individuals lacking the truth, having no eternal ability, foolishly claiming the body as theirs, wanting to rule over their selves. Herein, lay every temptation and downfall. Now in this ongoing battle, there are the individual distortions of belief, called the fog of war. So, among believers, we see mostly vague partial degrees of living GOD'S ways. After this battle is over, on the losing side we will see all the self-destroyed souls, filled with hells confusion, arrogance, deception, envy, insanity, and all individual divisive desires. All the victors of this battle are with GOD, where all do desire the created ruling omniscient truth of created sustained life, that endures through eternity. We know mankind's spoken word is *their* greatest power to influence the *temporary* course of things to come. So, consider the spoken word of GOD and what great power and ability must there be in those words to steer the *eternal* course of all created things. Scripture says, and we see it is true that for us, only FATHER can create *one* perfect physical environment through *their* desires. And in that FATHER has given *us* each a free-agency of force, we will desire to align to eternally sustain his creations, or we become contrary project forces against his creation. Now tell me, is all created physical complexity the true reality; or is the true reality its ruling creative conscious simplicity. Well, in

Dualism it is actually both, along with all the desires that arise within it all. By seeing all *subsequent* conscious souls are gifted in their lesser form and orders under GOD. And that *without all three* ascending facets of GOD, sin brings *irreparable* gaps of disconnection to each lost soul, losing all hope of being restored to FATHER. GOD'S law of love's value, and every desired choice for any imperfect separation, states that such created *gaps* between souls and love, can only be bridged at great cost. And the only true value known in a loving reality, is a sacrificial suffering painful perfect forgiveness of love, bringing the two sides back together. The omniscient truthful love of GOD, must have eternal continuity, there can be *no* gaps or separations in Heaven. Every separation is a weakness that allows the destructive elements of time. By free-willed evil arising below GOD, there are countless small gaps and endless great caverns opened up between the confused souls of mankind, and the perfect goodness of GOD. We see the perfect SON and HOLYSPIRIT holding *to*, and being held *by* FATHER. Can in suffering the horror and pain of being among and surrounded by sin, reach out to grab those few loving fearing hands, and span certain of these gaps. They are a most wondrous perfect bridge, to help us learn and cross back over to eternal life. But exactly how are these gaps and caverns bridged and crossed. We read FATHER gave to mankind all the love he has in his SON'S blood and death. For us, JESUS was asked to choose and bare unimaginable pain, as both a sign and price, showing how great is GOD'S love. We read how MESSIAH came down in love, at FATHER'S request. In CHRIST'S great love and trust in FATHER, he *literally* put his eternal soul at risk in separating and leaving behind his immortal body of eternal existence with FATHER. During his time of trying and sacrifice, CHRIST was not the fullness of an eternal GOD. Then being the mature soul of the only begotten SON, adjoined to the weak newborn flesh of mankind, as our souls are. This had to be for many Righteous and Holy reasons; one being to sense and deny all temptations of the flesh, making CHRIST our perfect example to emulate. And in this, being a Righteous Holy Judge over mankind's condition and experience. Prophecy says HOLYSPIRIT came down to overshadow and place the DNA from the loins of King David, within the egg of a most humble chosen virgin maiden. A maiden found of GOD to be pure and without defilement of worldly honor, wealth, or stature. The soul of the SON of GOD was bound to that body in all our physical weak restrictions. MESSIAH sensed and rejected every common, and all King David's inherent physical proclivities. One third of the mind of GOD was *literally* adjoined to and born into the lowly physical experiential estate of mankind. JESUS was for around 30 years, was minus his connection to HOLYSPIRIT and FATHER, just as we are at birth. He was however, watched over by many powerful Angels. Scripture says MESSIAH was born into a body that was uncomely; so, there was no physical beauty or strength in him, that any would desire him for his appearance, as that is vanity and lust. He was born without worldly advantage, and sent to accomplish a multitude of things. Including to live a lifelong sinless example, resisting all temptations against how we should desire to live. And to bring and restore as much as it might be heard, the original truth and love of GOD, which had been weakened, misinterpreted, and misused to allow the desires for this world. Third, was to suffer through the length of his harsh moment here, from being born in a manger where lowly animals fed, to walking the earth, every minute seeing and feeling all the continual rejection of the pure truth of GOD. CHRIST lived in the poverty and harsh physical existence of that day. And there were other reasons and purposes; but his last and most important, was he must be sinless, and pay an extraordinary painful price of love. MESSIAH had to suffer the harshest, most painful shameful public death, of any man that has ever drawn a breath in this world. This to adjoin all who truly desire to the HOLYSPIRIT. This unimaginable suffering before the world, was CHRIST'S Holy-offering of sacrificial, painful, responsible, faithful, perfect love. Being that required perfect payment for the required perfect forgiveness and strength that can perfectly bridge many certain gaps. In this, GOD gave everything possible, and the best of all they had to give. MESSIAH gave to the world his suffering life and precious blood, and all his precious words and wonders. And now, in and by that gift, the final Judgment of all continued rejection of GOD'S love, was thereby set and required. John 10:17-18, "*THEREFORE DOTH MY FATHER LOVE ME, BECAUSE I LAY DOWN MY LIFE, THAT I MIGHT TAKE IT AGAIN. NO MAN TAKETH IT FROM ME, BUT I LAY IT DOWN OF MYSELF. I HAVE POWER TO LAY IT DOWN, AND I HAVE POWER TO TAKE IT AGAIN. THIS COMMANDMENT HAVE I RECEIVED OF MY FATHER.*" FATHER told JESUS, he was the only one in a position of perfect value and of such perfect love to him, who could risk his soul, and assume and lay down his physical life in great pain, suffering, and love. All in hope, to break the cold hard hearts of many, and pay for and erase the evil of many. And *if* CHRIST *did not* succumb, to a single physical temptation to sin, he would be restored to his body as GOD. John 15:12-14, "*THIS IS MY COMMANDMENT, THAT YE LOVE ONE ANOTHER, AS I HAVE LOVED YOU. GREATER LOVE HATH NO MAN THAN THIS, THAT A MAN LAY DOWN HIS LIFE FOR HIS FRIENDS. YE ARE MY FRIENDS, IF YE DO WHATSOEVER I COMMAND YOU.*" FATHER could then see in this Holy Righteous suffering act of his SON, our MESSIAH, his taking and claiming the account of the sins of many others. But in bearing only *the price* for these sins upon himself, these could never be seen upon the SON as *his* sinful acts or thoughts; but only a perfect righteous loving payment. In this, many sins are not just washed from offenders, but from reality itself. CHRIST'S perfect gift, fills and closes these gaps for all who truly *repent and return* to GOD. A pure sacrificial responsible love is the only perfect force in all reality that can accomplish this wonder, called the restoring grace and mercy of GOD. We must always remember how CHRIST put his eternal soul at risk. Had CHRIST succumbed to one single temptation of the flesh, and sinned while in a mortal body, he would have been lost to FATHER, just like most souls. CHRIST would not have received his eternal body of GOD, or paid for our sins. If we deny any single aspect of this perfect understanding of what JESUS did, our sin and GOD'S sacrifice, love, and perfect forgiveness, lose all true and immutable meaning. After CHRIST'S sinless sacrifice of risking his eternal soul, in unimaginable pain and shame. He did rise to received again his eternal body, and again sat down victorious in his Throne. The price of forgiveness that enables our repentant indwelling of HOLYSPIRIT, was affirmed. We see by *not* one, but by the Trinity, a great cleansing perfect bridge was born. A great loving bridge of passage with HOLYSPIRIT as our guide. Only after MESSIAH'S victory over death and sin, was there a way for HOLYSPIRIT to connect, strengthen, teach, and lead us back over to FATHER. If we remove JESUS'S sinless experience as a man, or his suffering sacrifice as the SON, or our Baptism of the indwelling teaching HOLYSPIRIT, or FATHER being there above it all, to create, oversee, and accept this process as Holy and restorative. If we remove one perfect word or step, GOD'S equations of purpose, reason, love, and truth, to forgive and restore us as a true child, will fail. Without FATHER waiting for him in promise, MESSIAH who suffered and died as a man, could never be restored. I warn all, deny any of this at eternal peril. Yet unimaginably, in GOD giving all they have in this great sacrificial work of love, it means little to nothing for most souls. We see FATHER could not suffer and die, only his perfect SON could. Without all these acts of love, sacrifice, pain, and teaching, we see it leaves fallen mankind on the wrong side of their gaps with no possible bridge of hope. Now, all souls were gifted in their orders. We see the Second, and Third of GOD; then the Seven Arch Angels and the Four Seraphim; then the orders of Cherubim and lower Angels; and then Mankind who were gifted in their orders down to the least. All for accepting and living their gifted and created measure of eternal loving faithful responsibility. All being souls in the long sustaining order of perpetuating eternal life and love as one. We see in every gifted eternal free-will soul, that *our* own uncertainties, possibilities, probabilities, and our certainties, pertaining to our *free-will* and love, dwell entirely beyond GOD'S control. We see one true love, called faithful responsible eternal love is just that; and only its purity, responsibility, and faithfulness matter. We know the three souls of the Trinity perceive and create. And we know we and the angels perceive and create. We know that every soul's creating and indwelling our own perceptions in our awareness of them, is then our own existence. And by sharing and living the one common loving belief of GOD'S greater existence above us, is all that matters from one to another. And all we see fulfills this purpose, making our continued existence richly pleasing to both a once lonely FATHER, and all his Children. Now FATHER declared to JESUS that all FATHER has, and all that shall ever be, are in FATHER'S immeasurable love and trust, placed under the SON'S feet, as a gift of pure love, honor, and trust. So, if CHRIST had committed just one little imperfection of sin while dwelling in the flesh of mankind, he, we, and the whole of creation placed below his feet, would be surrendered in sin to Satan. Satan knew this, and the spiritual battle between CHRIST and Satan has been epic; and it is not over yet. Who do you want to help? Who do you want to win? Sin helps Satan. FATHER'S lower firmer realm would forever fall. This is the faithful responsibility to love and life that FATHER rested then and now upon our Holy King. How faithful, loving, beautiful, and strong he is above us all. JESUS has all our loving praise, honor, faith, hope, and glory forever, Amen. In both eternal body and soul, the SON again dwells in Heaven's Holy of Holies, as the HOLYSPIRIT exists everywhere GOD'S loving truth is found. And again, all our gifted souls, reside, exist, and react within the quantum firmament of GOD'S aether; yet most remain separate from GOD. To their desired degrees, most souls exist as close to and one with the lower physical firmament, as they desire to be. For this one great reason and purpose of souls choosing between GOD'S loving spiritual world, or ourselves in his created physical world, all reality was built, line upon line, precept upon precept, from the top down and from the outside in. If you can visualize this simplicity, it all came from HOLYSPIRIT, so give thanks. The Scriptures never once say GOD created all physical existence from nothing at all, as many falsely claim. It says that all perceived reality was formed from FATHER'S *breath* like forceful substantive ethereal projections that form his spoken words. We find by Scripture and all true scientific wisdom, that when either the conscious soul, or the physical, is studied alone and apart from the other, they never fully answer or prove anything. But taken together as one, through HOLYSPIRIT, they reveal all truth. Why is it that most cannot accept that our consciously projected forceful patterned catalyst, aligns and creates our particle-wave based perception. Or see that in studying all force contained and aligned physical points alone, can never unite the two different laws of consistency in the two realms. We see two realms coexist, and we see how the physical realm was in sequence, perceived and fashioned secondarily out from the quantum ethereal Heavenly realm above. Mankind has been at the door of this great feared knowledge for a long moment now, yet they refuse to step through that door until they see and hear a reflection of their own desires on the other side. Now around three thousand years ago, around a thousand years before his great sacrifice, YESHUA gave an iteration of his account of being a part of, and witnessing the creation of all we see. Spoken by him to King Solomon, who wrote it down for all who would desire to fully understand such things. In this chapter, JESUS speaks about his early loving relationship between him, HOLYSPIRIT, and FATHER. CHRIST declares his great love, and how he was brought forth and gifted out of FATHER'S soul. In this short chapter JESUS teaches us about his place, purpose, birthright, and responsibilities, in being the Firstborn of FATHER, and our faithful example to follow. Throughout Scripture, we see JESUS in his reverence, calls FATHER, "*My LORD*" and "*My GOD*" more often than FATHER, because FATHER is also still *his* LORD and GOD. In CHRIST'S words, he reveals his oneness with FATHER. And being between FATHER and us, we see our possible oneness in him. Now CHRIST speaks in this Chapter from his heavenly eternal spiritual perspective of existence, not from any worldly physical desires, fears, or pleasures. So, seek HOLYSPIRITS interpretation, and not your own. Read this poetic call of JESUS carefully and deeply, and please consider each word Spiritually. Imagine CHRIST speaking to you, saying these words slowly and softly in his greatest desire for you to find your eternal life in him. Prov. Chapter 8, "*DOTH NOT WISDOM CRY? AND UNDERSTANDING PUT FORTH HER VOICE? SHE STANDETH IN THE TOP OF THE HIGH PLACES, BY THE WAY IN THE PLACES OF THE PATHS. SHE CRIETH AT THE GATES, AT THE ENTRY OF THE CITY, AT THE COMING IN AT THE DOORS. UNTO YOU, O MEN, I CALL; AND MY VOICE IS TO THE SONS OF*

*MAN. O YE SIMPLE, UNDERSTAND WISDOM: AND YE FOOLS, BE YE OF AN UNDERSTANDING HEART. HEAR; FOR I WILL SPEAK OF EXCELLENT THINGS; AND THE OPENING OF MY LIPS SHALL BE RIGHT THINGS. FOR MY MOUTH SHALL SPEAK TRUTH; AND WICKEDNESS IS AN ABOMINATION TO MY LIPS. ALL THE WORDS OF MY MOUTH ARE IN RIGHTEOUSNESS; THERE IS NOTHING FROWARD OR PERVERSE IN THEM. THEY ARE ALL PLAIN TO HIM THAT UNDERSTANDETH, AND RIGHT TO THEM THAT FIND KNOWLEDGE. FOR WISDOM IS BETTER THAN RUBIES; AND ALL THINGS THAT MAY BE DESIRED ARE NOT TO BE COMPARED TO IT. I WISDOM DWELL WITH PRUDENCE, AND FIND OUT KNOWLEDGE OF WITTY INVENTIONS. THE FEAR OF THE* ***LORD*** *IS TO HATE EVIL: PRIDE, AND ARROGANCY, AND THE EVIL WAY, AND THE FROWARD MOUTH, DO I HATE. COUNSEL IS MINE, AND SOUND WISDOM: I AM UNDERSTANDING; I HAVE STRENGTH. BY ME KINGS REIGN, AND PRINCES DECREE JUSTICE. BY ME PRINCES RULE, AND NOBLES, EVEN ALL THE JUDGES OF THE EARTH. I LOVE THEM THAT LOVE ME; AND THOSE*

*THAT SEEK ME EARLY SHALL FIND ME. RICHES AND HONOR ARE WITH ME; YEA, DURABLE RICHES AND RIGHTEOUSNESS. MY FRUIT IS BETTER THAN GOLD, YEA, THAN FINE GOLD; AND MY REVENUE THAN CHOICE SILVER. I LEAD IN THE WAY OF RIGHTEOUSNESS, IN THE MIDST OF THE PATHS OF JUDGMENT: THAT I MAY CAUSE THOSE THAT LOVE ME TO INHERIT SUBSTANCE; AND I WILL FILL THEIR TREASURES. THE* ***LORD*** *POSSESSED ME IN THE BEGINNING OF HIS WAY, BEFORE HIS WORKS OF OLD. I WAS SET UP FROM EVERLASTING, FROM THE BEGINNING, OR EVER THE EARTH WAS. WHEN THERE WERE NO DEPTHS, I WAS BROUGHT FORTH; WHEN THERE WERE NO FOUNTAINS ABOUNDING WITH WATER. BEFORE THE MOUNTAINS WERE SETTLED, BEFORE THE HILLS WAS I BROUGHT FORTH; WHILE AS YET HE HAD NOT MADE THE EARTH, NOR THE FIELDS, NOR THE HIGHEST PART OF THE DUST OF THE WORLD. WHEN HE PREPARED THE HEAVENS, I WAS THERE: WHEN HE SET A COMPASS UPON THE FACE OF THE DEEP: WHEN HE GAVE TO THE SEA HIS DECREE, THAT THE WATERS SHOULD NOT PASS HIS*

*COMMANDMENTS: WHEN HE APPOINTED THE FOUNDATION OF THE EARTH: THEN I WAS BY HIM: AND I WAS DAILY HIS DELIGHT, REJOICING ALWAYS BEFORE HIM; REJOICING IN THE HABITABLE PART OF HIS EARTH; AND MY DELIGHTS WERE WITH THE SONS OF MEN. NOW THEREFOR HEARKEN UNTO ME, O YE CHILDREN: FOR BLESSED ARE THEY THAT KEEP MY WAYS. HEAR INSTRUCTION, AND BE WISE, AND REFUSE IT NOT. BLESSED IS THE MAN THAT HEARETH ME, WATCHING DAILY AT MY GATES, WAITING AT THE POST OF MY DOORS. FOR WHOSO FINDETH ME, FINDETH LIFE, AND SHALL OBTAIN FAVOUR OF THE* ***LORD***. *BUT HE THAT SINNETH AGAINST ME WRONGETH HIS OWN SOUL; ALL THEY THAT HATE ME LOVE DEATH.*" Many worldly Christian Preachers, in grave error, say these are Solomon's own metaphorical words, speaking of his self and worldly things. How can these be the words of a man. JESUS the SON, is by the greatest of loves, truly appointed the eternal King of Kings. Who appoints not worldly biased Kings, Princes, and Nobles *in* the earth. And not the imperfect biased Judges *in* the courts of *this world*. He speaks of the Kings

Princes and Nobles in Heaven, and those in Heaven who will be "Judges *of* the earth". These are the Judges that CHRIST appoints. We know that all the orders of Created Angels, being eternal free-will creatures, were created to dwell in the upper quantum spiritual realm of Heaven, serving and living within the proximity of GOD. So, the Angels have a clear, closer, knowing, greater perspective and measure of GOD'S existence. Not witnessing all physical creations, but much of it, and actually seeing GOD. Now in our knowing that when higher beings are sent into this world on a mission by FATHER, their vast ethereal body is translated, being condensed to *our* firmer physical size and sensations, and when they return to Heaven, they are inflated back to their ethereal state. So, why did JESUS tell Mary right after his resurrection, to not touch him, because he had not yet ascended to the FATHER. Was JESUS in a certain phase of transition from man, back to GOD. The full meaning of this has not been revealed to me as of yet. If it is revealed to anyone, I hope you will in turn share it with me and all. In Heaven we will feel the body, but more faintly; and in this fallen realm, most barely feel their soul. This latter condition is not our created purpose. We are created to feel the physical more than the Angels, yet meant to sense and know our soul in the presence of GOD above all physical life. Angels were given the simple true choice of desiring and sharing a great love for GOD'S perfect ways of life, or turning away. And by the Temptation to be individuals, one third *of* around three hundred million Cherubim and lower angels turned from GOD to follow Lucifer. We see Adam at his creation, was placed in a perfect physical Garden in a Perfect Earth covering several hundred square miles. Yet not having witnessed *any* creation events; he was even put in a deep sleep so he could not witness Eve's creation and gifting. Yet he did hear HOLYSPIRIT and was visited and taught daily by JESUS. So, Adam and then Eve, both saw and heard GODS in person teaching. GOD'S hope filled purpose for Adam and Eve, was for them to hear, learn, choose, and share a sacrificial faithful responsible eternal love with GOD and all. Now Adam and Eve were joined/married in the earth, to bear physical fruit after their kind, and fill the earth with the wonder of loving faithful responsible children for GOD. And in this, GOD would eternally love and keep them, giving every beat and breathe and force of physical life. In all this, giving to mankind that same choice of wanting or not wanting GOD'S perfect love, truth, and created ways of life. Their choice was to trust and want the created truth and pure love of eternal life that comes only from GOD, and be Spiritual Creatures like the Holy Angels; or to *not* trust or want that. Their *choice* of believing and wanting, or not believing and not wanting GOD'S spiritual and physical ways, was set in their being told to not eat of the fruit of one certain tree, or they would die (lose their sustaining connection to GOD). GOD named this tree of choice, the subsequent knowing of not only what is good, but if taken and eaten, then knowing what is evil by becoming evil. FATHER unavoidably creates a possibility of endless free-will evil choices by defining what is good. Without a defined choice, good and evil do not exist. All GOD physically creates, is meant to support every soul's existence and eternal good. And all who GOD consciously gift as souls, must choose GOD'S desired ways, or form their own individual desired ways. Gifted souls only know evil by choosing and becoming evil. Before mankind's fall, FATHER held his creation in perfect dynamic guided order. There was a perceived cycling of days, nights, and orbits of all things held in their true courses. The Earth was cooler and dryer then, and it never rained; yet a morning mist formed near the ground for vegetation. All perfect physical life was in eternal care. But now all things divide, inflate, and accelerate outward in hell's rising confusion. The earth now wobbles imperfect on its axis, and earth's orbit around the sun, is now imperfect. We see all the imperfect aspects of time, are the flaws that inflate and corrupt all creations in many ways. The physical creation now exists in unstable ways. And by reverse engineering all current disorder back, undoing all contrary physical corruptions, we see the moment where perfection was lost and the disorders of time entered and began to undo the perfect dynamic orders of reality. Materialist see these current imperfect physical conditions as being caused by a chaotic creation, and not by the sinful destroyers of it all. The first stretching of the entire creation was sudden and then it stopped, it being relative to Adam and Eve's first projected evil disorder. We now see all the damage evil projected forceful desires have had upon reality since that first destructive act of time was chosen. So, can one become so confused, as to no longer believe the one pure truth exists; then become unconfused and return back to believing the pure truth? I tend to doubt that. But before reaching that irreparable point called being a reprobate, or a liberal zombie in modern parlance. Before reaching that horizon, each lost soul's beliefs are malleable, partial, and often weak. So, some may escape their current beliefs, but their desires against being subject to GOD'S ways, is hard to overcome. Every soul who has let go of a loving trust and desire towards GOD, opens a door to endless confused possibilities. They become a soul wondering what about this or that and wander through endless godless choices. We must completely come back through that door and close it, knowing only confused division waits on the other side, and never open it again. Once that door is closed, all our former evil conscious effects and proclivities, with our continued effort can fade and be overcome. Now for any who *have* come back to truly know, love, and follow GOD'S ways; yet again *lose* that love and go back out that door. For these to return to the truth of GOD a second time is impossible. Read Heb. Chapter 6. Those souls had in great search and fervent desire, found and been found to receive great intimate things and love from GOD. Things given through a painful sacrifice beyond definable value and measure. This most precious gift, if set aside again as unwanted, such souls are without remedy, just as the fallen angels. Now consider that if a true choice for or against the eternal ways of GOD is set before all, but none ever choose against GOD, where is the evidence of true choice. Choice demands free-will, and free-will demands a true choice. These together, require and bring chance, being a real probability into the conscious equations of reality. Chance and choice can neither exist without the other. So, the existence of real choice, requires a real probability that all antithetical sides must at some point appear, or there was never a chance *or* a true choice. Again, this Free-Will Antithetical Principle, applies to any shared reality, having many-souls and a supremely defined eternal form of love and life, set beyond rewriting. Without antithetical free-will rebellion to it, a true choice for a certain well-defined eternal love can never exist. So, we see cost, loss, pain, and rejection, *must* be and are a real part of real love, as is found in each life. Choice and chance say all antithetical appearances, such as a defined love and its antithetical hate, must both be allowed, *and* must both appear, or neither exists. This is Reality's Painful Law of Defined Free-Will Love among many souls. *All* conscious abstract realities require the contrast of their antithesis for them to exist as something well defined. Big cannot exist without small or right without left. So it is, with good and love, that there must appear evil and hate. Even FATHER in once being all alone, was not considered the vast immense GOD he is now; not until we and all creations, were gifted and formed so small beneath him. It is our minuteness, that declares FATHER'S great size. These Principled Laws of Reality apply to all things beyond FATHER'S person. And they are all intimately defined by what GOD desires and creates. So, it is GOD'S sacrificial faithful responsible love alone that defines all eternal oneness, and they allow the free-willed desired divisions of all souls within all FATHER has created. We read Lucifer became prideful and envious of GOD. But pride and envy bring a strange antithetical blindness, being that once any soul envies another for being greater above them, they must believe those created below them, must envy them also, even if not true. So, prideful envy birthed all false perspective of individual power, privilege, and possession, which in all individuals, displaces FATHER'S created orders of loving faithful responsibility. Looking back at that first confused abstract choice, Lucifer presented to Eve in the Garden of GOD. Scriptures says that in hearing a subtle contradiction (untruth) from Lucifer, Eve felt her first ever envious selfish desire, in that she wanted that lie to be true. She actually felt her love for GOD weaken, leading her to choose her first individual lie of belief. We read how she considered her beliefs about what GOD had told her, but was drawn to her newborn selfish desire to have that one thing in that immense perfect abundant garden she was not given. Scripture says she saw the fruit of the one forbidden tree, and in her rising self-centered desire, it was pleasing to her eye and looked good to eat. She was losing her trust, thinking GOD was keeping something good from her. She considered, wanted, and then believed, at least parts of Satan's subtle lie, being that it would make her wiser like GOD. So, turning, forgetting, not loving, disobeying, and disbelieving what GOD said it would mean for her to eat and know both good and evil, and she would die in being separated from the eternal care of GOD. And like Lucifer, in wanting to rise up to be like GOD, Eve ate; but she only fell and became a far lesser individual. She turned from a pure *oneness* of faithful responsible love in GOD, family, and the ways of eternal life. By her own desire to have that one little thing called evil, that existed there just beyond all the perfect abundant good she was given by GOD, she was caught by envy and death. GOD needs to and does give *all* that is good and eternal to us, and keeps nothing good from their Children. So, Eve, then being evil herself, immediately did as evil does. Did she warn Adam? No. She sought to spread her confused individuality to the one to whom she was wed, who trusted and loved her. Adam's love and trust for GOD was then tempted by *his* love for her, and *her* lies, and in finding his own reasons, he chose to believe her above GOD. So, both parents of all mankind were turned from GOD'S perfect ways that sustained their eternal physical provision, and became part of the endless lies that define conscious individuals. The definition of Truth is all that is eternal, and the definition of a lie is all that is temporary. GOD'S perceived eternal provision, is all that is good, and this is beyond all honest spiritual debate. Adam's soul and body were the highest of all gifts and creations in the earth; and GOD had put the earth and its fullness under Adams dominion as a loving trusting gift. And by Adam then bowing to believe Satan, he surrendered his dominion to Satan. GOD hearing about this in the garden, brought that first repulsed recoil of FATHER'S hands that held the universe in perfect tune, balance, and purpose. A sudden stretching and stopping by FATHER'S hands, introduced a certain level of disorder, as reality fell into entropy and fission. There and then, the physical realm's process of increasing sorrow, pain, hell, and death, began. Eve and Adam made their own envious choices to sit upon the throne of defining what is good and true. They reversed their polarity of alignment with GOD and lost GOD'S eternal love and care, being set alone in their inability to sustain physical life. From that moment, this antithesis called sin, spread and grew in the earth, even as all FATHER'S desire for this fallen physical reality coincidingly fade. Scripture shows that at the very moment this creation *loses* the last of its purpose and potential, in that it cannot render even one more eternal love. GOD at that moment, will let it all succumb to fission and dissolve away from its perceived existence. 2nd Peter 3:10, "*BUT THE DAY OF THE LORD WILL COME AS A THIEF IN THE NIGHT; IN THE WHICH THE HEAVENS SHALL PASS AWAY WITH A GREAT NOISE, AND THE ELEMENTS SHALL MELT WITH FERVENT HEAT, THE EARTH ALSO AND THE WORKS THAT ARE THEREIN SHALL BE BURNED UP.*" GOD then judges all unjudged souls, and perceives a New Heaven and Earth for his found Children, all within view of a created Lake of Fire. Let's discuss *this* world's biased concepts of justice. Mankind's justice is biased to fit their individual beliefs and desires.

But GOD defines justice in his eternal perfection and self-reliance; so, who holds the true and final justice. If GOD'S understanding of belief is eternal, omniscient, omnipotent, omnipresent, and immutable above all others; will it not render pure perfect justice. For the eternal love and life of all, we surrender in joy to GOD'S perfect understanding above all. But like truth love and beauty, in *this* fallen world, justice is now falsely defined in the eye of the individual beholder. In this world, all who refuse to bow to GOD'S eternal truth, being small slow fallen souls, think they know the meaning of justice. *Eternal* Justice has zero to do with individual revenge, punishment, financial or reputational restoration, or worldly equity. FATHER'S justice, is about making his love whole. *Eternal* Justice has to do with our shame before CHRIST, and his forgiveness as we correct every error of belief. Eternal Justice makes JESUS, who has been greatly wronged, whole again. But justice in this world is seen as individual feelings of power, control, satisfaction, punishment, or coercion and nothing more. This world is incapable of knowing, interpreting, or weighting every consideration. And never consider the wrong done to GOD, or render eternal good in anything. We see tainted, biased, exaggerated, and false accusations and testimony. We see the guilty set free for insufficient evidence or political influence, and some sent to prison for years, only to be found innocent later. And this, in the U.S. where we spend great sums, and direct great process and resources to provide a high semblance of fairness. Out of all the individual, familial, institutional, or governmental judgments, that occur around this world daily; not *one* has ever been near to perfect. To degrees, they are all wrong, and *perfect* justice was never served. If we desire *perfect* justice, only GOD can render that. There soon comes a day when all perfect Judgment is set, as all evidence has been collected, and a Holy Righteous Judge sits. Such a fearful day, and I fear it greatly. Even if we are found sealed in the eternal mercy of GOD, there is no joy in that day; only fear and pain. But joy we are promised, follows for all who desire and perpetuate eternal love and life. But a different promise is made to those who don't desire GOD'S created ordered ways of life. Now after their fall, Adam and Eve procreated until chance occurrence and the certain level of entropy, or disorder in their day overcame the bodies to which they were adjoined. And all created bodies after Adam's fall, were born of and in their parent's physical imperfection, and a small slow physical fallen perspective.

But all born after the fall, still have that same true choice before them, for loving the ways of eternal life, or self-rule and death. But now, our choice is to desire to first return back to, and then eternally live in GOD'S ways of truth; but in this, how do we become unconfused. First, we regret every prior moment and see our every individual belief is corrupt. We turn, not partly, but completely in desire, fear, and shame, disavowing our prior understanding of belief. We wait for CHRIST to Judge if and when our heart is true. If he finds our process of repenting, meaning to turn from *all* sin, to be true and sincere. CHRIST can choose to pay, and erase our accountability by his great sacrifice. If and when he does, we are there and then seen as a brand-new clean soul, set towards the eternal perspective of HOLYSPIRIT. Deny all former understanding beyond simple math, reading, and spelling. Even modern word definitions are filled with corruptions to overcome. We must hear GOD'S definition of all words. We then have eternity before us to learn more and more without end. So again, just because we have humbly and shamefully asked, this does not mean we are instantly forgiven. Matthew 7:21-23, "*NOT EVERY ONE THAT SAITH UNTO ME, LORD, LORD, SHALL ENTER INTO THE KINGDOM*

*OF HEAVEN, BUT HE THAT DOETH THE WILL OF MY FATHER WHICH IS IN HEAVEN. MANY WILL SAY TO ME IN THAT DAY, LORD, LORD HAVE WE NOT PROPHESIED IN THY NAME? AND IN THY NAME CAST OUT DEVILS? AND IN THY NAME DONE MANY WONDERFUL WORKS? AND I WILL PROFESS UNTO THEM, I NEVER KNEW YOU; DEPART FROM ME, YE WHO WORK INIQUITY.*" Heaven is not for individuals who believe in GOD and do not want to be filled with hell, or be cast in a fiery prison. Heaven is for those who live as CHRIST in all oneness of love. There are many religious yet rebellious souls who believe they are saved, yet are not (according to GOD, not me). GOD truly wants all souls to strive to return in true love to find and live in the shared truthful ways of eternal life. Scripture says, among those early souls who re-found perfection, was a man named Enoch, who by his great love, did walk continually with GOD in perfection for many of his 365 years on earth. So, in his process of years, he was learned and found faithful, and was translated up to the quantum realm of Heaven. Enoch now lives as an elder of earth, and as a lower Holy Angel. GOD knew pride and rebellion, once loosed by envious free-will choice, *could* become an ever-increasing probability; yet these are the only proof that true free-will choice and chance exist. Now, in mankind's propagation of being born already in a fallen imperfect physical perspective, GOD justly did not close the door on their restoration. Our *inherited* and not chosen weakness towards desiring the fallen proclivities within mankind, could not be seen in GOD'S mercy as a final loss; so, it justly became a rescue mission. GOD saw souls being born into walking without an eternal perspective of choice, coming into existence already on the wrong side of his original created line of true choice. So, GOD held a just merciful hope to restore mankind's perspective and choice by increasing his own sacrifices of love, to build a painful way to return to GOD. To return to perfection requires great loving guidance from a rejected, and now cautious GOD. GOD'S last final sacrifice and effort, was their greatest possible gift of love. Now GOD'S earliest help, came as the distance between mankind, the Holy Angels and GOD, had not yet grown beyond hearing those Holy Angels. It was an era of Holy Angels above, calling all to repent, as HOLYSPIRIT felt for all opportunities. This offer of mercy to mankind, was not offered to the fallen Angels who fell from the heavenly presence of GOD.

Also, Angels were created only in the male semblance of GOD, having no propagation. Therefore, no angels were *born* into a fallen disadvantaged perspective, which may have qualified them for a chance to make the right choice. Now regarding those few Angels who were translated into a human body consistency on the earth, and went into women and bore children. GOD'S offer of mercy towards their offspring appears unclear to us, being far from GOD'S perfect plan. But these offspring between angels and women, *are* referred to in Scripture, as being mighty *men* in the earth, and not weaker angels in the earth. There seems no end to the confusion free-will souls bring to GOD that they must deal with. Remember, it's not *just* an unfaithful choice for imperfect ways, it is the contrary forceful disorder each sin projects into the physical creation of life itself. A faithful loving perfection that perpetuates eternal physical life, is a most important point. Our knowing, desiring, and choosing to perpetuate GOD'S perfect order, is one of the greatest aspects that separate all worldly religions from GOD'S true religion. In GOD'S perfection, forgiveness of our *past* imperfect destructive acts, cannot come with allowing our endless future destructive acts. There exists no reason, purpose, love, change of outcome, or eternal life, in that. GOD cannot change from perfection for us, and GOD has never once changed for us. Do you see Heaven as a place where souls can still sin. At its start, our newborn eternal love feels a strong emotional element of sacrificial forgiveness. It is a most powerful emotion in coming to understand we have only hurt and assaulted GOD'S love, projecting our destructive forces into the fabric of physical life. And for this, GOD came down to suffer our shame and pain before the world. I see it so clearly, and in our broken shame we lovingly reverse our desired beliefs to be a new Godly creature. This hope towards us all, resides within GOD, and it brings them joy at recovering each eternal soul that was once lost. The true value of an eternal soul, is only known to GOD. And as with the story of the prodigal son, the love for a lost soul's return, initially leaps high; but the love for the SON who responsibly stayed and faithfully loved throughout his existence, is forever set on high as FATHER'S far greater treasure. All pure love is true love, and to GOD it is worth seeking, finding, and saving. And that is all that really matters to GOD and to us. 2nd Cor. 6:18, "*AND WILL BE A* ***FATHER*** *UNTO YOU, AND YE SHALL BE MY SONS AND DAUGHTERS, SAITH THE*

***LORD*** *ALMIGHTY.*" That is one of the few verses that says both sons and daughters of GOD. This is because there were only a few times women were in attendance when these particular things were taught. Old languages were and *still* are formally worded and spoken towards the genders present. So, we know no women were present for the following: Romans 8:19, "*FOR THE EARNEST EXPECTATION OF THE CREATURE WAITETH FOR THE MANIFESTATION OF THE SONS OF* ***GOD***." John 1:12-13, "*BUT AS MANY AS RECEIVED HIM, TO THEM GAVE HE POWER TO BECOME THE SONS OF* ***GOD****, EVEN TO THEM THAT BELIEVE ON HIS NAME: WHICH WERE BORN, NOT OF BLOOD, NOR OF THE WILL OF THE FLESH, NOR OF THE WILL OF MAN, BUT OF* ***GOD***." ROMANS 8:14, "*FOR AS MANY AS ARE LED BY THE* ***SPIRIT*** *OF* ***GOD****, THEY ARE THE SONS OF* ***GOD***." Phil 2:15, "*THAT YE MAY BE BLAMELESS AND HARMLESS, THE SONS OF GOD, WITHOUT REBUKE, IN THE MIDST OF A CROOKED AND PERVERSE NATION, AMONG WHOM YE SHINE AS LIGHTS IN THE WORLD*;" 1st John 3:1-2, "*BEHOLD, WHAT MANNER OF LOVE THE* ***FATHER*** *HATH BESTOWED UPON US, THAT WE SHOULD BE*

*CALLED THE <u>SONS</u> OF* ***GOD**: THEREFORE THE WORLD KNOWETH US NOT, BECAUSE IT KNEW HIM NOT. BELOVED, NOW ARE WE THE SONS OF* ***GOD**, AND IT DOTH NOT YET APPEAR WHAT WE SHALL BE: BUT WE KNOW THAT, WHEN HE SHALL APPEAR, WE SHALL BE LIKE HIM; FOR WE SHALL SEE HIM AS HE IS.*" Women may have, but need not have been present for the following. Acts 2:17, "*AND IT SHALL COME TO PASS IN THE LAST DAYS, SAITH* ***GOD**, I WILL POUR OUT OF MY* ***SPIRIT*** *UPON ALL FLESH: AND <u>YOUR SONS AND YOUR DAUGHTERS</u> SHALL PROPHESY, AND YOUR YOUNG MEN SHALL SEE VISIONS, AND YOUR OLD MEN SHALL DREAM DREAMS.*" From their first desired acceptance of individual belief within Eve and then Adam, through the passing of just a few *thousand* years. We read that the entirety of mankind was found to be reprobate except for eight souls. GOD said the thoughts of all mankind at that moment were *only* on evil continually. So, a moment came with no eternal truth or love remaining in the earth outside the 8 souls of Noah and his family. And to save the love and truth they alone held, and restart hope for more love, FATHER in his grief for the second time, *greatly* stretched the creation, bringing a cleansing great flood upon the earth. This was not the destruction of salvageable eternal souls, but the removal of all perpetual destruction. The entire world at that time, was in what we might call a liberal reprobate zombie apocalypse. We see FATHER waited until the last possible moment to do this, showing how merciful GOD must be. Yet GOD must show the world they need *their* rule and care. Now, through earth's twelve or so thousand-year history to date, GOD instructed those near to him in this world, to chronicle the events of mankind's obedient and rebellious history, and enshrine the pure truth that GOD'S created ways of life are not to be compromised. These are known as Scripture, or the Holy Bible. These ancient writings tell us to avoid and reject the confused ways of individual hell. I have no experience or judgment of the Geneva Bible, so I can't recommend or dissuade you as to that translation; but all modern translations are weak and to be avoided. I know the Authorized King James Bible holds GOD'S perfect wisdom when it is interpreted by HOLYSPIRIT, bringing all to *one* common understanding and belief. Then GOD may commune with us in our shared belief of understanding. And from there, teach and lead us to all higher loving wisdom beyond the Scriptures. Surely, we cannot say our firm Scriptural foundation is the sum of all we are ever meant to receive from GOD for eternity. Yet without that proper pure Scriptural foundation, we can learn nothing at all. After the flood, and a substantial re-population, there came another point where mankind needed a strong correction to see all they had again blinded themselves to. So, GOD ordained the lesser animal blood sacrifice covenant for mankind's repentance of sin. This animal sacrifice was never in any spiritual way for GOD, but for breaking a sinner's heart. Back then, most souls held all life as sacred, giving reverence at taking the life of each animal for food. At first, this blood sacrifice held great meaning, as every time a soul sinned against eternal life, an offence worthy of death, in that it causes children to emulate. Yet, in their stead, they could offer and witness the life of an innocent animal die in their place, watching the force of life draining slowly and forever from the animal's eyes for no other reason than their sin against GOD'S ways of life. This sacrifice of life's force in their presence and place, was to break their heart and put them on their knees in tears of repentance; hopefully giving them strength to turn from the evil they had committed. But the meaning and strength were soon lost in mankind's rebellious over-use and abuse of the Blood Sacrifice. The hearts of most grew numb to their shame in taking an innocent life. It became a monetary ritual, as those today feel paying a traffic fine, feeling only monetary loss. We see false physical belief crept in, being if they just continue to physically pay for their sin with an animal's life, they are legally released from accountability without ever truly repenting. But this is not so. We read in Isaiah 1:11, "*TO WHAT PURPOSE IS THE MULTITUDE OF YOUR SACRIFICES UNTO ME? SAITH THE LORD: I AM FULL OF THE BURNT OFFERINGS OF RAMS, AND THE FAT OF FED BEASTS; AND I DELIGHT NOT IN THE BLOOD OF BULLOCKS, OR OF LAMBS, OR OF HE GOATS.*" Do any think FATHER finds pleasure in the blood of animals; or sacrificing life he himself created, could give him pleasure. Psalms 50:7-8, "*HEAR, O MY PEOPLE, AND I WILL SPEAK; O ISRAEL, AND I WILL TESTIFY AGAINST THEE: I AM GOD, EVEN THY GOD. I WILL NOT REPROVE THEE FOR THY SACRIFICES OR THY BURNT OFFERINGS, TO HAVE BEEN CONTINUALLY BEFORE ME.*" GOD never forgives any under false pretense. GOD does not ask a gift that is his already. Psalms 50:9-11, "*I WILL TAKE NO BULLOCK*

*OUT OF THY HOUSE, NOR HE GOATS OUT OF THY FOLDS. FOR EVERY BEAST OF THE FOREST IS MINE, AND THE CATTLE UPON A THOUSAND HILLS. I KNOW ALL THE FOWLS OF THE MOUNTAINS: AND THE WILD BEASTS OF THE FIELD ARE MINE*." If a soul believes a sacrificed animal in itself, satisfies a need in GOD, who and what must they believe GOD to be. It is never the flesh that defines a sacrifice, it is life, pain, and suffering, that gives sacrifice its meaning. Any soul who believes otherwise, does not know or understand, and does not believe in the one true GOD. Psalms 50:16-17, "*BUT UNTO THE WICKED GOD SAITH, WHAT HAST THOU TO DO TO DECLARE MY STATUTES, OR THAT THOU SHOULDEST TAKE MY COVENANT IN THY MOUTH? SEEING THOU HATEST INSTRUCTION, AND CASTEST MY WORDS BEHIND THEE*." Heb. 10:4-6, "*FOR IT IS NOT POSSIBLE THAT THE BLOOD OF BULLS AND OF GOATS SHOULD TAKE AWAY SINS. WHEREFORE WHEN HE COMETH INTO THE WORLD, HE SAITH, SACRIFICE AND OFFERING THOU WOULDEST NOT, BUT A BODY HAST THOU PREPARED ME: IN BURNT OFFERINGS AND SACRIFICES FOR SIN THOU HAST HAD NO*

*PLEASURE.*" MESSIAH says FATHER has no desire for him to physically suffer, yet he still comes as the greatest possible Sacrifice, in hope of reaching and breaking our cold hearts. But it must *never* be used in a legalistic way as before. So again, around 2,000 yrs. after the Great Flood, the earth had been replenished, and individualism had spread to reign again as vanity and religious self-rule, grew to be the greater proportion. After all prior help and examples had lost all their strength, it seems FATHER would have had enough, and had done, given, and shown all considerations and examples of his love. Yet he made that last foretold greatest effort before the end of all evil is determined. FATHER says in Isaiah 5:4, "*WHAT COULD HAVE BEEN DONE MORE TO MY VINEYARD, THAT I HAVE NOT DONE IN IT? WHEREFORE, WHEN I LOOKED THAT IT SHOULD BRING FORTH GRAPES, BROUGHT IT FORTH WILD GRAPES*?" Now can we see FATHER'S omniscience in a way to visualize, *not* just the summation of his knowledge, but just as importantly its domain. To know where GOD'S omniscience dwells, and can be found, and where it cannot dwell, or be found. We try… It is written, there are GOD'S eternal ways of truth that make us one, *and* the lies and ways of rebellion that form endless division between individual lost souls. There are also connections between gifted souls and the ethereal, or physical bodies to which we are adjoined. I must show how our souls are meant to be aligned with GOD, and how sin has entered into reality to divide us from GOD'S ways, with hell and death. We are spiritually called to become of one mind with CHRIST. Romans 8:9. "*BUT YE ARE NOT IN THE FLESH, BUT IN THE* ***SPIRIT****, IF SO BE THAT THE* ***SPIRIT*** *OF* ***GOD*** *DWELL IN YOU. NOW IF ANY MAN HAVE NOT THE* ***SPIRIT*** *OF* ***CHRIST****, HE IS NONE OF HIS. AND IF* ***CHRIST*** *BE IN YOU, THE BODY IS DEAD BECAUSE OF SIN; BUT THE* ***SPIRIT*** *IS LIFE BECAUSE OF RIGHTEOUSNESS.*" Before the angels and mankind were gifted and created, FATHER, SON, and HOLYSPIRIT, were the first three souls; yet they are of one mind in every way of sharing one pure perfect love, belief, and spiritual/scientific understanding. This Spiritual oneness of love, belief, and mind, through the indwelling HOLYSPIRIT, is found unaltered within the Seven Arch Angels, the Four Seraphim, and all the still *faithful* Holy Angels in their created orders. We also see a *re*unification in the saved among Mankind. All the above, live without blemish or confusion in the mind of GOD'S omniscience and faithful obedience. In all of this, *we* see we are part of the Sons, Daughters, and Family of GOD. *These* are the boundaries and domain of all omniscient truth and love, this being GOD'S Spiritual Kingdom. FATHER'S created truth and ways are found nowhere else. If any Angels or Mankind are not unified in love, belief, and understanding through GOD'S HOLYSPIRIT, they must exist in individual states of lust, and false belief. Lost souls desire, believe, and follow their own false beliefs. They believe they are good, but they are contrary and evil. Their individual ways and beliefs do *not* align with GOD. Pure eternal love is all powerful, and it is indisputable that this love moves only in GOD'S ways and purpose. Eternal love is the one cause GOD cannot resist. FATHER yields his will over all that is, to no other reason or purpose. GOD goes and comes for eternal love. GOD creates all things for eternal love; and separates all evil to protect his found love. GOD lives for love, and in his *SON*, GOD has died for love, and rose up again for love. Such a loving sacrifice by GOD to show us their love exists towards us for help and strength. There is no grander word than love, and when love exists in our eternal responsible faithfulness, no grander enduring meaning or connection between souls, can exist. Only a faithful responsible love is eternal, and JESUS is the epicenter and essence of FATHER'S love. In knowing this, we must in our greatest fear, know the evil it is to *not* regard, bow, and honor; or to disregard and dishonor in any form or manner, the SON and his suffered sacrifice for us. Until we each see *all* the profound spiritual efforts and meanings in both life and death. And feel the pain and love that was before and now and forever in this gift, and feel it in the spiritual way it is meant to endure. Until we intimately know all these in our core, *we* have not received the fullness of GOD'S spiritual gift. When I was lost, the true understanding of GOD did not flow within me. And after 45 years of learning, if I hear a soul take a name of GOD as an expletive, or exclamation, or qualifier of emphasis, or for any irreverent *vain* reason, or in any irreverent vain tone; I actually shudder inside, as it is spiritually alarming. And what must GOD feel. Please fear any such unrepented evil from your lips, should come before FATHER in your day of judgment. Saying the Holy names of GOD in vain, is *equal* to spitting in the face of GOD; don't *ever* do this. Now to know the lively ways of love and sacrifice, desired to be found in us. We examine CHRIST'S example, and in our strongest desire, we aspire to live this way. CHRIST speaks often of his loving eternal obedience to FATHER. Teaching us that these examples of his love and life eternal, are his everything. CHRIST never failed once from his first experience at being gifted his soul of self-awareness out from FATHER. Not once has CHRIST, having free-will, ever failed in his greatest of sacrificial love, to always hear and live in FATHER'S pure truth. And again, only FATHER can know and define eternal love and life. Does CHRIST'S loving obedience towards FATHER, reduce, diminish, or demean the place, or name of JESUS. Not at all. According to Scripture, by this fact, he is raised ever higher. HOLYSPIRIT comes to comfort us with love, and teach us the true interpretation of Scripture, which says all *individual* interpretations are weak and born of physical desires, and are not true. Truth taught by HOLYSPIRIT to all, is one single impenetrable shared true belief. Scripture interpreted by the many to themselves and taught to each other, will become endless individual doctrines, born of personal desires and cultures. HOLYSPIRIT'S true interpretation declares FATHER'S strong boundaries and singular perfect definitions set in all things; even FATHER'S love, truth, righteousness, and his ways. And if we *truly* know JESUS intimately, we know he dwells only within those same boundaries of love and truth. The disparaging or diminishing of CHRIST is not found in reiterating his Scriptural *self-described* attributes, but in denying those self-described attributes. Or to imply upon JESUS one's own individual contrary beliefs and desires. Or to say JESUS ever lived outside of FATHER'S perfect created ways, or fell to any individual imperfect desires. I tell you that is this world's true diminishing and disparaging of JESUS. An example of disparagement I often see, is when I reiterate GOD'S truth about the ways of life and love, most souls in confused blind pride and self-rule, say JESUS knows and accepts that in their heart, it is just their *individual* way to love, believe, and worship him. But it is written that all individual interpretations and all non-ordained individual ways to worship GOD, are *not* accepted. Self-ascribed ways of worship, breed confusion and separation among believers. In Scripture, FATHER clearly defined how we all shall worship him privately and corporately as his family, being in a way that pleases him. Not in the endless ways that please rebellious souls on an individual basis. So, let's read about JESUS'S obedience to FATHER'S pure truth. These verses are *our* example of CHRIST'S eternal life, lived within the defined ways set by and desired by FATHER. Most are directly spoken by CHRIST to his Disciples about his love and worship for FATHER. John 5:19-20, "*THEN ANSWERED* ***JESUS*** *AND SAID UNTO THEM, VERILY, VERILY I SAY UNTO YOU, THE* ***SON*** *CAN DO NOTHING OF HIMSELF, BUT WHAT HE SEETH THE* ***FATHER*** *DO: FOR WHAT THINGS SOEVER HE DOETH, THESE ALSO DOETH THE* ***SON*** *LIKEWISE. FOR THE* ***FATHER*** *LOVETH THE* ***SON****, AND SHEWETH HIM ALL THINGS THAT HIMSELF DOETH: AND HE WILL SHEW HIM GREATER WORKS THAN THESE, THAT YE MAY MARVEL.*" John 5:26, JESUS says, "*FOR AS THE* ***FATHER*** *HATH LIFE IN HIMSELF; SO HATH HE* <u>*GIVEN*</u> *TO THE* ***SON*** *TO HAVE LIFE IN HIMSELF;*" John 5:30, JESUS says, "*I CAN OF MINE OWN SELF DO NOTHING: AS I HEAR, I JUDGE: AND MY JUDGMENT IS JUST; BECAUSE I SEEK NOT MINE OWN WILL, BUT THE WILL OF THE* ***FATHER*** *WHICH HATH SENT ME.*" John 6:37-38, JESUS says, "*ALL THAT THE* ***FATHER*** *GIVETH ME SHALL COME TO ME; AND HIM THAT COMETH TO ME I WILL IN NO WISE CAST OUT. FOR I CAME DOWN FROM HEAVEN,* <u>*NOT*</u> *TO DO MINE OWN WILL, BUT THE WILL OF HIM THAT*

*SENT ME.*" John 8:25, "*THEN SAID THEY UNTO HIM, WHO ART THOU? AND* ***JESUS*** *SAITH UNTO THEM, EVEN THE SAME THAT I SAID UNTO YOU FROM THE BEGINNING.*" CHRIST has always described himself as existing according to FATHER'S desires. John 8:28-29, "*THEN SAID* ***JESUS*** *UNTO THEM, WHEN YE HAVE LIFTED UP THE* ***SON*** *OF MAN, THEN SHALL YE KNOW THAT I AM HE, AND THAT I DO NOTHING OF MYSELF; BUT AS THE* ***FATHER*** *HATH TAUGHT ME, I SPEAK THESE THINGS. AND HE THAT SENT ME IS WITH ME: THE* ***FATHER*** *HATH NOT LEFT ME ALONE; FOR I DO ALWAYS THOSE THINGS THAT PLEASE HIM.*" JESUS says that in that moment, they have rejected FATHER'S truth he came to speak. That they will Crucify him in rebellion, and he goes obediently to suffer, to show GOD'S love. John 9:31, "*NOW WE KNOW THAT* ***GOD*** *HEARETH NOT SINNERS: BUT IF ANY MAN BE A WORSHIPPER OF* ***GOD****, AND DOETH HIS WILL, HIM HE HEARETH.*" John 10:30, JESUS says, "*I AND MY* ***FATHER*** *ARE ONE.*" John 12:49-50, "*FOR I HAVE NOT SPOKEN OF MYSELF; BUT THE* ***FATHER*** *WHICH SENT ME, HE GAVE ME A COMMANDMENT WHAT I SHOULD SAY, AND WHAT*

*I SHOULD SPEAK. AND* <u>*I KNOW*</u> *THAT HIS COMMANDMENT IS LIFE EVERLASTING: WHATSOEVER I SPEAK THERFORE, EVEN AS THE* ***FATHER*** *SAID UNTO ME, SO I SPEAK.*" John 13:3, "***JESUS*** *KNOWING THAT THE* ***FATHER*** *HAD GIVEN ALL THINGS INTO HIS HANDS, AND THAT HE WAS COME FROM* ***GOD****, AND WENT TO* ***GOD****;*" John 13:15-16, JESUS says, "*FOR I HAVE GIVEN YOU AN EXAMPLE, THAT YE SHOULD DO AS I HAVE DONE TO YOU. VERILY, VERILY I SAY UNTO YOU. THE SERVANT IS NOT GREATER THAN HIS* ***LORD****;* <u>*NEITHER HAD HE THAT IS SENT GREATER THAN HE THAT SENT HIM*</u>." John 14:9-10, "***JESUS*** *SAITH UNTO HIM, HAVE I BEEN SO LONG TIME WITH YOU, AND YET HAST THOU NOT KNOWN ME PHILIP? HE THAT HATH SEEN ME HATH SEEN THE* ***FATHER****; AND HOW SAYEST THOU THEN, SHEW US THE* ***FATHER****? BELIEVEST THOU NOT THAT I AM IN THE* ***FATHER****, AND THE* ***FATHER*** *IN ME? THE WORDS THAT I SPEAK UNTO YOU I SPEAK NOT OF MYSELF, BUT* <u>*THE* ***FATHER*** *THAT DWELLETH IN ME, HE DOETH THE WORKS*</u>." John 14:15, JESUS says, "*IF YE LOVE ME,* <u>*KEEP MY COMMANDMENTS*</u>." The commands are not born of any selfish or arbitrary reasons. They are *not* for us to pick and choose, they are the ways of pure love, truth, and eternal life among many souls. In John 14:21, JESUS says, "*HE THAT HATH MY COMMANDMENTS, AND KEEPETH THEM, HE IT IS THAT LOVETH ME: AND HE THAT LOVETH ME SHALL BE LOVED OF MY FATHER, AND I WILL LOVE HIM, AND WILL MANIFEST MYSELF TO HIM.*" John 14:23-24, "*IF A MAN LOVE ME, HE WILL KEEP MY WORDS: AND MY FATHER WILL LOVE HIM, AND WE WILL COME UNTO HIM, AND MAKE OUR ABODE WITH HIM. HE THAT LOVETH ME NOT KEEPETH NOT MY SAYINGS: AND THE WORD WHICH YE HEAR IS NOT MINE, BUT THE FATHER'S WHICH SENT ME.*" JESUS says in John 14:28, "*YE HAVE HEARD HOW I SAID UNTO YOU, I GO AWAY, AND COME AGAIN UNTO YOU. IF YE LOVED ME, YE WOULD REJOICE, BECAUSE I SAID, I GO UNTO THE FATHER; FOR MY* ***FATHER*** *IS GREATER THAN I.*" JESUS clearly states he is not equal to FATHER, yet he is one GOD in truth, belief, and love. John 14:31, JESUS says, "*BUT THAT THE WORLD MAY KNOW THAT I LOVE THE FATHER; AND AS THE* ***FATHER*** *GAVE ME COMMANDMENT, EVER SO I DO.*

*ARISE, LET US GO HENCE.*" John 15:10, "*IF YE KEEP MY COMMANDMENTS, YE SHALL ABIDE IN MY LOVE; EVEN AS I HAVE KEPT MY **FATHERS** COMMANDMENTS, AND ABIDE IN HIS LOVE.*" John 16:27-28, JESUS says, "*FOR THE **FATHER** HIMSELF LOVETH YOU, BECAUSE YE HATH LOVED ME, AND HATH BELIEVED THAT I CAME OUT FROM **GOD**. I CAME FORTH FROM THE **FATHER**, AND AM COME INTO THE WORLD: AGAIN I LEAVE THE WORLD, AND GO TO THE **FATHER**.*" These words from JESUS, declare the souls of JESUS and HOLYSPIRIT, and all lesser souls are at their gifting, a lesser, to lesser sacrificed partitioned gift *out of* and coming forth *from* the Soul of FATHER. And that living by GOD'S commandments and sayings, are the ways of eternal love and life. John 17:7-9, JESUS prays to his FATHER, "*NOW THEY HAVE KNOWN THAT ALL THINGS WHATSOEVER THOU HAST GIVEN ME ARE OF THEE. FOR I HAVE GIVEN UNTO THEM THE WORDS WHICH THOU GAVEST ME; AND THEY HAVE RECEIVED THEM, AND HAVE KNOWN SURELY THAT I CAME OUT FROM THEE, AND THEY HAVE BELIEVED THAT THOU DIDST SEND ME. I PRAY FOR THEM: I PRAY NOT FOR THE*

*WORLD, BUT FOR THEM WHICH THOU HAST GIVEN ME; FOR THEY ARE THINE.*" John 17:16, "*THEY ARE NOT OF THE WORLD, EVEN AS I AM NOT OF THE WORLD.*" Speaking to Priests in John 8:42-43, "***JESUS*** *SAID UNTO THEM, IF* ***GOD*** *WERE YOUR* ***FATHER****, YE WOULD LOVE ME: FOR I PROCEEDED FORTH AND CAME FROM* ***GOD****; NEITHER CAME I OF MYSELF, BUT HE SENT ME. WHY DO YE NOT UNDERSTAND MY SPEECH? EVEN BECAUSE YE CANNOT HEAR MY WORD.*" There is wisdom in this last ask and answer from CHRIST. Being that by *their* slow small worldly perspective and life-long false beliefs born of worldly desires, they can no longer hear GOD'S true WORD. They hold tight to their own individual belief and have no desire to let it go in return for GOD'S pure truth. Sadly, these are lost reprobates. Another Prophetic verse, fore-telling of the first coming of MESSIAH is Psalms 40:7-8, in which, JESUS says, "*THEN SAID I, LO, I COME: IN THE VOLUME OF THE BOOK IT IS WRITTEN OF ME, I DELIGHT TO DO THY WILL, O MY* ***GOD****: YEA, THY LAW IS WITHIN MY HEART.*" The Churches of all false doctrines read around and avoid these Scriptures, as they are hard to misinterpret. These are but a few of the Scriptures that tell us our sacrificial love for FATHER dwells only in our knowing that all that exists beyond our soul, is physically perceived to exist by him, and so, it is his; all for the one purpose of bringing to him his desired loving family. We also know not one thing was perceived into existence to serve any soul's individual desires. CHRIST says even the very manner and steps of our search to find the truth of eternal love, are authored by FATHER. And all of this by JESUS, was and is eternally flawlessly followed and shared with us for our hearing, understanding, and emulating; but only if these things of GOD, we fervently desire. We must know we can never find GOD, unless by all pre-ordained steps, they lead us across a great distance to where they are. If we choose for ourselves the steps and path and ways we live in our search to find GOD? JESUS in his own perfect obedience, says we will never on our own find FATHER. Can the ways of finding GOD be any other, or any less, than the ways of living with GOD? No. Fallen Materialist try to find the truth of reality with both hands tied behind their back, and blinders on. And I cannot understand or see why. Now here, I must define the lesser required perfection of mankind, that GOD has declared in Scripture. I struggled over where in this book to put this understanding, as souls hold tight to beliefs that forgive all ongoing sin, even as no such forgiveness exists. All false doctrines to varying degrees allow for sinful living. But in believing a false doctrine, a soul cannot grow towards GOD, because false doctrines can't lead anyone anywhere good. New souls enter life void of knowledge, belief, or understanding the forces and values of omniscience, and only the few in this world, find these in any good measure. GOD describes sinful life, as being outside all omniscient ways, and as a path to destruction. Can GOD'S perfection ever exist in us? The perfection of GOD, as with all things is in the pureness and fullness of it; but not in its fullness for all lesser souls. Our perfection, is in living within GOD'S boundaries of created purpose. And our imperfection lies in living beyond those boundaries. If a soul yearns to live within GOD'S perfection, and never steps outside those boundaries, they are perfect in having no imperfections. Yet they are never the full omniscient perfection that resides in FATHER or all those above. Visualize how FATHER has gifted many lesser spheres, and if we live within his greater perfect sphere, without stepping out, we remain perfect. The perfection of mankind is to know that the omniscient truth comes only from GOD, and that GOD has their highest hopes and pure eternal desires for us all. So, we seek in our greatest desire and hope to learn and believe only what we have heard from GOD, avoiding all worldly and individual understanding and supposition. In living GOD'S way, we never hurt or destroy anything, and as we grow to learn of GOD'S perfection, we in true love, live by these. *Our* perfection lies in following and being partial and not the fullness of perfection. And as CHRIST commands, we live within the boundaries of the perfection of GOD. FATHER told Satan to behold Job, a perfect man in the Earth, meaning Job responsibly and faithfully lived in all GOD'S created ways. We are perfect to GOD, when they see our hope and desire are to follow in responsible sacrificial love, as we learn and live before them and each other. In living in GOD'S ways, we are a growing partial perfection in FATHER'S eyes. *Our* perfection rests in rejecting all imperfection, and our last perfection is not attained by any final occurrence, or at any point of our ever-expanding moment. As gifted souls, our perfection is always before us in following GOD. Even YESHUA, who has always been perfect, in that he has never sinned. Yet Scripture says JESUS was *perfected* in sacrificing himself for us. CHRIST freely *chose* out of love, and did that *next* most hard, yet *perfect* work FATHER asked of him. And HOLYSPIRIT leads us into our own continued perfection. Hear and follow HOLYSPIRIT, believing no imperfect sources, and you are within GOD'S boundaries. We can be seen by GOD as perfect in our current moment, yet our next moment remains before us as we choose to stay on the perfect path before us. Do any think a just loving GOD would require more, or *less*. JESUS said, "*BE YE PERFECT, EVEN AS YOUR FATHER IN HEAVEN IS PERFECT.*" If your preacher ever says we can never be perfect, walk out. Such preachers, even if unknowingly, preach the Spirit of Antichrist and deny the word of GOD. If you are still with us at this point, we love and bless your soul for your apparent loving spirit. Now having reiterated and revealed to a good extent, what we have been shown to be the relatable perimeters and parameters of GOD'S love, creation, belief, and our desire, to find our perfection with CHRIST. We hope by any and all true lasting change to your understanding and perspective, that together we win all our conscious battles to come. If you are moved to hold a *new* or *renewed* wondering for GOD'S fullest truth, we will suffer greatly if that is lost again. Remember, it is only a small slow interpretation of the physical aspects, sensed by the body, which can allow a soul to believe things appear "*as if*" they might be a far older godless way and age. That small slow perspective, blinds souls to the larger faster consciously deterministic truth. Please remain determined and steady within the creative conscious ways of GOD. Recognize and battle against every physical assumption, and expect that some false beliefs are more ingrained in you than you think. As an element of the truth, we see in the beginning a physical problem had to be overcome by FATHER to bring the existence of free-will choice and chance. This problem rested in the conscious and physical requirements that allow, enable, and grow love. Meaning physical chance and conscious free-will choice of belief must exist first for a true love towards GOD to arise. The choices made by souls in coming to love GOD, *cannot* be made under the control of GOD'S will. After creating the physical reality and life, FATHER, to an extent had to let go of control over each body's use, and most consciously initiated physical acts. GOD needs to consciously perceive and hold all quantum physical and ethereal reality and life in their consistencies all the time, and yet let us go to the extent that our free-will choice of motion and action exist. FATHER, being the creator and sustainer of all physical reality, allows all *true* free-willed physical action within it, bringing choice into play. Our freedom of true choice, brings to all souls the power to question and disobey GOD by our soul's desires and physical acts, being our choices born of our desires. I ask you to consider deeply, that there exists only one model for a created reality that could ever accomplish all of this. And see that FATHER has to see us simultaneously from both a quantum spiritual and classical physical approach, using his medium of singularities, being his aether. Again, the aether's points are unperceivable to mankind until they congeal into much larger structural complexities, then being large particle certainties, residing in layers of spin and orbits under FATHER'S fine tuning. By souls being aware through their own created perceptions, this allows for freely chosen interplay among the increasing fallen aspects of chance and intentional actions. In the perfection of the initial universe, physical interactions were lived out in GOD'S ordained *sustained* ways. But since angels and mankind fell, all quantum and physical aspects are losing their perfect tuning. The original problem for FATHER to overcome and bring a true allowance for choice between

GOD'S rule, and self-rule, is that all physical reality must appear to possibly stand on its own enough, to allow a soul to deny GOD'S forceful work. What levels and layers of complexity did GOD have to create to allow this? If you think it through, you see it was a most complex work. If FATHER created just the earth and the sun above, mankind would see irrefutable proof of GOD'S existence, as no false conjecture could be accepted. The true choice and ability to love GOD, can *only* appear in contrast to a true choice and ability to disbelieve in GOD. Reality has to look like it has two possibilities. GOD, for those who *desire* to see a great probability and reason, and no GOD for all who desire to see and believe in false possibilities. This is a firm scientific physical necessity for choice of belief to exist. Before FATHER'S external creation and gifting of souls, when FATHER existed all alone in the perfection of his internal consciously created dreaming reality, chance did not exist, and no uncertain occurrence was possible. As a possible uncertain forceful choice and influence of another free-will soul did not exist. Contrary beliefs, chance, and choice, came to exist only by desired debatable physical appearances, and more than one free-will physically sensing mobile desiring soul. FATHER projected varied four strengths of willful force of desire to sustain quantum particle entanglement, resonance, spin, and all larger mass rotation and orbits; and then lastly, his forceful patterned guidance over all ethereal and firmer life functions. But beyond all this, FATHER has and must greatly let go. FATHER set his two stronger forces of will upon the smallest particle aspects, and set a weaker more reaching force of electromagnetism in chemical reactions for life's ongoing mobile complexity. Then his weakest broadest force of will was set to entangle all the greater classical aspects, into an observed relative order of forced orbital influences. FATHER'S entangled projected force of attraction we call gravity, is theorized by science to exist and act by the building entangled force of mass pulling everything towards its center. Or by great mass, warping space/time, as that warping pushes or pulls all mass towards the center of that warp in space/time. But it is neither, both being purely physical conclusions. It is FATHER'S desired force of will he projects to entangle his created perception, just enough so physical life would be entangled to its sustaining environment, yet allowed to move through it in their desired yet limited courses. It is a strange paradox that the unimaginably strong nuclear forces, only reach out a minute distance. Yet the weakest force of gravity reaches across the universe. I clearly see GOD'S projected *desired* entangled purpose for doing this. But I can't see how any *godless* force could; or where a godless force could originate. Now light was created to carry patterns that describe GOD'S created reality, and allow us to perceive it in the created reality of our soul's chosen belief. Both Newton and Einstein's theories share insurmountable problems, a unified theory and non-local entanglement, are just two. During his creation process, FATHER projected and perceived *all* things as he told and willed the winds and the waves to not surpass his perfect entangled dynamic reasons of motion. That perfect model has been lost, as the winds and waves and all things now break past their once perfect dynamic courses. We see resonance, spin, rotation, and layers of orbits, and GOD'S four varied projections of willful desire over all physical aspects, are essential for FATHER to perceive and sustain a universe for free-will life and choice. Minus layers upon layers of spin, rotation, and orbit, you can search forever for another model for physical life, that under conscious, *or* physical primacy, provides the required complexity, chance, choice, and movement for love to arise, and you will never find one. Now much smaller faster particles than Quarks, form the angel's *vast* ethereal bodies, and photons are way too slow to enable their sight. FATHER is the eternally owner and sustainer of ethereal and firmer life. FATHER instituted two sets of quantum controls in each body. One is ours to control our physical movement, the other is FATHER'S, to sustain all life functions like cell mitosis, an immune system, and many more. Mankind has learned to have some effect over these, yet GOD owns and controls every particle in every atom of every cell. Now physical science knows they can never begin to identity what consciousness is, or where it is; but they do see these two distinct forms of control. Science puts their faith in a mystical form of physical orchestration they call the subconscious brain, guiding all functions of life. Which is like watching an orchestra play a symphony, minus musicians. Science sees these many things beyond physical cause and effect, so, they use their terms of a "knowing cause" saying this physical aspect knows to do this physical job. To know and control; is it GOD? Or a wet piece of meat in a skull? Is it a mystical meat that *knows* it exists, and knows how to work the miracles of life's functions behind our back? By their lack of desire for GOD, fallen mankind *must* believe this fairytale. And though they see not one provable physical reason to do so, they make their choice of belief. The only two choices for orchestrating the complexities of life, are GOD, or magic brain meat. And seeing they believe they own and are their body, their rebellion pushes GOD away, hurting his heart with their hate and envy. GOD will *never* alter or increase his patterns of sustaining force, to fix mankind's ever-increasing projected unrepentant damage. All rebellion pushes GOD away to allow physical corruption and death. When imperfections in a body grows too far *beyond* its created design to function, this brings the final separation of GOD'S physical work, called physical death. Mounting imperfections devolve all physical things from what once was, into their ongoing confused states. A body's *eternal* certainty, comes only by a Holy Covenant between GOD and us; it is a Holy contract between souls. By mankind breaking that contract, our physical half now exists in peril within a world of risk, chance, and evil physical choices. Reality can sever our soul's needed information at any moment. The body we use is always at risk of injury or illness that it cannot overcome at GOD'S created rate of cell creation. As we live, our souls form our chosen eternal conscious desires; and then we will be sealed as found or lost to GOD, as this physical body vanishes. Consider how a very strong imperfect worldly love, between imperfect individuals, might endure for a longer moment than other loves; yet imperfection must eventually win. This gives us our reasoned sense of how and why eternal love can only exist between a perfect GOD and their perfect children. GOD must hide so we can see the evil that fills this world, and see our desire and need for him. Finding and nurturing eternal life and love in this world is delicate work that can be suddenly destroyed. But there is good understanding in knowing an immense field of life's beautiful flowers, can endure eternal in the coming softest cool breeze, and most-gentle warm mist, under the SON'S warm loving light. Does this sound very hard, or impossible, or unpleasing? Why so; it can only be hard, impossible, or unpleasing, because of what one desires down in their core. If we love GOD'S ways of eternal life, they are a rewarding natural experience. Yet someone who abhors the thought of this way of life, cannot understand how *we* find our eternal peace and joy in these same ways. The probabilities and complexities of life's desires, between billions of souls, predicts all desires must exist; including our love for

GOD'S eternal ways. And to this we can attest. Are we saved by our obedient faithful love for GOD? Or can we live in rebellion till death is at hand, and with our last breath, say JESUS'S name and be saved? We must sense GOD intimately, we must desire to live in GOD'S ways and nurture our loving belief that GOD'S promise will be there. Love is our most unusual and powerful force. Love always has its inner elements of trust, hope, and fear. And our strongest desire will lead, pull, and push us by these elements. Whatever we desire most, at a point, must rule and over-power our lesser desires, and it must even interpret our understanding and belief. Judas Iscariot fought hard to overcome his worldly desires, wanting to love MESSIAH so much. But in great shame, he lost that fight at the very end to his greater love of self and worldly power, and was pulled down. He could not explain to any or himself why, or what he was doing, as he fell to who he truly was in his core desire. At his last, he wanted this world just a little more than he wanted JESUS. He was confused and ashamed at finding out who he truly was, so he hung himself. Can *you* imagine feeling such utter loss in turning from all *you* had found to be good. If we are too closely divided in our desires, free-will begins to appear circular in our waring efforts. We are all controlled and driven by our strongest desires, but to be pulled apart by two strong contrary desires is a hell to fear. In being controlled by waring desires, some lose sight, and deny free-will exists. One day GOD will show us each our core desires. I pray my eye is singular and not divided. Matt 6:22 says, "*THE LIGHT OF THE BODY IS THE EYE: IF THEREFORE THINE EYE BE SINGLE, THY WHOLE BODY SHALL BE FULL OF LIGHT.*" If one has a waring eye for both light and a denied or unrecognized looming darkness, can one knowingly equate this? Did Judas know he would turn on JESUS at the end? No. Did FATHER see a qualified quantified fatal core weakness in Judas, before he chose him? Did GOD equate in Judas that he loved JESUS so much, yet when push came to shove, he loved the world more, and would betray him for money. It was Prophesied that one of CHRIST'S own would betray him; so it was seen before he was chosen. We must hope and strive to not be a divided individual; not just for ourself, but for those we love. We must come out of this world to stand together and lift each other up in eternal love. Free-will is not the total control to alter our core desires that many wish it was. We love what we love. We can't say I will instantly love this, which I did not love before. Or I will hate this, that I did not hate before. But we can learn why we love a thing, and how it destroys all life. In this, our desires *can* be changed to hate a thing we once though was good and loved. Imagine several polished imperfect spheres, representing several imperfect loves. The first has one *irreparable* imperfection, the second has a thousand, and the third has a million. The one with a million flaws will be quickly destroyed, and the sphere with one flaw will last the longest. Yet by that one *irreparable* flaw, it too must eventually be destroyed. Desire to repair all your flaws, and accept no imperfections. To this world, eternal love sounds mechanistic and computational. But at its core we see all the eternal conceptual elements of desired reasoning, purpose, forgiveness, shame, and strength of hope to repair and sustain each other in a loving eternal experience. But we also see that the more often we have to forgive someone for hurting us in a same repetitive way, the less we feel any real love being returned from them. An eternal shared love may sound like something more than love to many, because it is not a worldly experience. Individuals from their worldly perspective, assume eternal love is giving and receiving continual eternal-forgiveness.

No. GOD'S forgiveness and mercy are their loving painful mechanisms for allowing our complete turn from all evil, not for our eternally committing and remaining evil. We see all our emotions are formed in good or evil polarities of projected force, as the effects of our own desires pull us towards the things we want. Some souls towards GOD, but most towards their selves and this world. Most all have said, I don't understand why I did that, or how could I do such a thing, or, I surely knew better. We see two nearly balanced yet conflicting desires, one for, and one against a thing known to be destructive to life. Yet a soul can be defeated by a slightly stronger desire for it, than against. Bad choices are *barely* being made, even after carefully weighing the consequences. Some souls infer and imply that by GOD'S existence and his creation appearing ambiguous to them, that they see a guise and a ruse. But remember, only mankind's desired chosen too small slow fallen physical perspective, sees any ambiguity in the first place. I see no ambiguity from the loving believing eternal perspective, GOD sees no ambiguity in all he has created, so there is none. But ambiguity arises from self-delusion of false beliefs. Consider how the universe could *not* exist in a way that does not look ambiguous from *every* false perspective, yet appear unambiguous from the one true correct perspective. If you ever see ambiguity in anything, it arises from how you desire to see all reality. Relative ambiguity grows from every subjective root that denies GOD'S true fullness. A guise perceived by the rebellious, is thought by them to be deception. So, these think it is GOD who has done evil to deceive them. These do not see that in GOD being there and being omniscient, that for them to from an imperfect confused perspective, assume a right to define evil from their small slow limited subjective understanding, requires they see their self as the Supreme Being. It is insane to say in a confused heart, I am the god of not only my life, but I will define and attribute, even to GOD, what *I* see as good and evil. There is *zero evil* in GOD'S perfect eternal plan; but there is every evidence of allowing every false choice of individual belief in all we see. To all who desire eternal love and life, the creation is not ambiguous, and the pure truth is not hidden, it is right there in the open for all to plainly see. GOD hides nothing pure and true, but most souls blind themselves to it all, burying it under their own beliefs. All choices made by all free-will souls, are made from either pure good desires, or pure evil desires at the far sides. And as great evidence of choice of belief, we see every manner of confused, lukewarm choices across the spectrum in between. Over two thirds of all mankind, confess belief in a god, yet most do not believe enough to live in GOD'S omniscient ways of eternal life. Belief in *a* god of your own desired design, is not GOD'S standard. Scripture says the fallen angels, who truly know and have seen the One True GOD, tremble at his name, having surrendered to their core desires to be individuals in belief. We see across the middle of the belief spectrum many believe in and want *a* god; yet they don't want the one true GOD. They want a heaven, but not the one true Heaven. So, they invent and try to climb the walls towards eternal life in delusional ways, filling reality with endless imperfect individual worldly vain religions. In all perfect justice, the suggestion of an evil guise on the part of GOD, is rightfully attributed as just more rebellion to GOD'S truth. Now all have heard the childhood story of a Prince who had to hide, not his good and loving soul, but his elevated honored status, appearance, and place. And taking on the semblance of a lowly pauper to travel far, searching for a bride who would see and love him for his goodness. Reason being, many knowing his place, falsely declare with their mouth their true love, wanting to marry him for his status. That child's tale like many, originated from Scripture. We saw the KING of all reality come into this world as the pure truth, dressed and coming as a lowly uncomely pauper. And for this, he was despised, rejected, and hated through the prideful eyes of most souls, because he was physically undesirable. But by a few he was heard, welcomed, and adored for his inner spiritual truth, love, and beauty. To see the true royal beauty of the eternal love of MESSIAH, is to see him standing on the highest peak, bathed in a purest radiant light, arms stretched out, his head bowed in sorrow, his entire appearance covered with deep cuts; being thin, frail, embattled, yet victorious and high above all, as his blood and love run down. This is *the* most *glorious* beauty any soul can ever behold. Most see this as horror and not beauty; just like they can't see GOD'S objective truth in their individual subjective ways. So, it is *our* desired perceived understanding of reality, that puts *our* finger on the scale, tipping the balance to create each soul's desired perception of GOD. Their own interpretation of reality created in their individual belief, is all most see. And I see that for any reality to introduce chance, and allow all choice of belief, and every soul's desire. Nothing less or more than this vast complexity we see, would accomplish this. If you look at reality properly, you see a Consciously Deterministic Anthropic Principle. The vainly religious can see GOD'S commands as just democratic cultural aspects of a bygone era. Most believe GOD'S truth is just something we know or tell, and not something more that GOD created for us to be, do, and live. The best definition of the *created* truth is, "reality existing and living as originally meant". To most souls, the truth means their believed testimony as to their experience. For those who do not want GOD'S omniscient truth, how easy it is in their own awareness to say the one pure truth does not exist, and this is what *I* add, subtract, and believe. Most professed believers in CHRIST, hear his truth in a broad, loose, far more accepting and inclusive way. Many believe that by admitting they sin, and that JESUS exists, they are saved. By this, they think GOD vain, and pleased by their simplistic profession of belief in his greatness, even as they live against his ways. Where is their love in that? Most believe that beyond their earnest profession of belief, that to differing degrees, physical desires, choices, and ways of living are to a great degree, their domain. Who knows what souls who say they don't believe in GOD are hanging on to. I wonder what do they see, what do they seek, and why seek it? Is it their hope for unaccountable eternal sleep at the last? Do they find their experience of physical life is not enough, or meaningless? We agree, the physical half of life alone is not enough; but we see all the wonders, just there beyond it. More and more, science admits the irreparable holes in all their theories say their standard model can't be right, or complete. We hear from many scientific fields that we need a paradigm (model) shift. They say the same things we say about their beliefs; being they can never provide a relative answer to all conscious/physical realities. Un-avoidably meaning they do not have one firm immutable answer for anything. Many in the sciences try to hold to integrity, but the tides of opinion turn strong, and most surrender to the political pressure of a false scientific consensus. Materialist do *not* see the glaring truth that their pure *physical* theories have one purpose, being to give souls enough comfort, courage, and strength, to deny the reality of GOD. Conscious force guidance is seen everywhere in everything, making GOD the only true probability. The world and heavens and their purpose, are open to us by Scripture and all true empirical evidence, bound together in peace. Before re-reading these Scriptural verses, please ask GOD to show you *why* he chronicled these scientific descriptions, millennia ago. Hebrews, 11:3, "*THROUGH FAITH WE UNDERSTAND THAT THE WORLDS WERE FRAMED BY THE WORD OF GOD, SO THAT THINGS WHICH ARE SEEN WERE NOT MADE OF THE THINGS WHICH DO APPEAR*." Romans 1:20-22, "*BECAUSE THAT WHICH MAY BE KNOWN OF **GOD** IS MANIFEST IN THEM; FOR **GOD** HATH SHEWED IT UNTO THEM. FOR THE INVISIBLE THINGS OF HIM FROM THE CREATION OF THE WORLD ARE CLEARLY SEEN, BEING UNDERSTOOD BY THE THINGS THAT ARE MADE, EVEN HIS ETERNAL POWER AND **GODHEAD**; SO THAT THEY ARE WITHOUT EXCUSE: BECAUSE THAT, WHEN THEY KNEW **GOD**, THEY GLORIFIED HIM NOT AS **GOD**, NEITHER WERE THANKFUL; BUT, BECAME VAIN IN THEIR IMAGINATIONS, AND THEIR FOOLISH HEART WAS DARKENED*." Psalms 33:6, "*BY THE **WORD** OF THE **LORD** WERE THE HEAVENS MADE; AND ALL THE HOST OF THEM BY THE BREATH OF HIS MOUTH*." 2nd Peter 3:10-13, "*BUT THE DAY OF THE **LORD** WILL COME AS A THEIF IN THE NIGHT; IN THE WHICH THE HEAVENS SHALL PASS AWAY WITH A GREAT NOISE, AND THE ELEMENTS*

*SHALL MELT WITH FERVENT HEAT, THE EARTH ALSO AND THE WORKS THAT ARE THEREIN SHALL BE BURNED UP. SEEING THAT ALL THESE THINGS SHALL BE DISSOLVED, WHAT MANNER OF PERSONS OUGHT YE TO BE IN ALL HOLY CONVERSATION AND GODLINESS, LOOKING FOR AND HASTING UNTO THE COMING OF THE DAY OF* ***GOD****, WHERIN THE HEAVENS BEING ON FIRE SHALL BE DISSOLVED, AND THE ELEMENTS SHALL MELT WITH FERVENT HEAT?*" Rev. 20:11, "*AND I SAW A GREAT WHITE THRONE, AND HIM THAT SAT ON IT, FROM WHOSE FACE THE EARTH AND THE HEAVEN FLED AWAY; AND THERE WAS FOUND NO PLACE FOR THEM.*" FATHER says we are *capable* of seeing *how* he has perceived all physical reality before his face. Yet many desperately see it another way that says GOD need not be there. I *know* FATHER projected his forceful patterns that created the reality we physically sense, and he will see its dissolution. All life functions require his pre-set constant forceful *patterned* guidance; but mankind says without a single proof, cause, or reason, that all these miraculous controls and our awareness, all arise by neural electromagnetic aspects within a wet magic meat relaying interface. The brains areas that correlate to our forms of perception, are like graphics and audio card correlations in a computer, in what goes to and comes from a speaker or monitor and nothing more. Visualize a light switch on a wall; but who pushes the switch, and why? The correlation of a light switch being flipped, and what occurs after, can't imply that switch is the creative source of the photons we then sense. GOD is the forceful sustainer and orchestrator of all functions in our body/interface. But *our* awareness, is distinct and other-worldly, being countless orders of magnitude more phenomenal and wondrous than neural-correlations can ever begin to describe. Do you know how many systems and functions in a human body, that even under mankind's limited understanding, now fall more and more under a most concise definition of "irreducible complexity". Which means we see an extremely complex, *multi*-component system, that cannot lose one component and have any useful function. It requires all its components to be what and where they are now. And there exists no evolving path of any lesser useful purpose along the way. So, the guidance of natural selection cannot be said or shown to explain its appearance. For example, a so-called evolution of an eye under natural selection, would require its many components, and their sub-components, to evolve for eons in unison, minus any help; being a contradiction to chance evolution. And again, strong emergence is a new word for projected forceful designed creations. Now the blending of two DNA's, can produce offspring having diminished and or enhanced pre-existing attributes. But we never see truly new information for any entirely new attribute, sense, or system, popping into existence. We see that through all genetic studies, and every experiment on cell mitosis, that every time the genetic information, remains the same, or information is damaged, broken off, or lost. But there has never once been a single spontaneous creation of any new information. So, we see only loss of perfection, which is not evolution, but *devolution*, being observed continually. Now, can evolution and devolution coexist? No. It is shown that RNA and DNA, have zero capacity or ability within itself to create any truly new information. And it is known that DNA is the only possible agent for bringing new information for any evolutionary change. It is also shown that the complexity of DNA itself, could never come into existence accept by GOD, or evolving or creating itself first and all alone. But how can the most complex component required for evolution, have first evolved its self? Could the most complex computer program ever, code itself before the first computer? If so, how, and why? Where could the vast understanding and immense physical information required to create DNA, come from minus a purposed design. Belief in evolution, lies in a blurred fury of false questions and false-answers. There is no purely physical way to get the unimaginable physical complexities and miracles of life, or conscious awareness, from undirected physical chance, engulfed in entropy. If we honestly step back and all at once view the complete body of mankind's knowledge, we see a huge odd *asymmetric* loose weak sagging incomplete cobweb. We see the glaring lack of evidence or proof in its many oddly shaped holes and caverns between their built web's threads, needing to be filled. We see many strands of supposition, unfounded leaps of faith, physical miracles, and tricks of the light, and numbers pasted together to oddly theoretically tell a lie. We see their underlying blind driving desire to build a consensus of belief that denies and runs from the truth. There is no other physical candidate to replace DNA as an agent of new information. So, if one believes in a purely physical Abiogenesis, and that the biological evolution of brand-new sensory systems and conscious senses evolved through DNA. Then evidence says DNA's magical evolutionary powers, died long ago. GOD filled the Earth with each species, using mostly a male female design to allow for the chanceful combining of their traits to be enhanced or reduced. But these chance enhancements and reductions don't equal evolution, nor deny the over-all guidance of GOD. Science *re*places GOD'S planned forceful guidance, with their ineffectual correlations and all chanceful cause and effect. All in their hope that our need for GOD as our creator, is no longer desired. I am unable to grasp such thoroughly confused thinking. These false science prophets, tell all to wait for a savior in some coming proof of earth bound, or other worldly Abiogenesis. Evolution means nothing less than a brand-new sense or system that never existed before, just appearing in one single body's DNA. Which according to natural selection, survives by being desired as a mate for its recognized advantage over other of its species. So, there must have always been a sentient sense within all creepy crawlers, that sensed advantage in another, even before any sense of sight was introduced. If a true appearance of a real leap of physical evolution occurred today, the only possible cause would be GOD. If today, one person was born with a brand-new sense, it must be the work of GOD. Show me how a brand-new sense, could be physically explained by the evolution of DNA. Consider what would have to occur for the first sense of sight in any creature to have strongly emerged. First, we must *assume* the miracle of physical life has by incalculable chance, come to exist and evolved up to a miraculous reproductive *blind* creepy crawly. And after overcoming all impossible odds, in just one *blind* creepy crawly, there just appears a brand-new complex cell, capable of sensing being hit by photons in a way that differentiate patterns and wavelengths of light. Which is one impossible physical occurrence! So, this one blind creepy crawly, was not handed this *new* DNA from its parent(s) who did not have it to hand down. So, this first creepy crawly, suddenly had brand-new complex DNA written within every cell of its body. But wait, this creepy crawly had to be born with *at least* three new complex interconnected systems. So, an *immense* amount of brand-new perfectly designed DNA, had to just appear. And each of these three new systems, required many coordinated functional sub-components, and *their* DNA, to also pop into existence. These parts must appear simultaneously in cooperation and coordination to provide any advantage of natural selection. Part one is the unimaginably complex cell, or cells, on or near its exterior surface. This new retina cell must come to exist, with all parts and functions being able to detect being struck by patterns and wavelengths of photons. Photons being quanta of energy that have zero mass, and travel at 670 million miles an hour. That is such a cool miracle to just appear by chance. Now this new retina cell had to have internal processing abilities and the ability to format and transfer this new information in a new formatted discernible pattered way. And so, at the same moment, this creepy crawly must be born with a new complex optic nerve to convey all this new formatted information *somewhere.* And, and, and, at this mystical moment, this new nerve must be connected to a new complex primal visual cortex. At least these three systems, which required a lot of new DNA just *appeared* to detect and sense light in a new way. Natural selection says this organism's new sense was favored to survive, because of a recognized advantage over others. So, a way of sensing this new advantage must have evolved in the other, still blind creepy crawlers, for other blind creepy crawlies to sense an advantage. Evolution of a new sense, is nothing like enhancement or reduction of pre-existing attributes.

Science says small changes to existing attributes is the evidence of evolution; but why? Yes, each offspring is different, but I see no evolution. A brand-new appearance of sight, needs a new complex receptor cell, a new optic nerve, and a new visual cortex all at once. *And* there must be a pre-existing sentient awareness, that can comprehend and use this new information to its advantage. Science simply says this happens, without showing a true process. This staggering amount of brand-new DNA to make these new complex systems, according to evolution, had to be written in the DNA of every cell at its conception, for it to have this advantage to pass on. How did this first organism get this new complex DNA from parents who did not have it to hand down. A new sense is the definition of creation, not evolution. *And* unless *two* creepy crawlies, both have this new sense, and mate, this new DNA in *one* creepy crawly alone, could not be successfully passed on. So, in only one bug, it may define them as a new form, unable to propagate. Evolutionist say that without knowing why, life just keeps genetically evolving and improving to create billions of improvements and other brand-new systems. Like the eye, with new lensing, new fluids, new muscles for turning and focusing, and greater information sharing between all these systems. And how did this creepy crawly create its first particle/wave-based perception of sight, and add its bright awareness of a created perception describing the dark exterior reality. The incomprehensible odds of the physical evolution of a visual system for perception, minus GOD'S gift of a soul's awareness, is beyond impossible. But in seeing *our* soul's knowing creative abilities, and the one true great creative GOD above, knowingly projecting and perceiving into existence the physical systems of light and sensing sight, and gifting us all our soul's sense of awareness, it all lies within perfect reason. We find the odds of GOD perceiving all things created, are a near certainty, and our only sound bet. While the odds *against* any pure chance, complex Physical Creation, Abiogenesis, and Evolution, can't even be calculated. Evolution asks every impossible faith from us, and lacks any reason for anyone to accept it, other than a strong desire and hope for it to be true to deny GOD'S existence. And the cost and loss of betting on the impossible is everything you now are and have, and would be or have for all eternity. Physical Creation by conscious patterned force projection, holds every sound reason and purpose, brought together with our bright conscious awareness to utilize it. A reality based upon conscious forceful creation, exhibits all soundness, seeing we do have *within us* this same ability to forcefully form particle-wave based perceptions. I have not heard or read of anyone living or dead, who ever give an alternate detailed scientific formula to qualify and quantify our created particle-wave based perceptions, or our awareness of them. Because there is no other possibility. They just say without a single reason or cause, that our awareness is an illusion. Does that in any way sound scientific, true, or honest? Now, let's revisit the reasoned difference between the two lively realms of the firmer and ethereal existence. First, we see mankind's standard model gives the physical impression of the quantum realm as being a dead-end road. Being only the smaller and smaller parts of each particle in *this* firm realm. The quantum higher realm of Heaven, is a woven *entangled* and relative continuum of existence, just like the woven firmer physical reality. It exists in its own qualities, and consistencies of continuity, everywhere and in every moment, under its own quantum realm laws. Mankind sees two sets of laws, but can't properly define them apart, so, they believe they are united. Being just another impossible unscientific belief and distraction that keeps them from seeing the truth. Through the quantum entangled realm, is how the bodies we use, receive all the forceful guidance from FATHER for all functions of physical life. This constant work is seen in all cell mitosis, the intelligent immune system, and the *countless* other aspects of designed life, that are all beyond physical chanceful occurrence. But, as to *our* choices of what we want or touch, or where we go, or what we outwardly and consciously do and believe out of desire, or what we feel, be it comfort, pleasure, pain, or anguish. These are *our* conscious creations, entangled to our soul alone. By all empirical evidence, reality requires projected patterns of conscious force that must be entangled over and upon the physical for creations to consist in complexity. And we see GOD'S required patterned force, comes down through his quantum gate. If we hear and answer GOD'S call, and wait daily at *his gate* to *his* quantum Kingdom, he deals with us there always. From Proverbs Chap. 8, "*NOW THEREFOR HEARKEN UNTO ME, O YE CHILDREN: FOR BLESSED ARE THEY THAT KEEP MY WAYS. HEAR INSTRUCTION, AND BE WISE, AND REFUSE IT NOT. BLESSED IS THE MAN THAT HEARETH ME, WATCHING DAILY AT MY GATES, WAITING AT THE POST OF MY DOORS. FOR WHOSO FINDETH ME,*

*FINDETH LIFE, AND SHALL OBTAIN FAVOUR OF THE* ***LORD****. BUT HE THAT SINNETH AGAINST ME WRONGETH HIS OWN SOUL; ALL THEY THAT HATE ME LOVE DEATH.*" Every revelation of truth, every non-conscious breath, and every heart is beating right now, by GOD'S entangled patterned desire. All for his great reason of bringing the aspects of free-will choice and chance, to allow free-will love to arise and exist. We see for now that to find love, GOD allows most rebellion and random acts of choice and chance that diminish a body beyond GOD'S current level of care. Meaning, if one is injured, or injures another beyond the body's ability to function and repair at GOD'S current set speed of cell replication; minus *great* reason, GOD stops sustaining that body. When GOD told Adam and Eve, that if they chose the abyss of individual imperfection, they would surely die. GOD meant that they would introduce a growing physical chance and disorder, and a growing need for repair beyond GOD'S desired pre-set patterned care. And GOD will not increase his care in all bodies. GOD can intervene, if moved for eternal cause; but as we see, this is most rare. All flesh is now at the mercy of ever-increasing disorder and chance injury. And within all devolving DNA, we pass down new defects born of chance and choice to our children. By all sinful choices throughout history, mankind out-runs all GOD'S repairs and ways of life. According to Scripture since the fall, mankind's life span has decreased from over 900 years, down to 120, then to 70 years. Why? Because they have increased their ways of hell and death. Yet for Righteous reasons beyond our purview, we at times see GOD intervene. We do hear doctors say once in a great while, they cannot explain why this one lived, or that one died; that for all that is known it just should not be. We see contradictions *against* all irreconcilable probabilities do occur. Their how and why, cannot and do not compute in any physical sense or understanding. And in any purely physical reality, these unaccountable variables in life and death should not occur. But in GOD'S model of conscious determinism, they are predicted and required. GOD'S gift of physical life in this corrupt fallen predatory reality, now sits on a knife's edge, far from its once perfect strength. After GOD'S rare healing, depending on belief, we hear those not believing in GOD, say it is an as yet unexplained physical anomaly, and some are. And we hear those who believe, say GOD had a reason to take one and leave another; and for some, GOD did. Again, in believing physical life is all there is, many ask why GOD would stop or allow harm to the physical body of a child? And we ask, if chance, or free-will choice, brings a child's body to deaths door, being found by GOD to be in an as yet un-sealed understanding of belief? Why would GOD force a child to stay in this world to be indoctrinated against him and dammed by unbelieving wolves? And if no child ever dies, all souls would know GOD must exist? Those who ask why GOD allows children to physically die or be hurt, believe deep down that a short physical life is all we get. So, who has a right to rail against GOD, seeing their false perspective of right and wrong, love and hate, life and death, truth and lies, and even the classical and quantum realms, is equal to their false understanding of GOD'S purposed creation. We are blessed to know of this gateway of communication between the heavenly quantum, and worldly classical realms, being a high traffic area. There is at this gate a connecting and a dividing and separating. This gateway is a line of accountability, and a line of hope and mercy all in one. If we visualized this gate correctly, we see this abstract line between the quantum and physical realms, literally runs down the center of every particle in this body we use, and every conscious desire, and every choice we make. We find the fallen physical world and the fallen soul's desires and choices *for* this physical world, on one side of reality. And the Godly soul's desires and choices for GOD'S Kingdom on the other side of reality. This *is* GOD'S gate, and *after* the fall of mankind from the truth, all souls were born fractured down the center, minus our spiritual connection on one side to GOD and being connected only to the physical world. The body is of this world, and is connected to us *and* to GOD for life. Yet the lost only exist as part of the physical world. While souls indwelled by HOLYSPIRIT, see, hear, and are seen and heard in both realms. At this hard to fully comprehend gate where the two realms come so near, yet go different ways; both coming from one source and purpose. We see most souls desiring and existing only on their side that faces this world. Now, all souls adjoined to HOLYSPIRIT are in no way static or identical. We are all quantifications of rising or falling loving perfection if you will. These connections in us are malleable, being taught and Judged by the perfect Justice of GOD'S love. Some will receive HOLYSPIRIT, but then leave GOD again, never able to return (The Judas equation). This difference between these two realms, can never be physically found, because it is spiritually and forcefully defined by GOD. And so, the never changing distance between GOD and every soul could not ever be less; yet every lost individual could not be farther from truly knowing or affectively communicate with GOD. It is only in both GOD knowing us, and we intimately knowing GOD that there can be true communication or communion of eternal life. For all rebellious believers also, GOD could never be closer or farther away in the same moment. All can see and know if GOD instantly heals a wound, calms a storm, or moves a body's location, being undeniable before mankind's physical senses. But most refuse to see GOD'S everyday wonders, they only see the rare and unusual. GOD reveals their selves to us in small personal private measures when we seek them in true eternal love. Now most souls falsely believe GOD *comes* to where they consciously and physically are in their sin to hear their prayers; but *we* know GOD, without moving, is always at that place that could never be farther from, or closer to us. We are either consciously one with GOD'S belief and truth, or we are individuals, being separated. There is no literal coming to or going from, between us and GOD. It is all in our truly turning to becoming one in love, truth, and polarity. There is no sensing GOD in any way, except purity of love and truth. Yes, metaphorically we say we go to GOD in steps; but these are spiritual steps. We go to GOD through spiritual steps of acceptance, change, and obedient emulation. In all essence, it is just our *growing* spiritual oneness of love and belief. Now GOD speaks of his Righteous anger, but what is anger that is Righteous. This anger arises, not because GOD in any arrogance or selfishness did not get his way. GOD is righteously angry because most souls refuse to see that without his guidance, they destroy others and the life GOD alone works to give us. GOD does not destroy in anger; he turns in rejection and pain from sustaining those living in destructive ways. Scripture says there is no way back to eternity for the defiled *physical* aspects we see; this world and all flesh are beyond redemption. Now, to erase our forgiven sins upon his creation, and remember them no more. GOD can just erase all our created perceptions of doing so, as he creates a New Heaven and Earth. GOD and we will not be able to recall those erased perceptions, we will only know that all our shame that was before, is gone forever. And we will look out upon those individuals in the Lake of Fire who held onto their evil. Scripture says, do not pray for healing this world, not until the coming Millennial Reign. Even then, there will only be a limited healing of this world. It is Scriptural that soon by mankind's own countless evil ways, a rapid increase in disease, famine, and the next *of* three more great wars, will bring death to one fourth of mankind. This world is gone mad through divisive beliefs and evil use of technology, as growing insanity unleashes more perverse aversions to GOD'S truth. How can we who believe and live according to GOD'S ways, compromise with anyone who does not. Most souls in the U.S. don't vote for presidents, they vote for gladiators to fight for their beliefs and ways in which they live. The standard in modern politics, is that a candidate's strengths outweigh their faults. Honesty, integrity, and morality are losing all meaning, and move down the list of a candidate's required qualities. There is the truth that GOD is real; so, we cannot redefine or remold GOD. All who do this abstractly, redefine GOD to themselves, but not in actuality. So, if ten individuals disagree on who GOD is, this is just individual lies. Only Scripture interpreted by HOLYSPIRIT can truly define GOD. We must understand, GOD makes all things in accordance with their perfection and desires to which they are eternally bound. If anyone thinks GOD changes or compromises, they don't understand the eternal perfect qualities and ways of omniscience. Most see GOD as having a compromising mind like mankind, thinking GOD wants to heard cats for eternity. Now it strikes us odd, that many souls look at the concept of living forever, and feel an aversion. The true cause for their aversion is personal, circular, yet unrecognized by them. It stems from pain and sorrow they accrue in rebellion to GOD'S ways of eternal life. There is no thought choice or act in existence that is benign; there exist no third, or grey area. We are by our desires in each moment, walking with and for GOD and eternal life, or in our individual contrary ways, moving from GOD. How I hope you see and hold to what GOD'S projected omniscience of creation truly is, means, and does. What it means for perpetual existence of created and sustained life. What it defines and brings, *and* what it eternally requires of us. Also, what its existence demands of its self, or it will cease to be; as this is GOD who is life. So, if the truth and omniscience of GOD is lost in us; our perfect existence and the source of eternal life disappears in us. To love GOD and life, is to protect GOD and life from all divisive imperfection. Or can we all in selfish denial, believe that the holder of the omniscient force of truth and eternal love, the perceiver and provider of all things, seeking a family in eternal peace and love. That instead, GOD'S ways and desires should be rejected, and GOD should be a slave to the rebellious as a provider of abundant eternal life and fulfiller of all *their* desires. That GOD should work eternally, perceiving the means and the repair of physical life for all who reject his ways and forcefully destroy themselves and all creation. GOD then being bound in the same hell mankind has created for themselves. GOD being hated and trying to satisfy the masses who want nothing less than to sit in his place. GOD then not being loved, or being GOD, but eternally and greatly abused as a servant to evil. Is this what most think our perfect omniscient loving GOD should be to a sinful world. Is this a just outcome for GOD after all their work and suffering to find eternal love. The question is, what do you want and need to make you eternally happy? If it is to be GOD, you are out of luck. Is it for GOD to forever provide you with existence and just leave you alone, strike two. If it is to live as you desire for a while, and then cease to exist, strike three; there exists no accountability in the latter. We are all at bat, and our only outcome is a home run, or a strike out. Within this spiritual conceptual line between the quantum and physical firmaments is the point and place of our required eternal choice, and our accountability to our choice. From this gate or line, GOD sees a soul's countenance, and only in his Children, can GOD see his bright countenance of HOLYSPIRIT; and so, these are watched closely by GOD and his Angels. We strive to learn, follow, and remain within GOD'S created ways. But all have stumbled before, and we may again. Though we *can* in a certain moment of our felt prayerful forgiveness, rest in our true salvation status, we cannot rest in our blessed final assurance till we are sealed. The world pulls at us from every side, using all worldly desires and our every weakness. Our perfection lies ever before us, and like Judas, we cannot see our last choice. Omniscience is of great importance to us, as this is where *our* desires for GOD, his Children, and Kingdom are understood, where all *true* belief unfolds and dwells. This is where all our faithful responsibility and accountability to GOD and each other arises. Now noting this fact that FATHER hears and sees *only* his children for their eternal love, causes weak believers who lack understanding to see an over-riding of free-will in some by GOD; believing they and others are forcefully chosen by GOD. Such belief reigns over the false doctrines of pre-destination, and once saved always saved. Just as within science, we see *false* assumptions are broadly accepted in all mixtures of false religious doctrine. Yet as to this line of our responsible accountability, we find understanding in Scripture as to where that line is set. Accountability and responsibility rest intact in between all souls, and GOD'S omniscient perfect mercy. I can't know the exact depths of GOD'S mercy, yet I will not tempt GOD or use my lack of knowing to allow myself an excuse to sin. GOD'S line of accountability to love itself, remains eternal, untouched, and at its created purposed place. GOD surely expresses his fading mercy upon the evil world at large. But GOD'S eternal forgiving mercy, fully appears for each soul at our truly turning from all sin. Some falsely believe if GOD to helps anyone it's an act of predestining. That is false, Scripture says in Heb. Chapter 6, that by eternal free-will choice, some receive the HOLYSPIRIT, yet by a new returning free-will desire for the world, they again reject GOD and fall away. Scripture says, these fall from a true oneness with GOD, and are lost eternal, just as the fallen Angels. I have seen this happen, and sense a risk in myself. We can say, if one grows in a loving connection to GOD, and *eternally* remains by their choice and desire, that these are by their love, and GOD'S returned love, on a predestined, or designed path to eternal life and joy with GOD. This is the only free-will predestined path to eternal life. Now Scriptures say in many ways, that all degrees of hate, are just degrees of lacking love, and that GOD can only eternally love, those who eternally love them. Secular science can never see the truth of the soul, or the origin of the force that creates life, which we see. Humanist see a godless reality, but all who deny the flesh and make our mind and heart one with GOD'S omniscient belief and understanding, are shown wondrous things. And until you communicate with FATHER in his language of belief and understanding, there can only be mis-communication. To most GOD'S truth is gibberish and *can*'t be heard. And by it contradicting their free-willed desires, they can refuse to believe it. We see creative conscious soul size relativity, and frame of reference, together with the understandings of UED and WPDT, disprove all godless beliefs. For years we have observed what we call the ever-present 50/50 rule of evidence for reality. Meaning once you see how we all create biased particle-wave-based perceptions, and see all the variable soul size observational frames of reference between mankind and GOD. Then all observed physical evidence, instantly appears to require GOD'S existence.

But if you don't see those, it appears to not require GOD'S existence. If GOD were not there, *all* evidence should say exactly that, and not give room or space for any alternate understanding. There should be one firm understanding, having no paradoxes or hidden variables or perturbations, suppositions, or physical miracles. The universe should not appear to be intelligently orchestrated and finely tuned unless it is; and it clearly does. Cosmic fine-tuning, places fallen mankind at *not* the physical center, but at the conscious center of all created firmer realities, with only two ways to explain this fine tuning. And not having any desire for GOD, many are forced against logic and all common sense to believe in those impossible materialist theories. I find it evidentiary that mankind has in these last days, found only two paths forward in science. One is Inflation Theory, which *cheats* to infinitely increase all possibilities to counter the fine-tuning problem, using a math error to found a theory of reality. The other is GOD. Multiverse is their unprovable godless god. How can they accept a theory founded on a miscalculation in a cosmic expansion equation. There are hundreds of errors that could have been made, but only this one false infinite answer allowed them to imagine a mystical belief in an endless runaway self-creating physical universes. Science hates and decries *every* infinite answer as being false and impossible, except for this one. Was this math error an act of chance, or an evil influence from beyond? Did Satan influence that error and whisper these temptations in their ear? How do these ragged theories in Physics, Cosmology, Anthropology, Climatology, and other fields of science, find credibility in this world? Their credibility comes not by their theories at all. They inherit their credibility from technological, engineering, and medical advancements that impact our lives; and even science fiction movies. Technical advancements, and all their wild theories, are packaged together as one science for credibility. If not for technology changing our life as we see it, and all super-human space and time travel movies, their false theories on reality would be laughed at by most. It is sad to see, in that you *cannot* run from the truth; yet most try. And you can't run from your soul; yet again, most try. It appears to me, the sciences no longer search for a physical truth, but run wildly from the conscious force truth that chases them. And as if by an enemy, when cornered or confounded by the evidence they find, they must grasp at any theoretical defense, no matter how far beyond belief or proof it is.

Their underlying desperate need they have and do not see and cannot admit, forces them to accept any false physical possibility that offers them comfort or strength in all their continued denials of GOD. All physical life on Earth, even in its dying state, needs every aspect of this dying universe to be what and where it is; as GOD reveals to us the great cause for this dying reality. Now not one fallen soul after a lifetime of study, can offer one proof *for* or *against* the existence of GOD. They can only debate the problems in their false theories, born of their faith, hope, and desire. Everyone accepts their own beliefs and each tiny piece of data they can spin to sustain their desires. Scientist admit they require physical miracles at the very beginning, and often throughout all their theories and models, and that to this day they have not proven one thing beyond doubt. How inconvenient and illogical it all is. Their slow small interpretations of intermediate data have a strange effect of imperfectly filling in tiny gaps here and now. But this practice only creates unanticipated larger irreparable gaps down the line. This is what perpetuates their process of continually building theories that crumble under their own weight, to be replaced again and again. Their building of crumbling physical theories, has occurred over and over;

yet they don't see it. For most souls, their ability to believe they are sane, rests in never finding any real truth. And for most, their never-ending search for *their own truth* is what comforts them. A belief in endless universes, leaves many with a false faith and hope in their ever-changing physical religious search. It reassures them that nothing is proven or disproven. By this, they can never be forced to abandon their individual beliefs. They are the church of the endless unknowing search, if you will. Do all souls fear and hope that GOD is either there, or not there? Do materialists seek a comfort derived from a consensus of belief effect, just like all congregant believers in false gods? Yes. We see a false perceived safety in numbers affect, even in matters of abstract belief. FATHER reveals that for true and real free-will choice to exist, he must allow all souls to see, believe, and accept the desires, and even the delusions of their heart. And we have overwhelming evidence of this being the case before us. Ten people can read or hear a quite simplistic understanding. Yet strangely, if discussed long enough in deep enough detail, all will by their own perspective and desired interpretation, see real differences here or there to at least some biased extent. Consider two lines starting at the same point long ago: one line being the truth of a conscious over physical primacy; and the other line defiled by denying the same. At the very beginning of this deviation, one-line veers ever so slightly away, yet as reality continually unfolds, the gap of belief between them grows. But there is only one line of emerging data, being viewed by souls from these two frames of reference and perspectives of reality. If there could be shown undeniable proof of GOD, all souls would try to follow GOD because of this proof. But most would fail by a growing animus and envy. We now see most have a desire for an open-ended interpretation of reality, formed through their small slow temporary physical perspective. From there, they can, if desired, believe this reality is billions of years old. One cannot say they are a true Child of GOD, yet believe in a physical reality, viewed through the fallen perspective of mankind. GOD perceived the finished universe in place, which is now being stretched. And our visual perceptions, are formed using slow photons. This is what the saved and the lost see though a differing interpretation. Now false beliefs start by seeing false possibilities. And by adding a few false possibilities together, false probabilities appear. And adding false probabilities together, births the ability and need to initiate false theories. This is how all fallen mankind builds individual beliefs. Are all conscious and physical realities mysterious, or miraculous to you. We see no mystery or miracle. Can GOD look at the universe as if it is strange, and wonder how and why he made it? It is inherent that until we lesser gifted souls, see and accept GOD'S perspective and belief, and know how and why we became aware, we cannot know anything truly existential. Why does reality appear vague from a fallen perspective? Because they have fallen so far into confusion, and away from the true perspective of GOD. They have fallen to be so small and slow consciously, that cavernous ambiguities and uncertainties as to what is true and real, fill them. By believing in a linear unfolding of time, they think they can grow in truth, as they literally shrink and die away from it. If a soul strongly does not desire GOD'S truth, they can never believe it. The truth is not a physical aspect that can be measured and weighed against lies. Truth can never be an empirical qualifiable quantifiable element. But GOD'S conscious truth shows us both how *and* why, and not just approximal physical measurements. If a soul can never accept the *eternal* how and why, GOD'S truth is forever beyond them, as his truth can never be *physically* known or found. The fallen no longer ask why, they are enamored with all the unanswerable physical questions, born of their unrestrained unguided imaginations and denial of GOD. The truth of reality goes so far beyond mankind's slow small observations of just 6% of the created physical aspects. GOD does not fit their narrative or desires; yet all they physically see doesn't properly fit all their desires or narratives either. They try so hard, using any and every means to make it fit, but it can't. So, they declare unseen variables and perturbations must exist, being a blind hope. Fallen mankind ignores the ardent problems found in all they try so hard to believe. They cannot grasp a Universe being held between the outstretched hands of FATHER. At this universe's creation, it was far more compact, with all aspects and orbits existing in a dynamic designed sustained stability. The original creation was the size of a sphere that maybe just over-filled FATHER'S two cupped hands. But at being rejected, his hands recoiled away in the pain of his love's rejection, and his creation is stretched from its perfection. So, it now exists in all its deteriorating consistencies, as FATHER still holds the quantum edges of it all. Are believers embarrassed to express GOD'S science to other souls? I hope not. I find only assured peace, joy, and love, in expressing GOD'S truths. We see fallen mankind's dystopian distress that comes from rejecting GOD'S truth. We fill with sorrow, as it is painful to see so many being lost in individual devolving states. And as the universe is being stretched more and more by FATHER, materialists from their slow small perspective, knowing only the limited aspects they sense, theorize and see a greatly distorted past. Psalms 104:1-4, "*O LORD MY GOD, THOU ART VERY GREAT; THOU ART CLOTHED WITH HONOUR AND MAJESTY. WHO COVEREST THYSELF WITH LIGHT AS WITH A GARMENT: WHO STRETCHEST OUT THE HEAVENS LIKE A CURTAIN: WHO LAYETH THE BEAMS OF HIS CHAMBERS IN THE WATERS: WHO MAKETH THE CLOUDS HIS CHARIOT: WHO WALKETH UPON THE WINGS OF THE WIND: WHO MAKETH HIS ANGELS SPIRITS; HIS MINISTERS A FLAMING FIRE:*" Isaiah 42:5, "*THUS SAITH GOD THE LORD, HE THAT CREATED THE HEAVENS*, AND *STRETCHED THEM OUT; HE THAT SPREAD FORTH THE EARTH, AND THAT WHICH COMETH OUT OF IT; HE THAT GIVETH BREATH UNTO THE PEOPLE UPON IT, AND SPIRIT TO THEM THAT WALK THEREIN:*" Many find the poetic beauty in Scripture to be human, and so they see that as annulling.

Mankind was brought in GOD'S reflection, and so can or should one such as GOD, speak in less than elegance, eloquence, wonder, and beauty, to convey their nuanced eternal truth and love. We see the word "*and*" in that last verse. Which we under-lined to point out the creation was first perceived in perfect stability, *and* stretched at a later point. For what possible reason, other than an empirical record, do these Scriptures written thousands of years before mankind knew reality is being stretched, say such an otherwise meaningless thing. We see so many facets of conscious creation to be studied. We need to evaluate and *compare* the state of all things in their current *rate* of entropy, fission, and expansion, to a once perfect reality, created in perfect dynamic conditions. Do you see how all force appears as a spherical aspect, and not linear? We can equate a size, value, and consistency for spherical forces, and all subsequent physical creations, before disorder and fission existed. Then just equate the path of two sudden stretches with full stops. The first being weaker and the second being stronger and flooding the earth. Then the one final sudden and continuous stretching. Three events all occurring between a perfect condition and the current condition, and occurring within around twelve thousand years. Many equations *can* be qualified and quantified, to build up and affirm GOD'S truth. Now when it comes to existential questions, for fallen mankind existing through their own slow small physical perspectives, they see only enough evidence to subjectively believe whatever they desire. Choose what you desire, see it, and believe it. Again, this is the insanity surrounding us, as everyone does just that. Have you ever seen a line-drawing which looks like a man, but flipped over, it resembles a woman with *ambiguities*; sound familiar. How can souls trust *only* in physical things they *believe* they see and touch, when they have *only* their secondary created perceptions of those experiences, being a second-hand here-say biased awareness. Our own secondary created dark particle-wave perceptions are at best trusted, and at worst wrong. And then our followed bright awareness of our perceptions, is other-worldly. My peace comes knowing the external dark silent reality was first perceived, not by us, but by the mind of FATHER. And I see all there is to life, from his eternal spiritual perspective. We all exist consciously, right at that divide of the real and unreal, the seen and unseen, the perceived and unperceived, the known and unknown, the believed and disbelieved. This is where all *true* spiritual and physical questions and answers are to be found. But, only if HOLYSPIRIT'S eternal perspective is held, and true questions are asked. Even if an individual stumbles upon a true eternal question and the answer is given to them in GOD'S Holy promise. That answer is still heard through evil desires, and a false interpretation will be created. By GOD'S omniscient design, his truth can only be heard by those who desire and find GOD. If you have held a secular or worldly religious belief, yet you are now sensing the truth, you are now experiencing a true mindful closeness to GOD. Again, see how you can willfully force your hand to dance in front of your eyes faster and slower, without planning each fluid motion; as you desire it, it just moves. Do you see a fluid conscious control over the physical, or a physical control over the conscious. Do you see the proof of your own created perceptions. The undeniable proof is everywhere for us to see. Please see that $E=MC^2$, *using photons*, does not apply to the realm of vast ethereal life perspectives. Yet, it does apply to a small slow physical realm perspective. Accept GOD'S greater UED model, and increase your comprehension by orders of magnitude. I ask, how do Materialist see energy as a first cause, and the material aspect as an affect? In UED, energy does not exist until force is applied to the purely physical aspect, energizing it. Is $E=MC^2$ inside out? Is it partial, purely physical, and incomplete? GOD'S truth of reality is like a vast ethereal tree hidden from fallen mankind, behind a thin physical blade of grass. Yes, by free-willed rebellious desires, most see only the blade of grass. Their founding miracles support all their physical desires, and visa-versa, as each lost wandering soul clings tightly to it all. When souls do not believe in GOD, it's not physical evidence alone, that supports or allows their belief, it's their founding miracles and purely physical interpretation of it all that allows all their faith, hope, and self-delusion. Destroy their founding miracle, and their eyes will glaze over in utter fear. And this is their looming fate, as that day of all Revelations approaches, our desired created perceptions, and what we believe, depends greatly upon what we want or don't want to see. This is not about seeing from the right or left, or from up or down. But from the perspective of our soul's eternal existence. Individuals, see endless tiny fragments of evidence, and by desired assumptions, they see, build, and affirm their godless beliefs. We can only see, if we open our mind to this far greater truth of reality, and reject all individual conflicting beliefs. We ask all to be honest and answer this question. Why should the tiny pieces of complex physicality, tell us that GOD is not there, instead of telling us GOD is there? Does each piece of this vast puzzle have a label on it, saying made by GOD, or not made by GOD? The proof of GOD is blurred by every small slow physical desire. But if a soul turns to see all things in the brightness of eternal life, and not the darkness of death, denying every belief that life is temporary; they can grow to see all the things they have seen their whole life in a dark misunderstanding, will shine more and more in new ways. Seeing something new, brighter, lifting us ever higher. Those who turn in a true desire for GOD through faith, prayer, study, and emulating CHRIST, at some point receive their undeniable Spiritual reconciliation. And that is a moment of such calm loving joy, which *can* grow eternally *as long* as we desire it above all things. Consider again, all the deeper spiritual purposes for the properties of physical light. Scripture says many times in many ways, that GOD is light. Meaning the properties of physical light, as with all external creations, are a part of the Holy purposed projected forceful desires of GOD. The properties of light have a forceful design that conveys the sentiments of our emotional facial expressions between souls. So, light was created to bring our lively loving visual experiential moments between each other. Though *all* size and speeds of light particles are of one spectrum, and none have any property of brightness; light is still said in Scripture to be the light, or life of souls. And again, GOD exists throughout and beyond our perceived physical dimensions. So, GOD surely move far faster than the speeds of photons. GOD is in all his ways, forcefully entangled to every physical aspect of his own created perceptions; being forcefully and physically right here and right there at every point of reality. GOD is everywhere that light is, or is not; and GOD could perceive all physical light to momentarily stand still, instituting utter darkness. If all light stood still, why would our reality become dark if photons are bright. Why must they hit our eyes in patterned reflected wavelengths for us to visualize reality. Physical light's broad *speed* and wavelength spectrum, allows each conscious size and speed perspective to see reality, relative to its self. We know the slow large photons of light fallen mankind relies on, no longer allow mankind to see GOD, the angels, or Heaven above. And for sin, FATHER pulls and stretches the very shell of reality outward, as mankind falls further and deeper behind. Yet for temporary comfort, mankind desires to see reality *as if* it has always expanded from a single point, and this contradicts Scripture. Even minus GOD as its creator, this stretching could have started with a stable universe, one seventh its current size. And if inflating, rather than being stretched, why do the universes far edges stretch faster by a pull, when they should be seen as the slower lagging indicator being pushed by inflation. This universe, can't even be truly measured or known by mankind, yet by their desire, they believe they do. Blinded by their small slow physical understanding, they cannot accept that GOD'S greater size relativity, rules over his creation. We know this corrupted physical reality began its fall into confusion, *after* Adam rejected GOD'S ways, and he lost his eternal perspective. We do not know how many days seasons and orbits around the sun, Adam and Eve lived in the Garden of GOD. But we know their first sin caused the first stretch that introduced enough disorder to initialize the process of physical death. That initial new low entropy, allowed the bodies of mankind to live for around 900+ years. The great flood was a second far greater sudden stretching and stopping. Scripture says that event greatly increased disorder, and reduced the average physical life span, from

900+ down to 120 years. And at CHRIST'S Crucifixion, the creation was stretched again, and set on its final continuous stretch towards dissolution. So then and there, a final judgment for the whole of creation was set. Scripture clearly shows us these three events occurred at around two-thousand-year intervals. A certain *secular* theory says that for two-thirds of this universe's existence, *after* a theorized inflation epoch, the universe dramatically slowed until an unknown variable caused it to accelerate faster and faster in this last third of its existence. This data may relate back to that mid-20th century cosmic expansion math error we spoke of earlier. Now in GOD'S true model, as GOD pulls at the outer edges of this spherical universe, this must cause a faster stretching dissolution to occur at its outer edges. Mankind does see the farther aspects of the universe moving away faster than the interior, and faster than the slowest speed of light. Yet their model can't give one reason for this to be so. Just another paradox in their model. GOD said around two thousand years ago in 2nd Peter 3:8, "*BUT, BELOVED, BE NOT IGNORANT OF THIS ONE THING, THAT ONE DAY IS WITH THE LORD AS A THOUSAND YEARS, AND A THOUSAND YEARS AS ONE DAY.*" Two thousand years ago, this mathematical snap-shot was given to us to consider. This Scripture means many things; one being that *not now*, but 2,000 years ago, the conscious effort and force required by FATHER to repair and maintain a perfect universe and all physical life, was not as when created. FATHER'S work to sustain perfection for *one day*, 2,000 years ago, would equal 365,000 times his desired chosen work that initially sustained his original perfect creation. FATHER can never choose or want to eternally work that hard for souls who hate him. This quantification revealed the over-all level of disorder in *that* day. This past value, can help estimate *when* the last perfection existed, and what has been lost. Remember, throughout this fallen creation's dissolution process, the original values of GOD'S forceful care, have remained constant, and so the need for more and more forceful repair has not been met. Before imperfect forceful rebellion and all subsequent stretching, a perfect universe was easily sustained. And FATHER could in peace and leisure, sustain, love, enjoy, and commune with everyone. FATHER will *not* be an abused repairman for eternity; would you want that? Now 2,000 years ago, there were far more than the original *two* people sinning; so, forcing far more disorder. But today we have billions of rebellious souls, projecting unimaginable levels of contrary force. So, all numbers and levels of forceful rebellious souls, are qualifiable and quantifiable considerations. We also must see and accept that the edges of the *quantum* realm, after GOD'S six eternal day firmer creation epoch, extended far beyond the edges of the physical realm. And we see this once much smaller physical realm, cannot stretch in size to anywhere near that of the quantum realm, or it will all fission and dissolve into an ethereal quantum state. Now finding and using the value of that (one to one thousand disorder equivalence) will require we consider all aspects. Time is one of mankind's most malleable unreliable consideration in physical science, and we see true science must consider all things from a vast eternal perspective. In falsely imagining reality's age and cause of disorder, and seeing no origin of force. They see linear time, where we see only our expanding spherical experiential moments, attacked by disorder, confusion, and tempting evil change. We must see reality, not just as it applies to all evil changes in *our* moment, but in GOD'S moment. All destructions are interwoven and increasing; and as it all expands faster, these destructive effects called time, must also increase. Around four thousand years ago, we do see FATHER'S second greatest sudden cosmological stretching event. A huge *sudden* pulling, that further weakened his then fallen entangled creation, and his desire for it. This greatly disrupted every celestial body's core and surface, causing a great flood on the earth. Scripture says FATHER saw all the thoughts of mankind were dark and on evil continually. Such a saturation of evil desire and belief, is not reparable. So, FATHER stretched every aspect of creation, down to each molecule, atom, and nucleus; then and there, starting or increasing fission. The crust *and* core of the earth, were violently stretched and all consistencies were weakened releasing great energy. The vast polar ice broke, scattered, and quickly melted to cool the oceans. While stretching the sun and the moon brought added heat, pressures, and gravitational forces, causing torrential storms to cover most of the earth's surface for forty days and nights. Great earthquakes and volcanoes occurred in huge numbers over a period of several months. This expansion of the univers, earth, moon, and sun, pulled all mountain ranges down *lower*, as the flooded face of the earth became the oceans floor. The earth then appeared as it did on its first day, being a watery sphere. There were great affects to earth's surface as the ground waters were forced out to saturate all dry rock. The appearance of the earth was greatly altered, as the hydraulic forces of oceans violently pouring over mountains, scouring, gouging, and braking up the surface for months. Bringing down here, heaving up there, washing back and forth from high to low places. Settling layers upon layers of sediment in new places over a most violent year. I imagine about now, the slow small physical perspective of isotope dating systems, are trying to pull many away. We must resist that small slow physical, timely, problematic interpretation. Now after a year, GOD had settled and cooled the Sun a bit, bringing more great winds, quakes, and drying, causing more great surface fractures and upheaval. The waters still moving violently back and forth, scarring the face of the earth even more, to be gradually forced back into the earth. And with rapid settling and cooling, the ice caps soon reformed. Now a drying out process occurred over several years under new higher pressures and temperatures. This second stretching of the universe resulted in a far greater insufficiency to FATHER'S conscious forceful work to sustain life. This causing physical life to fall into a far faster disordered confusion. By this new level of disorder, mankind's life expectancy dropped from over 900 to 120 years. And later,

GOD initiated that third, now *continuous* expansion, after MESSIAH'S Crucifixion. Scripture says the average life for mankind, was then reduced to 70 years. And though it is not written of, we know life expectancy dropped further, until mankind's technical and biological understanding grew enough to feebly fight back against disorders reach. Such qualifying quantifiable affects, seen chronologically are judiciously relative to entropy, fission, life expectancy, and earth's appearance of age. By mankind's increased knowledge in farming, diet, entropy, chemistry, biology, we see mankind wages their irrelevant war against death. Mankind may say they make great progress; but adding a few decades is a blink of a whisper of one meaningless breath in the eternal perspective. From a slow small temporary timely perspective, any tiny extended physical moment may be cherished in seeing something that is not truly there. Which all will realize, when it does disappear. Now in seeing love conveyed back and forth in great part through physical light, we know many aspects of GOD'S love ride upon light in its search. Yet it rides in its eternal measures in that faster light the fewer sense. Do you want to see by that fastest purest light? Scripture says so many times that GOD is light, meaning all speeds of physical light are a flowing gift of GOD. And *not* that GOD is only light and nothing more. Gen 1:3-4, "*AND GOD SAID, LET THERE BE LIGHT: AND THERE WAS LIGHT. AND GOD SAW THE LIGHT, THAT IT WAS GOOD:*" Meaning GOD *saw* it served its created purpose. Now the eyes, ears, skin, and every sensing aspect of a body, are a part of a neural interface, that sends and receives information to and from our soul directly. The brain also centralizes all sensed data in a second process for memorization. Now science thinks photons have wavelength but no mass, so from their perspective, they exist only as a quantum of information. A photon is huge, compared to particles that form Heaven and the ethereal bodies of the angels; yet it is the tiniest aspect seen from mankind's slow small perspective. But photons enable mankind's physical existence. A photon, being at mankind's edge of reasoned understanding, must be described in the dual manner of both a single defined quantum particle of energy, and also a blurred congealed energetic wave of countless photons moving through the aether. These sourced waves emanate from everywhere to everywhere, from every direction and in every direction. Light's *conscious* purpose is for all souls of all sizes and speeds to know and love each other through and within

GOD'S loving creation. Secondarily light is a source of energy for the physical half of our existence. Take away *all* light, and life disappears. These are the greater reasons FATHER said, "let there be light", and created all lesser eyes and bodies, perfectly fulfilling his purpose. Please see the simplicity of it all, and do not complicate or confuse this within your soul. Now, a faint residue of FATHER'S early, now dissolving spoken force, remains in space as an echo throughout the aether in every direction. Materialist say in error, this is a big-bang back-ground echo. We see *all* light particles, including the slow photon, are small and fast enough to escape the corrupting stretch of time; but where we see creation slow or expand, we see disorder of complexity increase. All experiments say these effects of disorder, or time, lose all affect as we near and reach the minimum speed of physical light. So, we see how and where *our* eternity enters, by growing our perspective and belief up to and beyond that speed to be one with GOD; from where mankind fell. We see the aether, and all the Prophesied coming new creations have only an eternal created purpose, and no divisive aspects. While slowing diminishing things are bound to a temporary perspective among time's divisive aspects of entropy, fission, hells confusion, and death. And in slowing and shrinking, they lose to entropy their once created omniscient physical state of complexity. There is a minimal conscious and physical speed, being relative to GOD'S perspective, required for FATHER to sustain any eternal reality. Why do entropy and fission, increase, as worldly *conscious* perspectives slow down. This slowing shrinking loss of mankind's omniscient information occurs as they spiritually separate from GOD, and fall and shrink, down into their own dark individual abyss, sensing only the dying physical realities around them. We consciously rise or fall, grow or shrink; and we forever rise *with* GOD, or fall *alone*. Time is a dissolving process, destroying the physical half of reality. But souls with GOD who spiritually reach and pass the minimum speed of eternity, leave those aspects behind with the body. The truth reveals all things of GOD and life beyond this dying physical facade. Our understanding of lights varied speeds for various conscious/physical sizes, and its *eternal* complex purpose, reveal the true meaning of eternal love and life. Again, the universe's far reaches, according to Einstein, paradoxically expand faster than all lesser light, making it impossible for corruption to reach the Holy of Holies. We see FATHER, must retreat from sin. These bodies we use, are a lost foregone conclusion; and to consciously and physically enter eternity at the end of our work here, we leave it behind. Now we are only consciously bound to imperfection, by imperfect desires. So, if we desire true change, we must change our desires. We can grow to feel safe spiritually, but never physically, until we receive our new angelic body. *Time* relates to all fallen *subjective* divisive physical aspects and is known and felt in so many malleable ways among the fallen. Do you see how the concepts and aspects of time can cease from your belief, yet the concept of eternity cannot. Eternity is not malleable, and can be accepted or denied. At its end, the universe will dissolve from its outer edges inward, to reveal the real illusion of all that is physical. But wherever we see a new created perception of FATHER, bathed in CHRIST'S love and light, there is our hope of eternal life. Now like light, our souls do not stop, turn, and *then* go again if we change conscious directions; we just instantly change. Yet without light, we could never be led to eternity. Remember that only from mankind's fallen slow small belief and perspective, is a photon the fastest known aspect of reality. And every witnessed action and reaction, must be the result of an equal or greater cause.

So, never look at the tiny physical aspects to form your perspective, but see the universe as a layered consciously entangled reaction, and FATHER as its force generator of cause, being far greater than the sum of all the effects seen in the universe combined. Never see the universe as big compared to us, but small compared to FATHER. Who can believe contrary to all we see, that the universe *self*-created, from a point singularity; and was not created by a great, loving, forceful, omniscient GOD. The enigma of forceful conscious awareness is seen everywhere. How can any soul believe it just appeared to exist. Conscious awareness has eternally existed, and mankind's slow small observation that time stops at the speed of photons is not a physical answer; it is a huge conscious question. What is there when time stops, if not eternity? Can you see a third possibility? If all divisive effects called time, fail at that slowest speed of light, this is where our conscious size and speed of eternal life begin. Time is the slowing evil affects that comes from corrupted information. Adam fell from his eternal larger faster soul's size and speed perspective of reality while in GOD'S perfect Garden. And mankind continues to cast aside GOD'S omniscient eternal ways, and descend into states of lesser existence. Scriptures use the word *fell* for a reason. To consciously fall is to let go of, and descend down from a far higher oneness of an omniscient loving understanding of GOD. Now GOD did not bring the aether's excitations of physical light, and *then*, try to find a purpose for it; and light still roams the fallen realms in GOD'S mission to find love. There was a pre-known, pre-formed, omniscient intent set upon every aspect at the start of everything's creation. Light is given such prominence in Scripture, and except for its great purpose, it would not be written about so often, so broadly, and boldly throughout Scripture. We know we never see any actual physical objects, but only their representative patterned information imbedded in light patterns, traveling throughout the aether. We find the true and holy purpose of light has everything to do with the quantum conscious and physical realms intersecting, as all speeds of light exist in all realms. 1st John 1:5, "*THIS THEN IS THE MESSAGE WHICH WE HAVE HEARD OF HIM, AND DECLARE UNTO YOU, THAT* ***GOD*** *<u>IS</u> LIGHT, AND IN HIM IS NO DARKNESS AT ALL.*" James 1:17, "*EVERY GOOD GIFT AND EVERY <u>PERFECT</u> GIFT IS FROM ABOVE, AND COMETH DOWN FROM THE* ***FATHER*** *OF LIGHT<u>**S**</u>, WITH WHOM IS NO VARIABLENESS, <u>NEITHER</u>*

*SHADOW OF TURNING.*" If the fastest light radiates out from FATHER in all directions, he can cast no shadow for his children to see. These verses say FATHER is supreme, as all below him cast a shadow in his presence. And our eternal love, as with JESUS, is seen by others in our loving obedience. We see a created hierarchy exists throughout reality, governing all conscious and physical experience. Eternal love, life, and light, are the highest elements that move throughout the aether, coming from, and returning to FATHER. Now each soul's created perceptions that are antithetical to GOD, are stored in the aether. So, nothing can ever be hidden in our day of accounting. After the Judgment, all created evil perceptions will be sealed in the aether within the lake of fire for the lost souls to recall and torment themselves. But our forgiven created perceptions, by CHRIST'S sacrifice, are cleansed from our aether; or put in the lake of fire's aether beyond our reach. And as written long ago, then GOD and all in Heaven, can never sense or recall any evil shameful perceptions ever again. But what is not cleansed, are our past painful memories of finding love. Remembered pains of love are essential to us in eternity, as these secured us to GOD and are part of our understanding the eternal value and cost of love. Do you think we will not remember our *pain* of knowing how JESUS died for us? The conscious and physical pain we felt or caused and will remember in Heaven, is beyond our current knowing. Finding eternal love will have been a painful process for GOD and us all to endure. But for GOD and all the saved, after the White Throne Judgment, there shall never be any new pain in GOD'S Kingdom. When it comes to memories, godless science implies that observed neural correlations are their source and cause. Science can see neural activity that correlates to our soul finding and retrieving our *created* and stored quantum particle records from our aether. And they can believe they see causation, but they can't show any causation. We are each a growing sphere of textured belief and desire, held and expressed as our souls ongoing growing summary of all prior created particle/wave-based perception. By our life's experiences existing in particles and waves, on *our* created quantum side of reality, we could say it is our created particle/wave-based quantum program that defines who we have become in our current moment. All coded, stored, and sensed as our operating system in our hard-drive of the aether. But in changing towards GOD, we see that though we can be forgiven and remain sinless; our shameful memories can still haunt, taunt, and tempt us, as they fade. Our beliefs and desires are not our soul, they *are* the current spirit of our lively existence. Again, Scripture says each soul is of either a good or evil spirit. The aether acts as our library, and this entire world's storage of digital information is insignificant in comparison to it. Imagine storing every streaming, three-dimensional, five sense perception of thought and act of every soul who ever lived in the entire universe. Each created particle-based image, sound, touch, taste, smell, and emotion, of every soul's ever-expanding created moment are all there in the aether. We are our soul, has a physical interface, and aether connections; yet if we lose our interface connection, our aether memory remains. Continuing with a computer analogy, our soul's spherical surface would be like our ram memory, holding the firm, vague, and nuanced information of our present moment. The human body's entire neural system is our interfacing relay, and the aether filling our soul's sphere is our hard drive. Our soul senses all our new and prior information through our limited, ever-diminishing interface. All our desires and hopes, being our spirit, can change and grow. A former pride, can now be our shame. A former denied truth, can be our only treasure. The Free-Will Uncertainty

Principle of Desire, found in all our moments of choosing, has built the former, the current, and will build the final spirit of our soul. GOD says he sees a new creature if we truly change our core beliefs. So, we are the same soul, having a brand-new spiritual definition. In CHRIST'S far greater ability to hold and see all *our* life's information on judgment day, he will know us in ways we cannot know ourselves. If, and *only* if JESUS says we are new, have we turned from our evil and found our new eternal purpose of love and life. Against the vastness of the aether, we see our soul's size and memory can eternally grow. Yet we must always exist through a functioning interfacing platform within the aether. If an ant had a tiny *eternal* soul, (it does not) and gained an expanding belief of understanding by examining and moving single grains of sand containing tiny aspects of truth from one pile of not knowing, into another pile of knowing. This ant's soul, could one day become the congealed belief of a mountain of wisdom. Each body/brain is just a physical processer of physical information; but the qualities of our gifted soul, are that we desire, know, love, choose, and generate our forceful patterns to create our perceived reality, all within GOD'S reality. And within GOD'S presence, we experience and acquire our eternal perspective of their truth and love. We see we have access to forever learn through GOD and know so much more than the fallen can ever see or know through their falling slowing experience. Do you see that to forever fall, brings no conscious end, just a separation. And each soul's slowing experience, will be felt as relative to their moment. Our capacity to learn and understand, can only end at the final edge of reality. We know FATHER'S promise comes not in our crying to him to keep forgiving us forever; but in our faithful responsible obedience to learn and live in his created sinless ways. We understand, and feel the ever-rising heights of true love. Souls living in vain religious beliefs do not want the one shared truth in Scripture that says all are not forgiven at their asking, or raised anew to commit the same destructions as before. Again, John 5:14, "*AFTERWARD JESUS FINDETH HIM IN THE TEMPLE, AND SAID UNTO HIM, BEHOLD, THOU ART MADE WHOLE: SIN NO MORE, LEST A WORSE THING COME UNTO THEE*." In desire to turn from sin, we may stumble in weakness, lacking wisdom and strength. But in shame at each sin, we grow in resolve of loving faithful obedience. Our spirit is seen, compared CHRIST. We are not given license to live as a sinful spirit if we ask CHRIST to save us. 2nd Peter 2:20-22, "*FOR IF AFTER THEY HAVE ESCAPED THE POLLUTIONS OF THE WORLD THROUGH THE KNOWLEDGE OF THE LORD AND SAVIOUR JESUS CHRIST, THEY ARE AGAIN ENTANGLED THEREIN, AND OVERCOME, THE LATTER END IS WORSE WITH THEM THAN THE BEGINNING. FOR IT HAD BEEN BETTER FOR THEM NOT TO HAVE KNOWN THE WAY OF RIGHTEOUSNESS, THAN, AFTER THEY HAVE KNOWN IT, TO TURN FROM THE HOLY COMMANDMENT DELIVERED UNTO THEM. BUT IT IS HAPPENED UNTO THEM ACCORDING TO THE TRUE PROVERB, THE DOG IS TURNED TO HIS OWN VOMIT AGAIN; AND THE SOW THAT WAS WASHED TO HER WALLOWING IN THE MIRE*." If we are GOD'S Children, we are called and set upon a new pure eternal path, not the same vile path of confused desires for this physical world. If lost in confused plural beliefs, professed Christians form a personal biblical belief to fit their own devolving understanding. Some do hear GOD'S truth, but by contrary desires they still cannot live it. So, they remain trapped in time, living in personal chosen ways. Many believers can't see that by their unfaithful weaker love for

GOD, and their worldly desires, GOD *must* reject them as the sinners they are to protect his Kingdom. GOD did not gift us a free-will soul, to choose whatever ways we want to live. GOD gave us free-will to see who and what we would love and choose to be. Worldly religious doctrines and godless scientific theories, come by false authority and a fallen purpose; being to allow sinners to believe they are saved. All these false coverings, have no affect or sway with GOD. How can we know GOD'S truth, from false truth? Only by intimately knowing GOD in all obedient biblical, spiritual, and physical, ways. Now, it being so important, we again raise our bright visualization as an intrinsic inherent aspect of our soul alone. Physical light is *only* a correlation, in its reflecting this physical reality's information, yet being no *cause* of bright awareness. There is no way we can create particle-wave-based perceptions, other than the way we have shown. If not by our projected force patterns of desire to exist, acting upon the aether, we could not create or visualize anything. A deteriorated vitreous gel in the eye, creates fluid traction and dragging upon the retina, and this stimulus, causes the soul to see little lightning strikes inside the eye. Such stimulated patterns of information sent to the soul, are not from physical light, but friction. So, a non-light stimulus of the retina, sends the same information as when it is hit by dark photons. There are many physical sources of information, received by our soul having nothing to do with light, that allow us to have visual experiences. Neural science talks about neural correlations as if they are something they are not. Many scientists knowingly or unknowingly imply false beliefs. They should show us a detailed process for how neurons project and create the patterns that define our differing senses. Show us from where comes our qualia of brightness, sound, smell, taste, and feelings of touch, pain, and pleasure. Describe what is our perceived brightness and other qualia. They can't, as they have no clue. A few more verses on the source and purpose of light. Ephesians 5:14, "*WHEREFORE HE SAITH, AWAKE THOU THAT SLEEPEST, AND ARISE FROM THE DEAD, AND CHRIST SHALL GIVE THEE LIGHT.*" Meaning after physical death, CHRIST'S projected information to form our bright perceptions will return to our soul. John 1:1-4, "*IN THE BEGINNING WAS THE WORD, AND THE WORD WAS WITH GOD, AND THE WORD WAS GOD. THE SAME WAS IN THE BEGINNING WITH GOD. ALL THINGS WERE MADE BY HIM; AND WITHOUT HIM*

*WAS NOT ANY THING MADE THAT WAS MADE. IN HIM WAS LIFE; AND THE LIFE WAS THE LIGHT OF MEN.*" John 8:12, "*THEN SPAKE JESUS AGAIN UNTO THEM, SAYING, I AM THE LIGHT OF THE WORLD: HE THAT FOLLOWETH ME SHALL NOT WALK IN DARKNESS, BUT SHALL HAVE THE LIGHT OF LIFE.*" Matt 6:22, "*THE LIGHT OF THE BODY IS THE EYE: IF THEREFORE THINE EYE BE SINGLE, THY WHOLE BODY SHALL BE FULL OF LIGHT.*" Meaning we can have a good spiritual eye alone, or the waring desires of both physical *and* spiritual eyes. Romans 13:12, "*THE NIGHT IS FAR SPENT, THE DAY IS AT HAND: LET US THEREFORE CAST OFF THE WORKS OF DARKNESS, AND LET US PUT ON THE ARMOUR OF LIGHT.*" 1st Tim 6:15-16, "*WHICH IN HIS TIMES HE SHALL SHEW, WHO IS THE BLESSED AND ONLY POTENTATE, THE KING OF KINGS, AND LORD OF LORDS; WHO ONLY HATH IMMORTALITY, DWELLING IN THE LIGHT WHICH NO MAN CAN APPROACH UNTO; WHOM NO MAN HATH SEEN, NOR CAN SEE: TO WHOM BE HONOUR AND POWER EVERLASTING.AMEN.*" These faster light particles and waves from FATHER, conveying his one truth, can only be seen by his Children. Rev 22:5,

"*AND THERE SHALL BE NO NIGHT THERE; AND THEY NEED NO CANDLE, NEITHER* <u>*LIGHT*</u> *OF THE SUN; FOR THE LORD GOD GIVETH THEM* <u>*LIGHT*</u>*: AND THEY SHALL REIGN FOR EVER AND EVER*." This is scientifically saying CHRIST Glory alone, will be our source of light in the New Earth and Heaven. Again, the requirements for an eye, or any complex multi-component system to *evolve* minus omniscient guidance, is just impossible. But the devolving of mankind's fallen eyes can be understood, as the eyes of mankind were never less, but once far greater than now. Before the fall, Adam and Eve saw more than all earthly creatures in clarity, distance, and color, far beyond our current use of photons. At his creation, Adam saw the force of life radiating past each body's surface. All physical abilities diminish as our physical half devolves. In Scripture, FATHER formed his outer appearance for his children, having *his* outward conscious perception facing forward, and *nothing* has ever been created behind FATHER. All gifts and creations eternally remain before his face, and it pleases FATHER upon his Throne, to rule over reality in this way. GOD, like us, perceives all free-willed things, only after they are created, or occur. And by CHRIST'S sacrifice, he will soon look upon all as we all fear and tremble, seeing each unforgiven sin as hate, rejection, and dishonoring his love. If FATHER in any moment of this world's billions of souls, were to look and see the sum of all occurring evil at once; he could never hold his Righteous anger that must instantly appear. And in that day FATHER does look upon it all, everything seen to be good and desired by CHRIST will eternally remain. So, we see in GOD'S shared perfect belief, after CHRIST'S Judgment of all eternal souls, only good souls remain freely and forcefully animated to live and create their shared free-will existence with GOD and others in Heaven. While the lost cast into their fiery prison, can barely move or feel anything under the lake's immense gravity and their new unfeeling indestructible body. In this world, fallen mankind's interpreted beliefs on reality have continually changed; and in each moment, what they now currently believe, is all they can know. So, they oddly see their own current truth of reality, though they have, and will forever alter it again and again. They constantly say, this is what we *now* know to be true, up until it no longer is what they know to be true. How do they not see this hellish merry-go-round they are on, and have always been on? Why can't they see their own perpetual confused self-deceit? This way of life is only lived by a temporary or permanent insanity; I see no other way for a continuous failure process to be embraced. Now, let's further consider the great lack of evidence, mankind calls dark matter and energy, and how their physics and model of reality provide no understanding that accounts for 94% of the firm reality. Yet these dark realities produce a gravitational lensing affect, which tells them something *is* there, raising their *profound* many questions about their standard model. Any modeled belief, that can*not* in any relative way, predict or account for 94% of physical reality, and 100% of conscious reality. And which has to constantly adjust and realign all its values, perimeters, and parameters of the six percent they think they sense, is a model born of insanity I cannot fathom. In all they espouse, I see nothing at all in their beliefs that makes the first common or scientific sense. They constantly have to try to understand and explain so many irregularities and paradoxes that arise, which they did not predict or expect. Such beliefs break my heart in seeing their profound unfounded hopes. Whereas UED and WCPDT, predict and account for it all; as it consciously, physically, and purposefully, predicts and requires reality's every tiniest aspect, to be just as it is. The minute points and grid of the dark aether, cannot interact with the far larger slower photons in a way that makes this 94% perceivable to them. But all the larger, still far too minute residual dissolving aspects, there within the aether, can cause that lensing of light. John 1:5, "*AND THE LIGHT SHINETH IN DARKNESS; AND THE DARKNESS COMPREHENDED IT NOT.*" This reference to light shining into the darkness, and the darkness "*comprehended it not*" in Dualism has both a spiritual and physical meaning. It physically means slow photons can't render comprehendible evidence of certain realities, such as dark matter, energy, *and* black holes. Once FATHER'S desire for his two substantive realms to exist in eternal *perfection*, was rejected by those he created it for. There remained no *eternal* desire within him for those same created great complexity to continue. Mankind's dark matter and energy, and their cosmological constant, are one in the same, being minute remnants and residual low-excitations of all the former complexities lingering throughout the aether. Which again, being too small, do not interact with large slow photons in very strong ways, observable to mankind. Again, being strong evidence of GOD'S fading desire for this firmer fallen reality to exist much longer. Do you see why FATHER is dissolving even the photons in his creation. Do you know that Nobel prize winners in physics, estimate and agree that in just the *last 2000 years*, the universe lost a huge percent of its photons. GOD'S vast aether grows less complex and colder by the moment, but what does this all mean spiritually? The most accepted physical interpretations of reality, no longer pose that light or reality requires a background medium to exist within or travel through. Using *their* standard model, they formed an experiment to reveal a predicted aether wind in reality that they could accept. But seeing their model is wrong, that Michaelson-Morley's experiment that found no aether wind, was also wrong. Now after dismissing an aether, I guess this lensing effect of light can't be an aether for them, as it does not fit in their theoretical model. All smaller faster particles moving through or existing within the aether, and all their affects, do exist below mankind's comprehension and perceptive ability. The aether's expansive grid design, allows all speed and size light particles their relative entangled existence and movement. There are so many aspects and interactions within reality, that fallen mankind cannot perceive. And their predictions using a 6% physical model are meaningless. Nothing is ever stopped by the aethers grid, and whatever affects it has on reality are unknowable to us. The aether facilitates all creations, and allows all movements through it, as this is what it was created to accomplish. Yet in traveling across the vastness of the universe, realities are affected by the aether *and* the complexity within it. Now we fear losing minds in this next point, not because of the point itself; but by worldly perspectives held towards this point. We read how the varying sizes of angelic souls and their ethereal bodies in heaven, are entangled to the orders of certain stars. Scripture also describes utterly dark stars, or bottomless pits (Black Holes) being inescapable prisons of powerful defeated fallen angels who warred against GOD, and lost their impossible battle. Scripture says these are imprisoned away from the greater creation around them by GOD. From a godless materialist view this may sound incredulous, because they cannot grasp or accept what the conscious truth and reality of GOD means. And you can't understand, until you look up and truly know that the vast Heavens, are the stage of FATHER'S Celestial Kingdom, Throne, and Holy Angelic Subjects, being currently fought over in a great battle. John 3:12, *"IF I HAVE TOLD YOU EARTHLY THINGS, AND YE BELIEVE NOT, HOW*

*SHALL YE BELIEVE, IF I TELL YOU OF HEAVENLY THINGS?"* GOD and immense powerful angels exist in a more forceful creative determinism; why is this ridiculous. Matt 25:30, "*AND CAST YE THE UNPROFITABLE SERVANT INTO OUTER DARKNESS: THERE SHALL BE WEEPING AND GNASHING OF TEETH.*" Rev 9:1, "*AND THE FIFTH ANGEL SOUNDED, AND I SAW A STAR FALL FROM HEAVEN UNTO THE EARTH: AND TO HIM WAS GIVEN THE KEY OF THE BOTTOMLESS PIT.*" One of many Scriptures equating angels to stars, and dark bottomless pits (Black Holes). Yet even most Church goers, turn away at hearing this, still holding a slow small godless scientific perspective. So, I won't argue it further; but if you are interested, read (KJV) Rev. 9:2 - 9:11, Rev. 11:7 - 11:8, Rev. 17:8, Rev. 20:1 - 20:3, and more. Please remember in the genesis creation model, that FATHER projected all light from himself on the first day, and then created the stars on the fourth day. So, if you believe the Bible, you must see that light is not just photons, and light does not come from stars alone. Yes, we receive true light particles coming from true created stars. But we can also receive projected holographic patterned wavelengths of light particles alone not coming from stars; and appearing as distant stars, galaxies, and more. Consider a chair two feet away and how by visual perception alone, we can't be as sure that chair is there. We know we receive lights patterns of information describing a chair. But we can also reach out, and in other ways, sense and accept the reality of a chair two feet away. But we can't do this with distant stars and Galaxies, which moves them into belief, faith, and an equated probability for mankind. If current means of our proving something is physically real, are diminished from five, down to one sense; to then believe it, our faith must increase. Mark 11:23, "*FOR VERILY I SAY UNTO YOU, THAT WHOSOEVER SHALL SAY UNTO THIS MOUNTAIN, BE THOU REMOVED, AND BE THOU CAST INTO THE SEA; AND SHALL NOT DOUBT IN HIS HEART, BUT SHALL BELIEVE THAT THOSE THINGS WHICH HE SAITH SHALL COME TO PASS; HE SHALL HAVE WHATSOEVER HE SAITH*." If you know GOD, that last line spiritually and scientifically says that mindful force over matter, creates the truth and lies that now fill souls and reality. It means mankind's fallen perspective says whatever you *see,* you shall *believe*. But a primacy of conscious projected perception says, that whatever GOD and we, consciously *believe* with all our heart, we shall *see*. This fact was known many thousands of years ago. Are there other Galaxies out there? There may be nothing beyond GOD'S first 300 million created stars. Mankind's small slow fallen perspective, sees the real and imagined extended universe accelerating outward and apart faster and faster. But they can't see that all so-called dark matter and energy, increase at a faster rate, consuming reality. The light in reality, required to find and know GOD from a fallen perspective is also fading at that faster rate. Satan, his army, and all fallen mankind, would want the end of GOD and all conscious existence and all reality forever; rather than receive what awaits them. Is Satan actually battling for the annihilation of all reality. Every desperate act of insanity and hate, will be waged in this futile war against the second coming of CHRIST. Rev. 12:7-12, "*AND THERE WAS WAR IN HEAVEN: MICHAEL AND HIS ANGELS FOUGHT AGAINST THE DRAGON; AND THE DRAGON FOUGHT AND HIS ANGELS, AND PREVAILED NOT; NEITHER WAS THEIR PLACE FOUND ANY MORE IN HEAVEN. AND THE GREAT DRAGON WAS CAST OUT, THAT OLD SERPENT, CALLED THE DEVIL, AND SATAN, WHICH DECEIVETH THE WHOLE WORLD: HE WAS CAST*

*OUT INTO THE EARTH, AND HIS ANGELS WERE CAST OUT WITH HIM. AND I HEARD A LOUD VOICE SAYING IN HEAVEN, NOW IS COME SALVATION, AND STRENGTH, AND THE KINGDOM OF OUR GOD, AND THE POWER OF HIS CHRIST: FOR THE ACCUSER OF OUR BRETHREN IS CAST DOWN, WHICH ACCUSED THEM BEFORE OUR GOD DAY AND NIGHT. AND THEY OVERCAME HIM BY THE BLOOD OF THE LAMB, AND BY THE WORD OF THEIR TESTIMONY; AND THEY LOVED NOT THEIR LIVES UNTO THE DEATH. THEREFORE REJOICE, YE HEAVENS, AND YE THAT DWELL IN THEM. WOE TO THE INHABITERS OF THE EARTH AND OF THE SEA! FOR THE DEVIL IS COME DOWN UNTO YOU, HAVING GREAT WRATH, BECAUSE HE KNOWETH THAT HE HATH BUT A SHORT TIME*." This is near, but Scripture says Satan and his angels are not defeated and cast down to earth, until the Fifth Trumpet of the coming Tribulation. A time in which young healthy people will die of heart attacks, out of fear for the insanity and devastation that comes upon the earth for about a ten-year period. JESUS said the beginning of sorrows is a period of turmoil, war, famine, and pestilence, prior to the Tribulation. And being brought by mankind's created hell, and the influence of the Four Horseman of the first four seals in Rev. Chap. 6-7. Who have been at work for probably over a century. We approach the end of that beginning of sorrows, which can't be compared to what comes next. No doomsday movie can compete with this coming reality. If you grasp FATHER'S creative force, can you imagine *his* wrath upon earths inhabitants, poured out for around ten years. Scripture tells about the truth, love, and light, of GOD. These three, that by a great price, do still come to us. These come to each soul, looking into their desires for GOD and their ways; as they leave no good loving soul behind. So, this battle now rages between all positives and negatives, as all light and dark forces, all values and choices, pull and push all souls and the creation apart. Science sees these physical effects, but not correctly. And they don't see a conscious forceful source, meaning, or purpose of it all. If we study all the past grand physical theories that have failed, we see every failure came as new evidence discredited their prior false interpretations. And every new theory has been based on, and interpreted through their same slow small physical assumptions. As they continually strive in the same wrong direction, looking for the *exact* same wrong answers. You can never hope to find an answer you desperately desire. That is biased science. You must search *all* possibilities in honesty, and accept what you find. How many unprovable assumptions and physical miracles does it take to believe in multi-verse theories? All such theories are imaginations of miracles born of a math error. Their own evidence and theories, allow that instead of starting from a singularity, we can start from a compact complex perfect creation, and still arrive here in the same fallen stretching reality. A once perfect creation, not inflating, resting in a lively minimal entropy equilibrium, under precise balanced forces. But then suddenly a conscious forceful anomaly forced a fatal imperfect disorder, stretching reality outward. The values of a far smaller universe, held in perfect dynamic-tuning for life around 12,000 yrs. ago, can be equated. And each new finding moves our awareness farther from seeing a purely physical origin. All problems arise for Materialist, from FATHER'S "Conscious Certainty and Uncertainty Principles" his "Desired Force Wave-Particle Creations" his "Expanded Observer and Zeno Effects" his "Conscious Soul Size Relativity of Perspective Principle" his "Six-Dimensional Reality" and more. These all show conscious awareness resides entirely in the quantum realm. How does this stand against their failing standard model, filled with paradoxes, ambiguities, a mind body problem, a non-unified theory, the fine-tuning problem, problematic forms of Synchronicity, the universes homogeneity problem, and so many other problems? These all arise in trying to define a purely physical reality. Reality echoes that our souls are all eternal forceful creative entities. All evidence rules out matter and energy as the creator of reality, leaving only GOD'S knowing forceful design. This all reveals the growing disarray within their godless physical model of reality. Yet many who deny GOD'S truth, are given worldly status by Satan. Where is the logic and reason to resist GOD'S truth. If consciousness evolved from the interactions of physical reality, it would be the trending direction in science; but it is not. Can they ever see the highest reasoned probability, that all physical complexity comes by forceful patterned design. So-many souls dwell in their *unreasoned* denials that both halves of reality, are needed to form all entangled complexities and then our awareness of it all. All created truth rests in FATHER'S omniscience, and we can interpret all realities through his understanding. The projected desired force in each soul's "Observer affect" required to create each certain single particle system, can be estimated. Then the total force in FATHER'S desires for his six days of creation, can also be estimated. Why do materialist think that the 6% of the physical *half* of reality they sense in a confused, too slow, too small way, is all they need, to define it all? Materialist are false preachers, politicians, and propogandists pushing false beliefs and immorality. The materialist doctrine says never ask about GOD, life, love, truth, or awareness of perceptions, just calculate physical data in blind faith. Doesn't this sound like a most strict cult? I see a godless loveless unreasoned religion; but what is their underlying eternal purpose? Do they know their purpose or reason? Do they see it is all to support their denial of GOD? Any data that falls near their modeled predictions holds great power to seduce all who say their loose interpretations of numbers can't lie. Yes, pure numbers don't lie, and we have no problem with pure true data. But from their small slow timely perspective, their subjective experiments and interpretations are relative *to* their size, speed, and biased perspective, and must always be a bit off. FATHER'S desire for his lighted creation to continue, is now only about 6% of his original desire. There is but a moment before the coming trouble. Our best *estimate*, as of this writing in Dec. 2012, is 15 to 35 years of rising chaos, before the start of around ten years of unimaginable destruction on earth. The Prophesied war and destruction of Iraq has past, and the Prophesied war and destruction of Iran is next. Then a third world-wide war, great famine, and disease, follow, just before FATHER'S Tribulation. This does not mean other major conflicts and events will not appear in between, but these three ordered events of Iraq, Iran, and a World war, bringing famine and disease, will occur in order. Scripture says the last 3 Tribulation Trumpets end at the Battle of Armageddon, and they bring the end of every reprobate life. Then CHRIST'S Reign on this fallen earth begins, as FATHER restores the destroyed earth to a certain extent for those who propagate in this Millennium. This will be the last period of teaching and trying by GOD. Over these 1000 years the love for GOD'S ways will once again fade in most. And after Satan is loosed after this thousand-years, hell spills forth again, bringing the very last war against CHRIST. The flesh of all CHRIST'S enemies, are then destroyed by fire sent down from Heaven. The universe then dissolves, but the aether and all resurrected souls remain as the White Throne Judgment begins. Then the final New Heaven and

Earth are perceived by FATHER, with the Lake of Fire being set outside the New City of Jerusalem's Gate. Currently, HOLYSPIRIT seeks loving souls through all good means, even reading this book is an opportunity. Do you know GOD sees your bright or dark desires right now, seeing what you desire to be true. What is there about death, hell, and eternal separate conscious loneliness, that attracts the many. Scripture speaks of a soul's backsliding, meaning a soul weakened, lost ground, and surrendered some control to a past imperfect desire they had overcome to a greater extent. GOD tells these, to see themselves, asking why has their countenance fallen. Gen 4:6, "*AND THE LORD SAID UNTO CAIN, WHY ART THOU WROTH? AND WHY IS THY COUNTENANCE FALLEN*?" Isaiah 3:9, "*THE SHEW OF THEIR COUNTENANCE DOTH WITNESS AGAINST THEM;*" HOLYSPIRIT is the brightness of truth coming into darkness. And our countenance of GOD'S truth and love, reveals some are brighter and closer to GOD than others. Proverbs, 4:18, "*BUT THE PATH OF THE JUST IS AS THE SHINING LIGHT, THAT SHINETH MORE AND MORE UNTO THE PERFECT DAY.*" The brighter and closer to GOD'S ways his Sons and Daughters are, is what allows GOD to work through them more to increase his family. The fewer among mankind may grow to be as bright as the lower Holy Angels. In understanding eternal love, and seeking to emulate JESUS, our countenance, or brightness, or spiritual appearance, equals our measure of the indwelling HOLYSPIRIT, and overcoming our evil proclivities. 2nd Kings 2:9-10, "*AND IT CAME TO PASS, WHEN THEY WERE GONE OVER, THAT ELIJAH SAID UNTO ELISHA, ASK WHAT I SHALL DO FOR THEE, BEFORE I BE TAKEN AWAY FROM THEE. AND ELISHA SAID, I PRAY THEE, LET A DOUBLE PORTION OF THY SPIRIT BE UPON ME. AND HE SAID, THOU HAST ASKED A HARD THING: NEVERTHELESS, IF THOU SEE ME WHEN I AM TAKEN FROM THEE, IT SHALL BE SO UNTO THEE; BUT IF NOT, IT SHALL NOT BE SO.*" Our measured indwelling of HOLYSPIRIT is equal to our righteousness and holiness, and it is different in all GOD'S growing learning Children. If anyone seeks to learn of, and love GOD, yet sees imperfection as GOD'S accepted eternal standard, that soul has already failed. When FATHER'S Kingdom and his Children are sealed after the Millennial reign, those who believe in CHRIST, yet continued in their sins will find themselves among all other sinners. Many can, and do interpret things in different ways, but none will ever disprove the truth that GOD sees this fallen earth as a fading dark sphere, with fewer and fewer bright to brighter souls upon it. And GOD will reap all bright children who return to them. So, to be and live more in the physical than the spiritual, is to find an ever-diminishing existence; and forever is so very cruel to these. Our ever-increasing, or ever-diminishing souls, can never cease to exist. Each lost soul is currently bound only to a fallen physical body that has no instincts or inclinations towards the ways of eternal life. But *we* are spiritually connected, raised, and drawn to live within GOD'S ways of life, that dwell high above death. In the coming New Jerusalem, a new body/interface will give us a less physical, far more spiritual life, formed within the new physical laws of Heaven. The current quantum and physical realms consistencies will no longer exist, but the aether is eternal. In Heaven, we will exist larger, faster, and closer to the aether. But there will still be an eternal divide between the lost and the saved. In Heaven, imagine you can feel much of what all others feel. Imagine feeling great joy in knowing only goodness, and feeling helped by helping, and loved by loving; feeling stronger in giving away your strength. FATHER forms all things perfect at their creation, and this universe was once perfect. But that first little sin set loose, all the evil desires that destroyed it. Yes, Eve sinned first, but it was Adam's sin that brought the fall and that first stretching of creation, *and* that very first measure of dark matter and energy. A devolving process began, that has run its course for around twelve thousand years. At least a third of the universe had reverted to dark energy and matter after the great flood, and around two thirds after MESSIAH'S Crucifixion. Did Satan expect the third great stretching would lead to a final disillusion and judgment of all corrupted realities. Equate your soul; how much of you truly wants GOD to be in, *and* over your life, and how much does not? CHRIST came to make us a way back to perfection. And at the rejecting desires of Israel's High Priests, the *Romans* tortured our LORD to death. Now we know, *no* subjective physically observed laws control reality. Yet they can be seen from a subjective perspective to vaguely describe the current state of the last remaining 6% of GOD'S once perfect creation, as it is observed in each devolving moment. Anyone trying to explain reality from a small slow temporary perspective, using just the remaining 6% of a devolving creation, has already failed. Our souls can exist, act, and communicate through firm, or less firm bodies. We know GOD spoke to Moses through a fiery bush, which did not burn. We know HOLYSPIRIT interprets all data correctly to reveal the science and truth of reality without fail. In these fading remnants of a once perfect dynamic creation, we see proof of our gifted soul and created body, brought together in the image of GOD. Science continually finds and interprets bits of subjective data, yet never comes near a conclusion; they only grow farther away. Their false physical answers, breed only-larger unanswerable false questions. They can't see how their godless beliefs, push them into insane untestable theories. A pure and true path of inquiry could never lead to dead-end unprovable theories. Now should we say GOD is alive? No, not in a current understanding. It is best to say GOD eternally exists, or is all existence; but not that they are simply alive. Life, now being adjoined to death, is seen as a physical process on the physical side of reality, while our soul eternally exists. FATHER creates and orchestrates every aspect of physical life in giving it all from within himself. So, he exists beyond all current definitions of being alive. JESUS *was* alive for around 33 yrs., so he could die for us. Alive, implies being born in the flesh, and all that now lives, also dies. Yet CHRIST and HOLYSPIRIT did in their way, come to be newly born to exist, but not in the flesh as we did. Every truth of reality mankind runs from, is down the path of understanding who and all GOD is, and what FATHER defines as good and evil. See all the divisions and connections of the two current realms, see all conscious force of creative desire, see the aether, see our created perceptions, and see our awareness of them. Under GOD, every aspect of quantum mechanics and the forceful design for Abiogenesis, the required guidance for the continuation of life, healing, and death; everything, down to the beginning and the end of this universe, and every question in between, above, below and beside. They are all answered by GOD'S one pure truth; even all the questions of why. Not one true question is not answered. Now, we turn to show which fields of secular science are the most damaging to souls. Speaking about the world population as a whole, this is easy. The fields that hurt most souls, are Evolution, Neurology, Cosmology, and Geology. Their godless theoretical views of Abiogenesis and Evolution of Species, their belief in a magical self-awareness of brain meat, and an extremely old universe *and* earth. These lies are all propagated to naïve children so Satan can steal their souls from GOD. Teachers, maybe unknowingly, teach many things known in science to not only be questioned, but discredited. Yet, they still indoctrinate all young children, and it is being allowed and desired by many parents. I would tremble in my sleep for destroying the hope of these children with all these spiritual poisons. Such teachers hold a bloody sword of powerful confusion, because the earth and the body are touchable evidence, and the source of physical pleasure to feed one's evil desires. While Heaven is the untouchable mystery most *subjectively* resist. The Earth and the body are things we perceive, touch, need, and strongly sense, and have a most certain relationship with. But teach your children to see the supposed geological record in the truth that time is an abstract malleable illusion, born of a small slow temporary perspective. And that the fossil record is 90+% short of being credible. For me, the fallen elements of confusion, entropy, and fission, drive me to see GOD who perceived all things. Show your children how GOD, in hiding their hand to such an extent, is required for our Godly loving need to arise, seeing we are born fallen. And why a belief in scientific falsehoods, pushes souls beyond the reach of the truth. Teach them the seen probabilities of a conscious force beginning, and the impossibility of a non-conscious beginning. Show them how starting any physical life from non-life, minus GOD, requires a perfect environment, a design, and a designer. Show your children how all the *great voids* found in all the false theories for Abiogenesis, show they can never begin to prove it. Show them how the four forces, minus conscious guidance, seen in all particle spin, rotation, and all layered orbits, could never create or account for all the great complexity we see. Teach them that a moment of relative dynamic physical perfection and profound stability under GOD'S strong fusing Principles, created the conditioned platform to then create all the greater complexities of life. Teach your children the entire creation was once a tightly entangled united system of fine tuning and upward pressures designed to sustain all that complexity. And an unknown probability for rebellion in mankind's prideful envious curious desires, became a certainty as individuals began to interject their own endless physical disorders, starting a Butterfly-Effect of Chaos. Every contrary choice made, introduces a force of uncertainty, bringing disorder to the original forceful order. This new entropy shot inward from every direction, as FATHER'S hands pulled away at the far edges of his once perfect order. All Geological and Evolutionary studies are currently filled with as many wild assumptions, imperfect answers, and truly unanswerable questions, as all other theoretical fields. Never look in any *wanton* desire, at their declared fossil record, or false methods of fossil dating. Their fossil record is not there, and their dating system is a collaboration of 20+ confused witnesses, being interpreted through a small slow physical perspective. Stare *intensely* at the great lack of a *required* fossil record. Stare at their many assumptions and proven falsifications. Do not hear what they say is there, focus and ponder on all that is missing. Ask yourself how and why should ninety-plus percent of the fossil record, required to prove evolution, be missing. Could the only honest answer be, you can't find something that was never there? Around the mid-20$^{th}$ century, evidence turned against earth-bound evolution in a fierce way, as they found the required fossil record did not exist. And found fossils their isotope dating methods, placed out of a required evolutionary order. This according to their own theory *cannot* happen. The study of Abiogenesis and the Evolution of Species, long ago, painted these fields into an inescapable corner. All new evidence rarely supports, and often hurt the theories of

Evolution and Abiogenesis. So, with no way strait up the middle, they tried an end run they called the Cambrian event. Positing a sudden appearance of many entirely new, *vastly* different species. This fell apart under all the thick smoke, heavy mirrors, mysticisms, and illusions that formed this misaligned supposition. All their physical Abiogenesis and Evolution theories are caverns of pure impossibility. So, as reality set in, they were forced to find a new faith; like when the fine tuning of the universe forced them to conjure up abstract Multiverse theories. So, as earth-bound Abiogenesis and Evolution failed, leaving only two ways out, they turned their gaze upward. No, not to the greatest probabilities of GOD, but to an even more impossible illusion. They now try to believe that a basic organism of life, came to earth from space. And so, the godless MSM has now been saturated with a decades long search for life out there, ignoring the fact that such beliefs only multiply their problems. Yet it is a misdirection that cannot be disproved, turning public scrutiny away from their failure of an earthbound Abiogenesis. They ask us to overlook the great failures of *all* their former impossible assumptions, and now have faith and hope in their more impossible assumption that a messianic nugget of highly interpretable physical data will fall from space to resurrect Abiogenesis and Evolution from the grave. All while they still indoctrinate little children with their godless religion. They raise an arrogant hostile indignation at any proffer of allowing a conscious forceful design to be taught as even a secondary consideration. The only souls who believe in Evolution and Abiogenesis, are souls who need and want a belief that denies GOD and the Bible. Their problems are not with any personal god, but with the one true GOD. And so, they hate the Bible that declares them. And guess what, they are now turning to push Islam, which is the most antichrist religion in existence. If you believe just one thing contrary to GOD or the true Bible, you *deny* the entire Bible. Do not try to choose to believe the parts of the Bible you like and agree with. I *assure* you all, you can find *nothing* of GOD that way. What do you truly believe? Please think about, ask, and consider all things together. See the big picture, as we all will soon stand before GOD to account for our desired chosen beliefs in our life. Can you imagine standing before GOD, found to be Satan's child, having encouraged or allowed so many godless lies to be taught to eternally destroy so many little naïve children. Scripture warns, it is a *fearful* thing to fall into the hands of an angry GOD. I feel both my righteous anger and my fear at seeing this abomination. And anger is an emotion I only feel towards sin. Scripture says there are both evil and righteous angers. I know GOD hates many things and is righteously angry. So, I hold and do the same. I know in GOD'S day of judgment, that all who taught, or demanded, or allowed the teaching of these lies to GOD'S children, will see their hands, lips, and clothes, soaked in the blood of countless children. Such evil souls will then and there, forever wish they could cease to exist before GOD. But this will be just one of their endless forever-unmet desires. The scientific fossil record has far more missing eons, than fossils. It is a vessel that can't hold water. Who gives them a right to teach GOD'S children that the earth appears to them *as if* it is a certain age, when it *is,* and appears to be a far less age to GOD and us. Their false *unanswerable* questions, outnumber their many false answers. There are not just small gaps in the fossil record to prove an evolution of species, there are not just vast caverns, there are not just millennia missing. There are endless eons of fossil changes missing between all species. They hold up highly interpretive fragments of bone they cling to for faith. Fragments that at best, are just desired biased interpretations. We see upon earth today, people of every skeletal shape, size, feature, stature, build, and bent. We can survey all the photographs in the world and see people and animals alive today, who present much like those given for their proposed early forms of mankind and other species. There is zero true evidence of evolution. Now by the inflation (stretching) of GOD'S Creation *away* from perfection, atomic entropy, or fission, first appeared among the heaviest element, and on down to isotopes like carbon. But current data, shows all fission is slowing, making fission an unreliable variable. *And* by all decaying physical complexity being contaminated in many known and unknown ways, these and many other factors show carbon dating is a far more subjective biased tool than claimed. They admit that of their twenty+ isotope, visual, and other dating methods, they can never use one method alone, as they all give different results. So, they most often rely on averaging their findings of several methods, that still produce an ambiguous result. Which for reasons that allude me, they then trust. And this process becomes more convoluted and unreliable with the actual age of the object being tested. How can their hearing 3 or 4 biased blind witnesses, giving different accounts, render reliable data.

Can we take unreliable witnesses having different stories, into a courtroom and form an average of their accounts to convict someone of a crime? Why not, it is the exact same reasoning? We see disease, diet, and conditions, severely alter the appearances of standing skeletal features; some even slow or increase the aging process of the living. Tell me why I should allow a biased godless soul to take a fossilized bone or fragment from the ground, and convince me they see something ancient, and distinct. And from that, imply it is a distinct evolutionary change in a species. How can anyone declare what a fossilized bone is, by variations from a so-called norm, when a *norm* for each bone in each species has never been defined. An average within great current deviations, is all that can be accepted concerning bone size and shape. In life, bones deteriorate and shrink, and if buried in wet soil peat or rock, under differing pressures, bones will bend deform and shrink, or expand before and during the indeterminate processes in each stage of bone fossilization. Mankind disregards and overlooks so much truth to fulfill their desired faith of belief. Do you see the desire and faith it takes to hold such impossible beliefs as Abiogenesis, Evolution of Species, and Carbon Dating, as a possibility, let alone a probability or certainty. The calculated odds of the many different right and left enzymes found among nature, coming and bonding together into just one *purposeful* protein needed to begin to form one simple cell, are beyond probability. Odds as close to impossible as can ever be. And a cell, requires 150-300 different protein formations. I have no doubt many want and try, and do believe in Abiogenesis; but their belief does not make it true. If the entire world believes a lie, does that make it true? The one truth is not subjective or created by consensus, but lies of conspiracy are. If I have a steel penny in my closed hand, and scientist ponder all evidence, and say to the world, all data suggests I have something else, this will convince many. But again, when it comes to defining reality, any consensus among fallen mankind means nothing. We must consider their conscious reason and purpose, *and* see all the true physical evidence for ourself. It turns out that the fallen, in general, are eagerly deceived. The world is largely become an adult version of the childhood game, follow the ungodly leader. By chance and probability, it is impossible for every soul to be wrong about every single thing. But nearly the entire world, can easily be deceived about many things, once they have a desire for such lies to be true. And with the ardent propagation of lies through biased education, news, and entertainment, they indoctrinate the young and old with their evil desired faith. Do the media care if *everyone* believes their lie? No. They only care about the consensus that do believe them. How do you turn the world to a lie, you turn the young. Darwin said the fossil record must grow to fill in all the many vast gaps, or evolution is a lie. So, in that this filling, never even began to happen, Darwin must wail and gnash his teeth and tremble, seeing the countless children he has helped to destroy. There has not been one physical theory proven in any way. Yet they thrive on hyperbole, conjecture, and faith. Do we see any ongoing evolving in any aspects of physical life? Why not, did it just stop? Why aren't new complex proteins forming in the wild? Why do we not see Abiogenesis still occurring in the wild? Now we find many complete distinct dinosaur and animal skeletons; but just as with mankind, nothing in the immense gaps between all species. Evolutionist say the vast missing fossil record is a real and grave concern, and they know no reason why this highly searched for required evidence, is non-existent. Yet they still hold their godless faith, and force their godless religion upon naïve children. Consider the following seriously, and ask why would this be written in Scripture around four thousand years ago? FATHER speaking in Job, 40:15-24, "*BEHOLD NOW BEHEMOTH, WHICH I MADE WITH THEE; HE EATETH GRASS AS AN OX. LO NOW, HIS STRENGTH IS IN HIS LOINS, AND HIS FORCE IS IN THE NAVEL OF HIS BELLY. HE MOVETH HIS TAIL LIKE A CEDAR: THE SINEWS OF HIS STONE'S ARE WRAPPED TOGETHER. HIS BONE'S ARE LIKE BARS OF IRON. HE IS CHIEF OF THE WAYS OF* ***GOD****: HE THAT MADE HIM CAN MAKE HIS SWORD TO APPROACH UNTO HIM. SURELY THE MOUNTAINS BRING HIM FORTH FOOD, WHERE ALL THE BEAST OF THE FIELD PLAY, HE LIETH UNDER THE SHADY TREES, IN THE COVERT OF THE REED, AND FENS. THE SHADY TREES COVER HIM WITH THEIR SHADOW; THE* WILLOWS *OF THE BROOK COMPASS HIM ABOUT. BEHOLD HE DRINKETH UP A RIVER, AND HASTETH NOT: HE TRUSTETH THAT HE CAN DRAW UP JORDAN INTO HIS MOUTH. HE TAKETH IT WITH HIS EYES: HIS NOSE PIERCETH THROUGH SNARES.*" FATHER spoke to Job saying "behold now" and "made with thee" referring to Job *seeing* at least one of the fewer brontosaurus type mammals still alive, before following many other species into extinction by entropy, physical imperfections, changing environment, predation, and the great flood. And we see all species dying out in their order of vulnerability. This description of a brontosaurus, cannot apply to any other animal that walked the earth. Is it just impossible for most to believe these creatures still existed around 4-5000 years ago, because of carbon dating? We know that after the creation fell into physical confusion. Creatures began to turn more and more on one another in predation. We read a description of a brontosaurus, written at a moment that according to evolutionist, a brontosaurus could not exist. Yet one is perfectly described to have lived at that time. There is zero evidence that men of that era had un-earthed, assembled, or studied fossils. They did not dig up, deduce, and broadly publish the great length, breadth, scope, and size, of this animal's neck and tail. A tail being thick and long as a cedar tree. Or the great body and long neck of this beast, appearing as though it could, poetically, drink up the Jordan river. So, this animal Job had seen, must have been among the last of these giants. Now we ask, do you feel this cannot be? Do you feel pulled back towards the slow small impossible physical perspective of Evolution, and those many obscure carbon-decay dating methods? Are you bound by your physical desires and perspective towards discredited godless beliefs? Or can you truly and fully let them go, seeing only the determined facts and truth of GOD? Dating a year that anything died or lived so long ago, is based on hope and faith alone. A temporary physical perspective will create a problematic appearance in a soul's perceptions, including their desired perceptions of quantum and classical measurements. So why would *I* accept their interpretations of their biased perceived measurements, when I see all their deceptive perspectives, and the evil reasons that they see and accept them. The fullest understanding of all differing soul size relativity demands, that to see a true measurement, the vast fastest perspective of GOD must be held by that conscious observer. And all lesser slower perspectives, *must* err in their biased interpretation. We see a Supreme conscious force at work, yet he allows every soul to *create*, see, and believe, the desires of their heart. Fallen scientist, having differing interpretations and beliefs are my irrefutable evidence of this. Do you see how all biased line of sight perception, allows one to subjectively see another soul as being wrong or evil, when they themselves may be the far greater evil before GOD. I do not align with any divisive so-called Christians or Jews who claim GOD understands, allows, or accepts, any ongoing sin. Yet it is still the far-left godless liberal who perpetuate the far more imperfect hateful divisiveness in this world. Yes, greater to lesser evil exist, yet all are divisive in their measure. By their individual desire, many see their evil ways of life, as being good, or not divisive to temporary life. Yet from an eternal perspective, GOD says they are divisive to eternal life. Eternal life has higher standards and requirements, just like eternal love. We need to understand that this ability to blind one's self to GOD'S truth, rests as much in what one physically can't see or know, as in what they do physically see, and think they know. Relying most upon that which is externally seen or felt, perpetuates a physical perspective. And what is seen and felt, may then mold and define a soul's spirit. All questions that define life, truth, and love, are asked and answered truly or falsely by each soul. How do *we* step out of a temporary perspective, and into GOD'S eternal reason and purpose? How do we escape individual false religious reasons, purposes, and our individual truths of life, love, and desire, for this physical illusion? In those falling far behind, there is no understanding of eternal life, only a blindness of not knowing the true spiritual purpose of life. Their interpretation of all seen evidence and life combined, shows they cannot find the truth, from the 6% of physical reality they see and study in their fallen limited senses and biased desires. The truth of reality can never be known by those who refuse to go to GOD with true loving questions. Now fallen mankind senses and interprets speed size and radiations from their touchable and untouchable sources. And this being all they can detect; it is all they can believe. Geologist accept possible past catastrophic asteroid events, but deny a past worldwide flood event, when there is ample evidence and enough water to do so. FATHER could change the face of the earth by his simple forceful perception of it being so. He could also, perceive thousands of asteroids impacting the earth. At FATHER'S desire, the Sun can be hotter or cooler, changing rain forest to deserts and back again. FATHER perceiving massive volcano eruptions and low solar activity, could bring a sudden ice age. Do a search for *images* of a 20,000-year bar-graph of the "Younger, Dryas event" and take a screen shot to preserve it. See the sudden drastic changes in sea levels and temperatures found in geological data, which dwarf the last hundred years they panic over. What caused *those* huge changes? The northern and the southern thirds of the world were quickly frozen over in deep ice, and in short order, it was gone, leaving its scarring evidence. We read how FATHER during earth's first six rotations, force perceived every mountain, valley, river, ocean, desert, forest, and all life, to appear in a substantive semblance that pleased him. But GOD also *allows* it to be seen as fulfilling the desired chosen beliefs of mankind. We know individuals perceive and believe whatever they desire. So, we trust only GOD'S Scriptural history. Yes, six evenings and mornings were seen in eternity, during FATHER'S original layered creation events of earth by his perception process. And between six to twelve thousand or more earths orbits around the sun, have occurred to date. We see an *undisclosed* duration of Earth's perfect moment in eternity, where Adam was alone; and another *undisclosed* duration of the moment where Adam and Eve lived in the Garden of GOD, before their rebellion to the truth of life. Each could have been tens to thousands of earth orbits around the sun. *We* do find every path leading to anything, even the *current* depth of mankind's fall, will in obvious ways be relative to and reflective of a path back. So, for several reasons, six thousand+ years before Adam's fall, plus six thousand+ after is just a reference, we consider as the most-likely age of the creation. Of course, there are many consciously forced events to consider and equate. Now it takes us a substantial moment of living, to gain an adequate measure of anything truly good; and around the same duration, to turn, leave, and lose *all* of that good. The perfect reason purpose and truth of reality are found by spatial spiritual discerning and observing the appcarance of *all* realities as bound together in one work, created in GOD'S forceful desired way. Otherwise, one can only see, know, and be confused by the near small slow fragments of reality they physically sense. And now, in all rising confusion and rebellion, this world runs into dark insane liberalism, leading the lost down a *hallway* called secular socialism that leads all from the great room of religious liberty, to room of forceful communistic denials of GOD'S ways. Socialism and Communism are the strongest forms of godless control and indoctrination. This world is now falling into a far-left zombie insanity, forcing its Socialist-Marxist ideals into every aspect of entertainment, religion, education, science, healthcare, sports, work, finance, even relaxation and sleep. They work to mold our every thought and belief, being godless authoritarianism. Imagine two societies: one has low taxes and great self-reliant freedoms of investing and enterprise that allow its many millions of people to be all they can be, to the best of their ability. But the socialist/communist society, greatly limits its people in what they can achieve, as all means of production are controlled by just several thousand politicians. In the free society, millions of souls, having great ideas, dreams, and goals, create greater products, companies, jobs, and great wealth within their nation. So, even at a lower tax rate, they bring in far more money for the common good of the people than a socialist-communist society. The socialist-communist society, whose leaders are at work controlling and socially molding their people to be drones, are ill equipped to have great ideas and dreams to build and run great companies, resulting in new products jobs and wealth. So, even at much higher tax rates, they bring in far less money for the common good of the people and nation as they choke out all free enterprise. Imagine one free soul with a great dream and drive, who builds several great companies worth hundreds of billions of dollars, that in extended ways, create 400,000 good paying jobs. And see this one soul, paying between 4 and 10 billion dollars in tax each year. The only way to see this soul as a great evil, is through envy and hate. Now most established religions, have been pushed into an ineffectiveness that allows and embraces much evil. It can mean nothing to the lost, that reality makes perfect sense through the truth of a qualified quantified patterned projected conscious force variable, creating every particle-wave based reality. Liberals are the self-declared highest order of the animal kingdom, being a purely physical creatures in a purely physical realm. All such reprobates, have lost all ability and desire for eternal spiritual-understanding. Now, by the existence of the word "*omniscience*" it follows that omniscience must exist. But who defines its form? Well, only the omniscient can define it; being FATHER, having all understanding and purpose in all his highest possible accounts. These, with all hopeful compassion and faithful responsibility, define FATHER'S eternal love. We see GOD'S enduring effort and suffering, trying to get mankind to choose their loving omniscience, and let go of *all* individual divisive confused beliefs. And we see mankind's endless efforts, asking GOD to alter and compromise his eternal created ways, to fit not GOD'S desires, but mankind's temporary physical desires. It is not even possible for one physical reality to fulfill each soul's individual desires. That would require billions of different worlds, created to match the desires of each fallen soul.

Yet there are the fewer who see, hear, and only desire the loving beauty seen in GOD'S ways, in one created reality. All of fallen mankind's desires and efforts are spent trying to deny or redefine GOD in their own image. GOD does not change, ever. Many can't see how the long-lost created truth, means their current belief is wrong. If lost souls cannot see GOD'S truth, should they be allowed to defend their contrary desires, standing before GOD? No soul can look at GOD and say their own contrary desires and beliefs do not make them evil. Yet Scripture says many *will* ask why? Now, if you say you have *no* belief in GOD, without doubt you have a formed concept of a GOD you see others believing in, that being your reason not to believe. And usually, an atheist's concept of GOD is filled with the injustice of imposing superstitious rituals upon the weak minded. If this should be remotely close to a soul's formed conception, I say in all their desires and assumptions, they do not know GOD in the least; and no atheist does. The term "*GOD*" meaning Highest Supreme Soul over Reality, from whom all lesser souls and realities have come forth. And the modern term "FATHER" comes from an ancient word meaning creator. And FATHER is the Creator of all that is good. In their confusion, the secular world believes mankind to be the highest life form. And many materialists and atheist think *they* are among the higher examples of what mankind can, and are meant to be. They imply that if gods do exist, the educated liberals are the best example. Now, I hope you find and feel great interest in this next hard to receive part. We find great solace, and hold no fear of mathematical truths in life when correctly applied. We find comfort that GOD'S math and love, hold no *imperfect* bias. Yet, GOD is *perfectly* biased against any imperfect understanding of any and all created realities. Now, among mankind, math at its farthest edges and deepest interiors is being cleverly abused; we see that at math's unrestrained unreasoned limits, it gets quite subjective and interpretive. As we said earlier, there is no path of human endeavor, founded on any lesser imperfect good, or obedience to any lesser love, truth, and understanding, that leads to eternal life; save for lovingly emulating CHRIST'S eternal ways. By all slow small perspectives, and devised mathematical operations and principles, mankind defines the current values of what still remains of FATHER'S consciously projected forces and creations. We see mankind's efforts to qualify and quantify all physical aspects and laws, and the entangled relevance between all physical things. So, consider how with much information about all conscious choice being made, statisticians can quantify the dynamic beliefs and motives among the masses. So, why is it then beyond reach, to go farther? Do you not see the predictable conscious and physical destruction that comes from all the endless divisive, confused, conflicting, individual and consensus beliefs. Do you see how worshipping every individual's right to live in any conflicting way, makes it impossible for mankind to form any functioning societal norm? Such ways must end in uprising or war, as it always has. Again, our gift of free-will is not about our right to believe anything we desire; it is only about our choosing GOD'S eternal ways, or our own temporary godless ways. We need the world to take a dual, equal approach, in each scientific field. On the one side, researching the primacy of conscious force projection over all physical creations. And the other side researching a physical primacy over creation, as it does now. This race to the truth and proof of reality would stimulate much debate, needing checks and balances of sharing all data, analysis, and interpretation between both sides. By one side equating and applying Conscious Force Principles to all aspects of reality. We could begin to qualify, quantify, formulate, and equate the actual conscious force projections that over-ride chance to create choice and all three-dimensional certainties. We can calculate the odds of hundreds of things written thousands of years ago in Scripture about GOD'S layered creation process, being set in a true order for rising environmental and biological complexity. Being a new understanding not known to mankind in the days of the author. There are so many facts and observances to be brought together to show the near certainty of GOD'S Dualist Model being true. Using every consideration to show how the current fallen reality is an intentioned result of GOD'S and mankind's contrary forceful desires. And why on the Materialist's side, we see a diminishing probability that it could all be a chain of purely physical chanceful events, started by an unexplainable physical miracle. What could it mean to mankind? What change could come to science, religion, and the whole world, to see two calculated odds set side by side, day by day. Two approaches working in a search for knowledge and truth. Being under public scrutiny and fear of being shown to cheat before the whole world, that signifying a humiliating scientific failure. Ponder the exclamations from around the world in hearing that all material sciences, compete and share funds, platforms, and forums in a scientific pursuit of studying the models of both a conscious and physical determinism. To those who say you cannot allow GOD in science. We say you are terrified of being shown how wrong you are. Beyond all doubt, *all* scientific inquiry is born of our desires and perspectives, which forms hope in as yet unproven things being true. Each soul's perspective-and-belief on reality, is that soul's religion. Materialism is a material religious belief. If any religion must be kept at bay in science, then all religions (unproven beliefs) must be kept at bay, and all theoretical science disappears. *Albert Einstein*, once wrote "Science without Religion is lame, and Religion without Science is blind". What all these fearful materialists and secularists demand, is that only their religion be allowed. I see their ever-rising *reasoned* fear of losing this debate. They know, the only way they can win, is to keep GOD'S voice silenced. This is the way all the dark liberal fascist work, in all aspects of life. Our side does not want *them* silenced; we actually want the world to see their confused contrasting failing beliefs alongside ours, to illuminate and affirm GOD'S truth. Liberalism breaks down morality, and leads to insane totalitarianism. There are only a few materialists of courage and mature patience who may not fear a reordering of the scientific world. Yet they do fear the relentless abuse they would incur to their last breath from their peers in publicly saying so. True debate and research are dead under liberal control. The zombie liberal way, is a forced godless order upon the world. But by his eternal supremacy, JESUS wins this war. Liberals from their temporary perspective, can't see an eternal end game. For devils it is about how many souls they confuse in the process. They refuse to see how a dual track of pursuing science will bring broader understanding and new fields of research. It will bring new interest, courses, teachers, and economic benefit for all institutions of honest study. The honest approach is the sound societal business model. Yet their unenlightened core desires, deny and hide the truth that *their* beliefs are evil. If we honestly equate and set all statistical values of probability, we can first, at the very least, publish a probability of each primacy being correct. By seeing set probabilities side by side, no matter how small or large, if at first 20% conscious primacy, to 80% physical primacy. Can you imagine a more honest debate coming to life, based on the evidence we currently have. Right out of the gate, we could see an equated current 1 in 5 chance that projected forceful creative conscious desire, rules, creates, and affects, all perceived physical reality. Yet such an honest *purely scientific* debate in itself, could not establish one supreme religion. Imagine all evidence found on either side must be open sourced, published, debated and examined by the world. This would unite souls in a search for the truth of life, whatever it should be. And yes, scientific, political, religious, and commercial interests, would exploit this as this world learns to use and abuse it, like they do with all knowledge. But this would be far more uniting and growing across the center, than the growing division we have now. The search for truth itself, can't be pre-defined or restricted; and the only alternative is to suppress the truth, being a most vile act. Suppression of Godly belief and expression, is the liberals most ignoble authoritarian demonic trait. This debate is here to stay, and seeing each sides strengths and insufficiencies, both sides must acknowledge there are questions they can never ask or answer. Like what was before a Big Bang, or was there a beginning to GOD. But the spherical eternal entity of GOD is at least knowable, while all undirected purely physical happenstance leading to great complexity and conscious awareness is not. The only answer is that all the changes we now see, being consciously and physically complex, must come by these two basic eternal aspects. In faith, we all must accept a currently unprovable foundation to support our desired belief; but we all see that the conscious and physical sides have proven connections. And by the entanglement of force, together they create all complexity. Whether one is secular or believing, neither has a yet proven foundation to force upon others. Two contrary beliefs, both rest in their desired accepted probabilities, both having their motivational and empirical *probabilities*, but no proven certainties as to primacy. So, there are question each side cannot ask or answer. These will be the last questions each side to answers. This is the true way to seek knowledge and answers. We all find our souls within FATHER'S perceived reality, sensing and interpreting the complexities of his and our consciously forced physical creations interacting. All our *eternal* souls have the capability, but not all have or find a desire to adopt GOD'S spiritual perspective, beliefs, and defined ways of eternal life. And all individual beliefs, coupled to all individual physical ways of life, form and bring all mankind's hell and death. Leaving a physical perspective, grows our spiritual perspective and conscious speed with each step. But falling back into the physical again, recorrupts and slows our perspective at each step. We can choose to deny all this, and be only half of who we are meant to be, But if we desire to understand it all, we can come out of darkness into GOD'S fastest light. Think of your learning and belief as your chosen path of desire. I pray GOD is your *greatest* desire, and not just a real, yet lesser desire. All fallen souls believe and share destructive lies, yet they think they share truth, by believing they are more-good than evil. Please consider the secular variable *they* call superior intelligence, and what might it truly be. Is cognitive ability, simply the wet network of flowing electro-chemical reactions in neurons. Or is the first cause and foundation of cognition, born in the core desires of each soul needing, desiring, choosing, and coming to believe that certain probabilities are true. We know there are variables within each brain/interface structure that affect our processing the limited physical information our souls receive and send. But that varying amount of *patterned* information, passing through our interface, does not translate into our ability to choose the truth. Truth is desired and believed within our soul's love; our beliefs are not born of electro-chemical reactions in wet fatty brain meat. We know computers process data, while having zero awareness of that data. We can burn every computer and human brain, and GOD'S truth has not been affected in the least. To know the truth, the need for a better brain is a false premise, as it is all a choice of faith. No matter our physical interface, we can, in our relative perspective, love GOD just as intimately as anyone else. And for many with a superior brain/interface structure, believing GOD'S truth may be impossible for them to see. The eternal perspective and desire for GOD we hold, are the factors that matter. It all rests in our faithful loving vantage point to see the reality before us. As a visual exercise, see this world as a conscious landscape, and the omniscient truth of GOD'S understanding and belief of life and reality, are a most-clear pure living water. But by flowing down into the low valley of confusion, we see endless individual false beliefs of understanding life and reality, all swirling together. So, the slowed valley waters are blurred and muddied. Only the water at its highest fastest source renders a clear eternal perspective, the low valley only holds a small slow blurred confused perspective. Living waters of truth come from the purest highest source, so we must climb to the top of this spiritual world, where we see only GOD. Many souls find comfort and prefer the silted uncertain waters of the valley, where incomplete imperfect beliefs pool together. For these, the effect of many accepted and allowed beliefs, means no one has the truth. This gives them the comfort of allowance to the rebellious, who don't want GOD'S one truth to exist. If one is in, and wishes to climb out of a confused valley perspective, they must for that choice to exist, first see *two* far distant high mountains, being a choice for an eternal spiritual knowledge, or a temporary physical knowledge. A soul who climbs out of the valley, to a higher plateau, has found a more focused conscious vantage point, but they have gained no greater cognitive ability. Each distant mountain's rivers below, represent all believed, yet incomplete defined understandings of all fervent religious, political, and scientific factions. If they desire to see more and more, they see the two distant mountains, feeding those strong rivers, coming from two opposing perspectives. The far higher mountain of GOD'S truth, offers *eternal* understanding of life. The other lower mountain top, offers a false physical understanding of reality. At the top of the waterfall on GOD'S Mountain, resides the purest dualist perspective of desire truth and reality a soul can ever receive. Every fervent driven soul, desires to climb to the top of one of these mountain-perspectives. Which one is determined by the desire and belief for or against GOD each soul has chosen. Yet, *every* answer received on top of GOD'S higher mountain, as all wisdom, rains down from high above, and from before and above the clouds. We must forever look up and forever climb and soar for all the higher truths to come. We see few souls have a desire for an eternal spiritual perspective, which is currently a lonely perspective in this world. Now turning back to our created particle-based memories; it should strike all as odd, that neuroscientist say we should *not* think of brain memory, as brain memory storage; but in ways of brain recall and processing of memory. They say this because the brain is known to be a processor of many functions, but lacks the analog capacity to store all our memories. *And* memory processing is all they actually see in the brain. The brain is a quantum analog information interface processor and relay, not a digital hard drive or ram storage. There is evidence, that a single entangled neural branch and neuron within *each* brain hemisphere, is our quantum link to each new created quantum memory pattern within the aether filling our spherical soul. In the most severe cases of epilepsy, surgeons have to sever the two hemispheres of the brain; and at times, *removed* half of a brain. Each half has areas and specific neurons that can do certain functions; including those single neurons that connect to our quantum memories. So, souls who have lost half their brain processing, do not lose half of their memory, or half of who they are. Yet strangely many die from a mild blow to one side of the brain, while others survive a steel rod run through the entire brain, never losing consciousness. Memories can never be destroyed by brain damage, only our interface ability to retrieve and process them. Science has never found any step-by-step process in a brain to form visual, audio, and other particle-wave-based perceptions that enable our memorialized awareness and understanding. Their godless faith says that within complex neural interactions, that like rubbing a magic lamp, a genie of knowing believing awareness just pops out. Their high standards of rigorous scientific proof, amaze me. No one can ever physically measure our soul's understanding potential. Mankind can measure different degrees of *desired* learned belief in certain things; and they know that even this is skewed by one's *desire* to be tested. Yet the world says those who get higher scores on IQ tests are superior in understanding. I know my understanding of GOD, comes only from GOD, channeled through my great desired need to know, it is not by my brain structure or neural count. So, *my* IQ does not mean what they imply. It is beyond all doubt that we receive things new to us, from two sources. One being from without, as in our reading, observing, or audible instruction; the other is from within. This *we* call our inspired revelations from GOD, and the Materialist call it godless inspiration of wet brain-meat. Yes, I know brains are far more fat than meat, but they are flesh, and people and animals do eat brains. Now Godly revelation, or wet brain fat inspiration, must both come from someone or somewhere. They can't come from no-one nowhere or nothing, as no-one nowhere and nothing are aspects of non-existence. You cannot get reality from nothing and non-existence, that is an insane belief to the utmost degree. Now mankind can in no way, measure our untapped learning ability; so, they measure the tapped and call it our IQ. Which only measures our remembrance and visualization of what they try to teach us all, comparing us against the tested population. They use this score to imply a certain ability and capacity; but among average interface structures, it measures only what one has desired to know. Most souls who score high on IQ tests, feel pride, drawing them deeper into their smaller slower temporary physical perspective. Are some unable to understand, because they lack intellect? No. But they can lack a desire or reason to know or believe what is being taught. Our fervent desire to know GOD, is our need in our broadest measure to not just know *how*, but *why* all created things are entangled and relative, and interact with every other conscious and physical cause and effect. Now on my first administered IQ test, I scored the highest for my school grade in Miami-Dade County. Which means no more than that I had a very *strong* desire to know the meaning and relevance of all things. IQ test identify the knowledge they search for; they seek out all *useful* square peg physical understanding that fits their secular square peg holes. A liberal darkness has controlled the sciences for centuries, seeking ways to get souls to deny GOD. Does any of this mean round pegs of desire lack the ability to learn? Or just that they lack a desire for the square peg beliefs, liberal fascist push all to accept? Square peg materialists do not like Metaphysical round peg understanding and hate such things, not because of *their* intellect, but because of *their* desires. These will score very low on an IQ test based on knowing GOD'S reality. To Materialist, consciousness has many unknown aspects and affects, yet out of evil desires they refute any theory approaching conscious primacy. They can never know reality through their ways which are designed to fail; they are trapped in a loop process. They exist in a prison of never-ending evil imaginings. Please see that FATHER created the universe in a way that reveals his purpose and who he is to all who desire to see; but this can never be forced on anyone. Since mankind's fall, we see increasing genetic errors passed down and occurring in DNA and systems that limit, alter, and eventually *reject* the flow of life's patterned forces. Short of physical death, the worst of these states is called persistent vegetative state. In such a state, all brain activity associated with awareness is not detected. But the rest of the body's nerves are still part of the interface, and so the soul dwells unhurt in the quantum realm in a state of lesser to non-experiential existence, as it pertains to interacting with this world. In a persistent vegetative state, a soul is denied from *strongly* existing and interacting here with us. So, we see in these souls, a line between their lively existence, and non-existence is straddled, bringing confused debate to fallen mankind's interpretations as to what is life. Yet, despite any neural limits or damage, it is our soul's desires which determine what we do and do not learn. Body traits and inherited proclivities, can *if allowed,* play a big controlling part in our learning. Consider how people are driven to learn by certain desires and fascinations for this world and beyond, yet others are not. And what of savants? Do they have damaged neural connections to this reality, or just a few very forceful pure desires. Are they so averted from any broader physical perspective, that they expend all creative ability on a few most desired aspects, and lock out others in a phobic repulsion. If you choose to believe complex neural activity equates to intelligence and our awareness, this shackles you in an extremely narrow search focus, and severely limits the evidence you may consider. This world holds a huge spectrum of false religious, cultural, political, and moral beliefs. Do you believe all souls are influenced to believe and know what they do, because of how their brain grew? Are there learned conservatives, and learned liberals? Is belief about who gets to the children first? Are there learned atheist, and the learned who know there is a GOD? There are learned people who believe in the one pure truth; and learned people who believe in endless false truths. What is their goal in measuring the knowledge in people filled with endless desired beliefs? Is it as liberal atheists and materialists say, that believers in the True

GOD are afflicted with ignorant superstition? Do some have a godless belief because of their brain structure? In being honest, we all know we see no causation between so-called intellect, and belief. But we do see clear links between one's desired beliefs and their IQ. So, which is the *evident* definer of how we interpret reality? Is it our brain, or our fervent desire? It is the one that defines what we intently study and remember. It is not a neural switch; it is a switch of desire. There is no *honesty* in saying a belief in GOD is born of desire, but a disbelief in GOD is born of intellect. And in between these two opposing pure beliefs, we see a full spectrum of confusion, ambivalence, indecision, and great denial. And to date, the primacy, or true cause, of *our* conscious desired awareness *and* the physical reality, are no closer to being understood by fallen mankind, than the first time they considered them. Now, if you are moved towards GOD in our discussion in any way. Please know that it was not a neural learned movement; it was a desire in your soul for what you read heard and saw, to be true. You heard and saw the truth that sparked your desired eternal perspective, and only then were you moved. And if you are moved, we hope you will be consumed with responsible faithful love for GOD, and hold a great fear for what it means to cast eternal life away and lose all love. All aspects of physical life and reality, were born of FATHER'S forceful desired fusion. So, these creations can only remain by his same enduring projected forceful desire. The fusion of creation, brought and enables all life; but fission came only after the fall, and as part of evil, hell, death, and destruction. GOD holds fusion in his right hand for his eternal gifts of love and life, and fission in his left hand in allowing every free-will choice for hell, death, and dissolution. Col 1:16-17, "*FOR BY HIM WERE ALL THINGS CREATED, THAT ARE IN HEAVEN, AND THAT ARE IN EARTH, VISIBLE AND INVISIBLE, WHETHER THEY BE THRONE'S, OR DOMINIONS, OR PRINCIPLEITIES, OR POWERS: ALL THINGS WERE CREATED BY HIM, AND FOR HIM: AND HE IS BEFORE ALL THINGS, AND BY HIM ALL THINGS CONSIST.*" Consist, meaning to form or hold together under GOD'S projected desires, and be bound in having a certain formed consistency. Millennia before mankind knew about atoms, fusion, or fission. Scripture said, that at the very last, there will be no more souls that can be saved to FATHER'S eternal purpose, turning his purpose from hope, fusion, and creation, to find love, to what Scripture calls the fission of his wrath. Zech. 14:12, "*AND THIS SHALL BE THE PLAGUE WHEREWITH THE LORD WILL SMITE ALL THE PEOPLE THAT HAVE FOUGHT AGAINST JERUSALEM; THEIR FLESH SHALL CONSUME AWAY WHILE THEY STAND UPON THEIR FEET, AND THEIR EYES SHALL CONSUME AWAY IN THEIR HOLES, AND THEIR TONGUE SHALL CONSUME AWAY IN THEIR MOUTH*." 2nd Peter 3:12, "*THE HEAVENS BEING ON FIRE SHALL BE DISSOLVED, AND THE ELEMENTS SHALL MELT WITH FERVENT HEAT*." This is obviously describing runaway fission, *millennia* before an atom or its nucleus or the creative forces were known to mankind. By these and many Scriptures, and what we see in this world; we see as part of GOD'S wrath, that fission is a final agent of cleansing. Scripture also says, by growing division, a nuclear, chemical, and biological, third world war is near. Now did mankind overcome the projected force of GOD and split all those atoms in nuclear tests. Or did GOD let go and weaken his will upon those less stable materials in past nuclear explosions, allowing mankind's delusion that they can rule the forces of creation. Also allowing the world to see what comes, once they destroy

GOD'S last forceful desire for it to consist any longer; just as foretold. We know FATHER stretching his creation out of its perfect order, allowed some fission to begin long ago. And hearing whispered inspiration from dark sources, of dark knowledge, mankind followed a dark path of dark power. Yet at the instant of every runaway fission event, seeing FATHER causes all things to consist, it was his will that allowed the dark delusions of their heart to see their final damnation. So, FATHER did let go his desire over those certain fissionable materials for them to consist. Now after seeing that first runaway fission event, we heard a Prophetic Judgment from GOD, that was whispered to Oppenheimer for this world, saying "*I* am become death, the destroyer of worlds." These words, wrongly attributed to Oppenheimer and mankind, actually declare FATHER has become the destroyer of *all* worlds. All evil pushes FATHER towards the dissolution of his Creation. Back in the 1950's, 60's, and 70s, Churches fought hard against lies, but lost the battle to keep prayer in school and stop abortion as two 6-3 and 7-2 liberal majority Supreme Courts denied our country's faith and love of life. We must never stop fighting to over-turn those and other godless decisions. Now our emotional feelings and beliefs are not immutable in our moments of choice, they can change, grow, and wane. In Scripture, we hear of a man, coming to his first weak miniscule belief. JESUS says in Mark 9:23-24, "*IF THOU CANST BELIEVE, ALL THINGS ARE POSSIBLE TO THEM THAT BELIEVE.*" The man replied to JESUS, "help thou mine unbelief." What do these words mean. The statement from JESUS meant, the omniscient truth of eternal love and life, comes in all hope into this abyss of confusion, to lead all truly desirous souls back to eternal perfection. The response from the man meant he had just sensed his first true desire for belief in what he heard that held great love and promise; but he held a *new* impure, weak, ignorant belief, burdened and set at war against all his prior beliefs. His new belief was there, born just then by desire alone, but it lacked the strength gained through exercising and understanding what was his new desired belief. He had just then, truly turned in his core to desire something new, but needed to know all the greater strengths of what he had just come to desire. How can we grow and gain a pure strong belief in something we do not have at first. By our desired work, we search and learn. Jer. 29:13, "*AND YE SHALL SEEK ME, AND FIND ME, WHEN YE SHALL SEARCH FOR ME WITH ALL YOUR*

*HEART.*" Consider that if you barely see in the far, far distant desert, a tiny dim flickering light, as maybe a tiny candle far away on a high desert hill. You can scarcely perceive its flickering, yet you see and believe that faint twinkling light is real. But you don't know in any depth or strength of understanding, what is this light you see and believe in. If you have no greater desire born of love, and your weakness is to stay where you are; you can never know any more than your weakest of belief, that a certain unknown light is there. But if you are driven by over-powering love, and spend the rest of your life *running* to that light, keeping your sight focused intently in wonder; growing and being strengthened daily. With every mile traversed, you see more and more light, till one day after years of running, you come so close, close enough to touch; but in reverent fear you dare not touch. You see a great glowing loving form of a pure immense living light. A light that can be heard speaking the truth, and felt as pure love. Looking up and down and side to side, you try to take in its great size and beauty. Appearing as you see it to be an immense pure crystal, having billions of brilliant different colored cut polished facets of the one created omniscient truth of life. Shimmering, shining, sparkling, speaking, singing, and moving in every blinding unknown color and resonance. Speaking each word like a corralling of many loud voices moving through deep clear waters. A light so new bright and beautiful, it can never be received all at once by diminished mortal eyes; or be declared or described by the greatest of oratory skills. This is sensing a spectrum of warm loving qualia closeness and brightness in so many new ways, that it can never be related to others; this must be experienced to know it. So, now your soul sees, hears, knows, and is touched by much more than that faint distant twinkle in which you first believed. Our once, so weak belief, through all our love, patience, and faithful responsible striving endurance, is rewarded with our great strong understanding of what is the source and way of life eternal. GOD has promised all, if we search for more and more of them all our life, as we would a desired treasure, and go to them, we will find GOD'S light of eternal life and be with them in our own place as FATHER'S beloved Children. There are so many greater and lesser degrees of desire and effort within those who search, and so many false interpretations taught by worldly preachers. Many teach that GOD comes to us in all our imperfection, and a perfect Holy GOD, eternally stays with us to hear all our selfish prayers, and forever love us all in our unfaithful irresponsible individual ways. GOD'S love does reach deep into this world; yet never to stay with, but to lead us out. In professed believers, there are certain things in this world they desire and can never let go of. By these desires they are slowed, stayed, and turned back in their search. All our truly repentant prayers, desires, and efforts to find, hold, and understand the love required to eternally live with GOD, will always be answered at their proper place and moment. Never think our perfect moment for asking, is also the perfect moment for GOD'S answer. It can be, but far more often, this is not the case. A true loving eternal prayer, may require years to be righteously answered. *Our* loving need for the good, can only be born in *us* by our desire. GOD cannot instill or impart anything good to us without our desire to have them. Learning, including every understanding of GOD, is a process of our discarding all former lies, and hearing and building a pure Godly love and belief. GOD pleads with us to search them out, and they want to reveal all things of their eternal love to all who find them. Now let's talk more about the Angels. We are talking about gifted souls, and created bodies, on a far higher quantum ethereal order than we. FATHER is far greater in size than the entire universe, and the ethereal heavenly bodies and souls of Angels are infinitely smaller than FATHER, yet immensely larger and more forceful than we are. The truth sounds like reality greatly eclipses this world's genre of science-fiction super heroes and villains. And it literally does by many orders of magnitude. Scripture describes Angels, as being created across a broad spectrum of size, strength, and spiritual closeness and purpose to GOD. Having varied gifted real projected conscious forceful affects that make ours seem benign. Yet *these* Beings, are real ethereal quantum realm entities, that will soon be revealed to us. *All* light was first projected in FATHER'S love, to convey and share his reality, ways, and omniscient loving beliefs. Imagine space as not a dark background, but as a somewhat more lit background. This is how much light FATHER first created. But we see most of FATHER'S original light is dissolved back into the aether, and the light from the stars we now see, comes by vastly larger Angels. We see in Scripture repeatedly, how the angelic bodies of stars are created for signs wonders and many good works from GOD. Angels, being the size and force of stars, are so big and forceful, that we struggle to comprehend them compared to us. Now Materialists say our conscious awareness emerges from within complex activity, and that it may emerge within the activity of the programs and the internet. Yet strangely, they do not accept that awareness can exists within vast spheres of quantum elemental fusion, having incalculable numbers of quantum equations and complex quantum processes occurring all at once. A star on a far higher level, can equate to a quantum computer. If awareness emerges from energetic physical complexity, it would emerge within the fusion of stars and in many other ways, long before ever emerging in a brain, or the internet. Now the internet is constantly utilized, changing, infected, and controlled by external sources. In each moment, it can only be described as a ball of organized white noise. Might *our* souls be connected to the internet, or to stars. The internet is filled with digital information, while brains, our souls, and the stars utilize quantum analog information. If a soul were connected to the internet, they would instantly go insane, flooded with all forms of bundled fragmented information, beyond any comprehension. Consciously existing in the internet, would be a pure insanity. Scriptures often say and intimate the forceful souls of Holy Angels, are entangled to the stars of Heaven. Judges 5:20, "*THEY FOUGHT*

*FROM HEAVEN; THE STARS IN THEIR COURSES FOUGHT AGAINST SISERA.*" Dan. 8:10, "*AND IT WAXED GREAT, EVEN TO THE HOST OF HEAVEN; AND IT CAST DOWN SOME OF THE HOST AND OF THE STARS TO THE GROUND, AND STAMPED UPON THEM.*" Rev 12:4, "*AND HIS TAIL DREW THE THIRD PART OF THE STARS OF HEAVEN, AND DID CAST THEM TO THE EARTH:*" The forceful desires of angels being entangled to the stars, raises a question and an answer about the fallen Angels. Scripture says clearly, the SON came to the fallen world over 2000 years ago, and by his Baptism of obedient suffering and sacrifice, the feared determined end of evil, was set. Matt 11:12, "*AND FROM THE DAYS OF JOHN THE BAPTIST UNTIL NOW THE KINGDOM OF HEAVEN SUFFERETH VIOLENCE, AND THE VIOLENT TAKE IT BY FORCE.*" Rev 12:7-8, "*AND THERE WAS WAR IN HEAVEN: MICHAEL AND HIS ANGELS FOUGHT AGAINST THE DRAGON; AND THE DRAGON FOUGHT AND HIS ANGELS, AND PREVAILED NOT; NEITHER WAS THEIR PLACE FOUND ANY MORE IN HEAVEN.*" Scriptures say, this literal conscious, spiritual, and physical war, between the faithful and fallen Angels, has raged for over 2000 years, and is ongoing. It will continue until FATHER ends it at the Fifth Trumpet of the Tribulation. Is there evidence of this ongoing battle? Scriptures say when a fallen Angel, or star, is subdued, they are chained in utter darkness in a bottomless pit, falling forever and cut off completely from the outer aether of reality. We see this utterly dark fallen prison, perfectly describes a black hole star. There are now almost 100 *candidates* and confirmed Black Holes named by science. Never consider their estimates, only the facts. I don't know how many out of 100 are confirmed. But by an angel being imprisoned in their black cold shrinking awareness, and the aether of the outer physical existence being warped around them. It must have a negative affect that can draw its physical surroundings into its darkness. These angels inside can't see anything outside their prison boundary, because without a free aether, light does not move, and love cannot exist. Their dark hatred of GOD, is a bottomless pit of individual belief. There is this very interesting thing about light in Scripture that occurred in the days of Moses. There was a plague of darkness, sent for a moment by GOD; but only upon the followers of Pharaoh. The Israelites were not affected. And we see in CHRIST'S Revelation to John, this temporary plague comes again at the fifth vial, towards the very end of the Tribulation; and again, only upon those following the last godless kingdom. There are many GOD sent events in the Old Testament, that have their mirrored events in the last days Prophecy. This first plague of darkness on Pharos Kingdom, was said to be physically felt, causing great fear and distress to one's reality itself. This was all information of physical light, love, and life in the aether, being turned around and withheld from all godless individuals; yet not GOD'S Children next to them. This plague speaks not of physical blindness due to ocular change. This is where not one new aspect of light, love, or life's force, touches any evil person's body. For a moment, these souls experience reality at the very edge, as each soul became individual cold stumbling grasping human black holes. Scripture also says the Sun's surface, for *a short moment*, will raise from its current 10,000, to 70,000 degrees Fahrenheit. And will for that same short moment, scorch and darken the bodies of any lost souls who don't hide from it. How much this will affect the Sun's coronal radiation and solar wind upon the earth is up for study. The Sun has two great mysteries: its solar winds, being particles other than light, that *beyond* all physical laws, hit the earth at a million miles an hour.

And its corona, or outer atmosphere, which is *mysteriously* millions of degrees hotter than the sun's surface. Now Scripture declares GOD in ways that can appear circular to a linear mindset, because they are spherical, and eternal. Circular, being the start of everything new, releasing from the end of everything old, and so on. Some may ask is GOD an irresistible force and immovable object. Well, all creations combined cannot move FATHER, as only sacrificial eternal love, lowers his gaze to take notice of our good desires. *We* all started at our unknowing gifted point of almost zero understanding and belief. From there we have all gone in our chosen directions of desire. I recall in fifth grade, a liberal teacher who tried to *indoctrinate* the class on the wonder of being individuals, using her out of country vacation. Saying we should seek and embrace a variety of sights and every differing cultural experience. Now I use the one-sided term indoctrination, because she gave no example of the evil contrary affects that arise from living life that way. In our current moment of choosing between good and evil, surrounded by chance, remember that everything that exists here is a two-edged sword. Everything, for yet a moment, still cuts both ways, and from both sides. That teacher told us that to experience such varieties, was the joy that gave life meaning and worth. I strongly sensed something imperfect in her words, yet they felt enticing. I did later in seeing this world, come to know that what she said was not true. And that living as individuals without barriers, brought all strife, envy, and every war. Living a worldly way divides people from all Godly love for each other. And without eternal grounding Godly beliefs, we are no more than dust in untamed winds. That teacher's belief raises divisions of desires between those who have, and have not experienced these things. Her beliefs were more than simple desires for physical experiences, they were her desires for abundant physical pleasures, security, beauty, comfort, wealth, power, and her pride in bragging about her travel experience. Desired physical experiences in spiritual ignorance, at the last, leave only feelings of un-fulfillment in most who have had all these things, and resentment and depression in most who desired, yet never had them. Through my existence, I came to see that teacher's temporary joy in experiencing the things of this world, was illusory. I came to see physical possessions and experiences were her master. In their abundance, all such experiences build the walls of a personal prison of desire and dependence for more. Any long-continued exercising of such physical experiences, eventually leaves one with a worn diminished desire for the same. I see all one-sided arguments as forms of indoctrination. I managed at a naïve age to reject that teacher's ideology out of my shear desire to do so. If you now ask why *I* do not detail both sides of *every* point in this book? Well, I do continually point out there are *always* contrary sides to all things. And I know souls can't always give an alternate view to everything said; but in school, you rarely hear an alternative life view given. I know the pure truth of life can never be found down that teacher's path. I wonder, might I have become forever confused if I had believed her. I believe in that moment, in error she believed she was right, her desire for GOD was negligible, and her desire to turn each child towards worldly physical experiences was strong. Every physical action and consequence of life I have ever seen, was consciously driven by desires. If you feel you are detecting any political faith in me, I say no. I see the lesser of two evils, but socialist, marxist, and liberals are the greater evil. I hold no faith or hope that any government formed in this world, can bring world peace. World Governments are driven by mankind's desires, confusion, pride, and greed.

Forms of Governance are not evil in themselves, the evil lies in centralized power, privilege, possessions, and the governing and governed people's desires. I love and vow to follow the King of Heaven; so, for government, I am a Spiritual Monarchist. In this world, all Godly Scriptural instruction is the only governance that can hold back hell's confusion. Scriptural values and truth are lost everywhere; and if you focus on *their* evil, *all* countries have enough atrocities in their past to turn the head and stomach of us all. But when we look at the loving charity and inspiration given by each country, Israel and the U.S. are at the top. And for governance that allows religious liberty, reduces poverty, and conscious misery, these two shine bright. From within and without, the endless demagogues railing at Israel and the U.S., exaggerate, amplify, and focus only on the imperfections that exist equally, or more so in every *other* country and form of government. Look at Russia whose poor are poorer than ours; yet Vladimir Putin who came to power of modest means, now has hundreds of billions. Venezuela's socialist leader came in lacking, but is now worth a billion. Communist Cuba, whose people are very poor, yet the Castro's are worth about a billion. Kim Jong-Un of North Korea, who executed his own family who did not satisfy his needs, lives in splendor, taking his country's young women as he desires. China is a godless evil trying to control the whole world, and stop all Godly religion. All the radical Muslim countries, want to control and kill all who do not conform to their belief. Europe is mired in zombie liberal multi-cultural fighting. Look at all the current slavery, genocide, and corruption, in African Countries. Systemic rampant Baltic corruption, and so much more. No liberals here in the U.S. focus on those current greater horrors. All the great evil forces in this world, focus on and exaggerate the imperfections of the U.S. and Israel. GOD hates this and all who do it. The Physical Sciences and Materialists, are all in political concert with secular government and all secular funding. The sciences have always been right up there in corruption with Government, the Media, Education, and Religion. It has never been the big four, it has always been the big five in world politics. And since liberals control the institutions and money for research that acquire most experimental data, they feel they alone hold a patent to interpret that data for us all. Yet these theorists forever bicker and debate between each other about their own interpretations. When asked about their internal divisions, they claim to hold an unbiased scientific pureness of heart and mind. I have listened to many souls across the entire spectrum of knowledge, and find a firm declared *dis*belief in any god, to be a minority across that spectrum. But it is not because any are learned or unlearned. It is their desires towards GOD that tip their scale of evidence this way or that. It is never the evidence alone, but always the finger of desire on the scale that tips the balance. All souls must believe what they desire is good, even if they are evil. How does one easily accept they are wrong and evil; they must be convinced. Now the path of desire in some can still be changed as they find in a convincing measure, a new understanding they truly desire. Imagine a person searching in great desire for a huge diamond in a great mountain of sand. But let's add that this strange sand, unknowingly to them can give eternal joyful life. They will never in joy, find the sand that slips through their fingers, being thrown aside, as they hunt a valueless gem. Evil perspectives and desires blind souls to eternal truth and value. FATHER'S first greatest gifted soul, was solely for finding a reasoned purpose and hope of freely desired choice of love. If we look deep enough into the underlying reason and purpose, we see FATHER seeks all who agree in his eternal loving spiritual physical perfection. These loving senses, liberties, restrictions, and ways of eternal spiritual life, are *not* desired by most souls; so again, eternal love is rare and costly. There is no way-round this painful fact within any shared reality of free-will choice and a defined love. Now across the millennia, mankind has offered so many physical theories that define the physical reality alone, but they all denied where and what we consciously are, and why and how we are consciously here. Most souls *want* to believe we are only physical, and here to physically sense, enjoy, and interact with this world. And nothing short of shared omniscience, can concur such belief. We aspire to much more than can be sensed through our body-interface, being the conscious spiritual forceful mystery of all true realities. What is the source of force on the physical side, name it? There is none, it is all an illusion. When you think of yourself, do you only see yourself in the mirror? Do you think of the body as yourself, do you feel you are your brain and body? Or do you know you are an eternal soul, loosely aligned to perceive things through the information sensed by a now temporary dying, physical interfacing body? If you knew your self-awareness had been transferred into a *sensing*, consciously controlled android, that looked and felt like a body, would you still see that new body in the mirror as yourself? If we perceive the body-brain interface to be who we are, we are enslaved to serve it, as we must believe it is our existence. Look into a mirror and know that what you see, has no control over your soul that you do not first relinquish. Physical desired sensations are powerful under the fear of physical death and losing all worldly love. And knowing the true conscious realities of Dualism, brings us to a scary uncomfortable reality most don't care to know. Being that *all* we are in each moment, eternally given us by GOD as ours, is our eternal spatial spherical estate of free-willed desired belief and awareness, existing only through our projected patterns of forceful desire into the aether to create the particle/wave perceptions we indwell to experience our existence. And we can never be more or less than this, here, or in another world. If you think the body you sense, is yours, then keep it for eternity. You can't, it is not yours; but your soul is yours forever. Something on loan, or lease, means temporary control, not ownership. This body is not ours, never was, never will be. I see physical death as the loss of my first loaned physical operating platform. And for lack of a better term, at that loss, my created awareness will be *temporarily* closed, awaiting my next upgraded operating platform. And for the truly saved, their good created perceptions of Godly understanding remain in the aether to be known to us through our new platform. But until our soul's separation from this body, we strongly sense our existence in this realm; a realm mankind has also assumed to be theirs to define in their own image of evil decisions and profoundly confused beliefs. Considering all things, we see FATHER can provide unto himself all things but two. And *we* were, and are, gifted and created in the hope of our providing these things he desires. And for GOD, all things coming from us but these two, have no eternal value. If GOD should ask of us these two things they need, work, and suffer to find; being their *giving* and *receiving* of faithful responsible eternal love, through our responsible faithful reverent observant alignment. These being the only two things we can offer that are eternally pleasing to all souls, including GOD above. Do you then offer GOD a worldly misaligned contrary love instead? Or something physical they have always had, which is already theirs, and not ours, and is now corrupted? And should GOD be pleased, because we thought our worldly physical gift would please them, because it pleased us to give it? If we offer to GOD such things, do we stand before GOD in vain fruitless religion, just as Cain did? Cain's gift was not given in responsible faithful love, but given as his best physical tribute. While Abel chose to give *his* best, in his best love, to whom he loved most. GOD sees not the gift, but the eternal love with which it was given. See reality from the perspective of GOD who is investing so much in a search for *eternal* love. The things GOD cannot have alone, the things GOD cannot make alone, the thing GOD *needs* and desires is an *eternal* loving family. I often have to dry my eyes when writing of GOD'S endless pain of unmet need. Do you think of him as *a* God, or as your loving FATHER? Should not our loving FATHER have the loving righteous holy child he longs to find, treasure, protect, and give all good things? Does GOD not deserve the loving family that would please them? Or does GOD deserve a family of unfaithful sinful children they must work endlessly to sustain in their contrary waring ways, receiving only endless disobedience and dying love? The Law of *Eternal* Flowing Love states each giver and receiver of *eternal* love, must be *eternally* good, or that love must eventually fail. There can only be one common eternal way, there can never be a give and take of contrary ways between billions of differing souls. This fallen world is drowning in all such ways of hell and death, it is selfish, depraved, and wicked towards GOD. Can we consider the coming day that GOD stops allowing all new evil? That FATHER must bring an eternal separation of all who reject his eternal ways and love. If FATHER sees a soul has *eternally* rejected his ways; yet that soul says I have not *eternally* rejected his ways; who is right? What has occurred, is this lost soul has redefined in their belief, what GOD wants and believes. This soul wants FATHER to accept the way they want to accept him, and allow *them* to define him to be as they desire. The omniscience of FATHER is right in all things, it cannot fail. If I search and strive for GOD my whole life on earth, and at my Judgment, GOD tells me to depart, I missed the mark. I know beyond doubt, GOD is eternally loving and true, and it was *I* who failed, and my love was not eternal. When we look deep into this creation, down at the cause and reason, we find we are gifted and created to choose between GOD'S ways, or our own ways. GOD is the cause, and this is our free-will reason. What powerful influences the many sensed forms of physical information can have over us if we ever forget their place, purpose, and true value. We must rule over them, and never be ruled by them; ever. Or we have placed and loved the Creation above its Creator. Many only know the things they see, hear, taste, smell, touch, and are touched by, and these rule their life. From a fear and agony of being hurt and rejected, to comfort and ecstasy, it is by these physical senses that most are blinded to GOD and are slaves to a fallen world. It is all amazing, fragile, and tempting to the flesh and the free-will soul, to say play a god if you want, go, enjoy, build, consume, hurt, kill, and destroy. Feel and desire all ruling physical sensations, forget FATHER'S responsible faithful loving ways and commands. This while beyond the universe, FATHER'S will and love for it all fades. A desire to control all the ways of our life and be an individual, is the true root of all evil. And in this fallen world, the power and ability to have this in all its grand splendor of privilege, power, and possessions, is found through wealth, being the love of money. Wealth is no more than having all the available physical securities, privileges, comforts, and powers, one desires. By desiring more than is given, we see eternal love was destroyed in a third of the Angels, and is destroyed in most all of mankind on earth. This, for all eternal reason and purpose is for a short moment, left beyond the will of GOD. Now the sum of true omniscience is as follows: we see FATHER knows who he is, and he alone knows the unbearable loneliness of once being the only soul to exist. FATHER knows his control, his ability, and the detailed once perfect structure of all he created. FATHER knows what is eternally good by his desires, and that only death and divisiveness exist beyond his desires. FATHER knows contrary beliefs bring divisions that destroy souls and reality. FATHER knows his desires and needs for family, and every detail of reality existing under his measured projected forces. He knows every qualifiable and quantifiable aspect of all the physical forms of life he sustains, and life's very reason and purpose. FATHER knows that by vreating chance and gifting free-will choice, that in each moment beyond his sight, there exists the near endless bounds of evil physical desires, thoughts, and actions of hell and death by all evil souls. FATHER knows the strengths of all his ordained and given willful forces. He knows the high place, rank, and honor he has given his SON and HOLYSPIRIT, and Holy Angels and Mankind, all whom he has set over his creations. FATHER knows his coming final plans and what they bring at the last. So, in that this is all that can or does matter in FATHER'S eternal picture, this is then the omniscience of FATHER we are called to learn and live by. Omniscience stops at the outermost edge of all created perfection. And just beyond FATHER'S omniscience, we see endless individual imperfect free-will rebellion and confusion. One might ask, who could not desire the life that comes in knowing the truth and love of GOD, and embrace it? First, we have to look at the cost of GOD'S truth, for it is the cost they reject, not the reward. Now *we* do not hold a worldly understanding of the word truth, that *most* souls are accustomed to. Truth to most is something they tell, hear, or believe. Yet the pure truth of reality is a created way of life, to be faithfully lived in our desired responsibility without fail. FATHER'S omniscient truth, defines our faithful responsibility to help sustain a gifted created eternal life. Truth is in every point, contrary to this world's endlessly changing individual beliefs. What does our living in GOD'S truth cost? Simple, we leave behind every false individual belief, and every desire and act of imperfection; and this cost starts now by choice, or it is imposed later in our total loss. This cost is not just telling and knowing the truth forever, but living the truth now and forever. But rejecting or forestalling the pure truth, has a far greater cost. We need eternal sight, and not a timely temporary sight. Eternal sight sees all temporal things of this world, as the fading illusions they are, knowing they will soon be gone forever. Seeing every powerful *physical* emotion for *this* world, every strong desire for *this* world, and every selfish *physical* pleasure *of this* world, are all temporary and void of eternal value or meaning. What would such a conscious turn lead you to? Do you see this bringing the complete loss of all you physically desire and know? Do you see it as an unwanted way of eternal life, lacking all things that physically please you? Then you are a strong *physical* creature, and living the *spiritual* truth of eternal life with GOD is currently not for you in it being misaligned with your current beliefs; so, work very hard to change your beliefs. The temporary illusion, may be your current great desire above GOD and spiritual life; yet souls here and there repent and find GOD does become their greatest desire in this world. But that is not our final greatest experience, for we can't comprehend the *spiritual* elation awaiting us on the other side. For me, my glimpse of the new has tainted everything here. Every once bright color on this side turns to dullness. All desires for this side sound like growing confusion, and having a smell of hell, death, and destruction. On this side we see a world filled with rotten covetous, pride, perversion, and the endless stress of division, sickness, and insanity. If you see this, and want to leave this horrid world for the spiritual; you must surrender now, sight unseen. We ask *you* to consider, see, and accept, that living here and now through love and faith in GOD'S truthful ways, is our only conveyance from this life to the next. It requires we turn to, and emulating CHRIST. We are promised in Scripture, that in choosing CHRIST'S pure life, our eternal joy and happiness beyond understanding comes at the last. Who can and do *you* trust? I trust only GOD'S promise to their Children. Yet we do not find mankind lining up around the block, as the world seems strongly repulsed at this offer. Which is not just the offer of a lifetime, but the deal of eternity. Through all worldly politics, science, religion, and desire, most form their individual variations of truth, all having their greater degrees of untruth. The cost of eternal life is losing our evil and all we now temporarily are physically. But our cost of refusal is giving up everything we will ever truly have or be for all eternity with a loving GOD. Do we want the now, or the eternal? Don't say both. Sadly, free-will dictates that we will either desire eternal peaceful love above all things, at its good reasoned cost. Or we won't. As to the one pure truth of GOD, there is nothing new to be told or heard, as it eternal and perfectly remains. It has been said and shown to us; so, hear and see it, and be found. Since the fall of mankind, a pure good desire is no longer there at first within us, it was lost long ago. In this world, GOD'S truth must be sought externally. And if spoken in this world, it must be CHRIST'S word for word true *inflection*, through HOLYSPIRIT. If not, it has been misinterpreted and nuanced by a fallen soul's individual desired bias. We see at reality's beginning, the omniscient truth was shared daily with all, as it grew in Heaven and the Garden of GOD. And GOD in that garden, did set out the one tiny line of choice for mankind to believe and live in GOD'S omniscient truth. Being to avoid any belief or desire to choose the fruit of one certain tree. Meaning this choice was set between remaining and living as one in the love, life, and truth of GOD. Or choosing their own belief, and thereby become an individual. This was set in front of them as the only *choice* to avoid, in a vast garden filled with the many perfect choices of truthful eternal life. To choose this one tiny denied choice, to turn inward to a new divisive divided individual understanding, was to rebel against the perfect continuity of GOD'S perceived reality. On one side, the choice of *not* eating the fruit of that one tree, was not just believing GOD, but living, loving, and trusting the truth of GOD'S Love. On the other side, in *eating* that tree's fruit, was the confused turning to believe one's self, or another; thereby not living GOD'S created truth. So, in seeing that fruit as good for them, *after* GOD said it was bad for them, was their new individual belief that GOD had lied to them. In believing that, they became individuals who knew the conscious confusion of doing, becoming, and now being evil (meaning contrary). This brought them separation from their conscious oneness with GOD. The anointed Cherub Lucifer, was the first to deceive himself and conceived in himself the first lust and lie ever; and he quickly spread his confusion of plural beliefs. So, he is called in Scripture, the *father* of all lies; meaning the creator of false beliefs. Lucifer did subtly and politically, in partial truth mixed with tempting lies, tell Eve that if she did not believe GOD'S warnings, her eyes would be opened, and she would be like GOD, knowing the difference between good and evil. Before considering Lucifer's deceptively worded choice, Eve had always lived true, never doubting or disobeying what she had been taught by GOD. But in that moment, by her diminishing loving trust and fear of losing GOD, she did believe GOD deceived her. She believed her first confusion of hell's individual desires, thinking GOD was keeping something good from her; and she ate. We now see by the choices for sin growing to reign in this world, GOD and his love are pushed out further and farther. Yet for a short moment, GOD provides a merciful way out of hell's confusion and death's grip upon any who want his ways. But this gate soon closes, as the entire creation is in a death spiral of its final free-will choices for the vile and abominable. Most souls are now foregone and destined by their own forged ardent dark desires within. Scripture from its first word to its last, says many times to seek, hold tight, and live by GOD'S pure omniscient commands, and do *not* lean upon our own understanding. 1st John 1:6, "*IF WE SAY THAT WE HAVE FELLOWSHIP WITH HIM, AND WALK IN DARKNESS, WE LIE, AND DO NOT THE TRUTH.*" We lie, not only when we do not know or tell the truth. But if we do not *do* the truth. 1st John, 2:4-6, "*HE THAT SAITH, I KNOW HIM, AND KEEPETH NOT HIS COMMANDMENTS, IS A LIAR, AND THE TRUTH IS NOT IN HIM. BUT WHOSO KEEPETH HIS WORD, IN*

*HIM VERILY IS THE LOVE OF* ***GOD*** *PERFECTED: HEREBY KNOW WE THAT WE ARE IN HIM*." 1st John 2:15-17, "*LOVE NOT THE WORLD, NEITHER THE THINGS THAT ARE IN THE WORLD. IF ANY MAN LOVE THE WORLD, THE LOVE OF THE* ***FATHER*** *IS NOT IN HIM. FOR ALL THAT IS IN THE WORLD, THE LUST OF THE FLESH, AND THE LUST OF THE EYES, AND THE PRIDE OF LIFE, IS NOT OF THE* ***FATHER****, BUT IS OF THE WORLD. AND THE WORLD PASSETH AWAY, AND THE LUST THEREOF: BUT HE THAT DOETH THE WILL OF* ***GOD*** *ABIDETH FOR EVER*." 1st John 3:2-9, "*BELOVED, NOW ARE WE THE SONS OF* ***GOD****, AND IT DOTH NOT YET APPEAR WHAT WE SHALL BE: BUT WE KNOW THAT, WHEN HE SHALL APPEAR, WE SHALL BE LIKE HIM; FOR WE SHALL SEE HIM AS HE IS. AND EVERY MAN THAT HATH THIS HOPE IN HIM PURIFIETH HIMSELF, EVEN AS HE IS PURE. WHOSO COMMITETH SIN TRANSGRESSETH THE LAW: FOR SIN IS THE TRANSGRESSION OF THE LAW. AND YE KNOW THAT HE WAS MANIFEST TO TAKE AWAY OUR SINS; AND IN HIM IS NO SIN. WHOSO ABIDETH IN HIM SINNETH NOT: WHOSO SINNETH HATH NOT SEEN*

*HIM, NEITHER KNOWN HIM. LITTLE CHILDREN, LET NO MAN DECEIVE YOU: HE THAT DOETH RIGHTEOUSNESS IS RIGHTEOUS, EVEN AS HE IS RIGHTEOUS. HE THAT COMMITETH SIN IS OF THE DEVIL; FOR THE DEVIL SINNETH FROM THE BEGINNING. FOR THIS PURPOSE THE* ***SON*** *OF* ***GOD*** *WAS MANIFESTED, THAT HE MIGHT DESTROY THE WORKS OF THE DEVIL. WHOSOEVER IS BORN OF* ***GOD*** *DOTH NOT COMMIT SIN; FOR HIS SEED REMAINETH IN HIM: AND HE CANNOT SIN, BECAUSE HE IS BORN OF* ***GOD***." 2nd John 1:1-4, "*THE ELDER UNTO THE ELECT LADY AND HER CHILDREN, WHOM I LOVE IN THE TRUTH; AND NOT I ONLY, BUT ALSO THEY THAT HAVE KNOWN THE TRUTH; FOR THE TRUTHS SAKE, WHICH DWELLETH IN US, AND SHALL BE WITH US FOR EVER, GRACE BE WITH YOU, MERCY, AND PEACE, FROM* ***GOD*** *THE* ***FATHER****, AND FROM THE* ***LORD JESUS CHRIST****, THE* ***SON*** *OF THE* ***FATHER****, IN TRUTH AND LOVE. I REJOICE GREATLY THAT I FOUND OF THY CHILDREN WALKING IN TRUTH, AS WE HAVE RECEIVED A COMMANDMENT FROM THE* ***FATHER***." Walking in GOD'S truth, is only in living

GOD'S eternal ways. The omniscient eternal ways of life, are spiritually too wide, and physically too narrow for most souls to desire and live by. Yet GOD'S truth and commands, by their nature, call *us* to our lively adherence. And still, many souls hope GOD will draw *his* line on the far side of *their* worldly physical ways. Scriptures describe a fiery lake and its purpose. It being a secure complete eternal separation and confinement of lost souls away from the Children of GOD. By the whole of Scripture through HOLYSPIRIT, we find this lake of fire is a never-ending place and chasm of separation, whose immense gravity resists most physical movement. All lost souls there, are surrounded by this confining lake of fire, filled with all the evil the souls therein, did themselves cause and create. Please see that only by the pure truth and love of GOD, can all things exist eternally. So, GOD'S truth and love *must* be seen and valued as the highest beauty in all things. If our responsible accountability to GOD'S love and truth were not a prerequisite, that would deny the pure eternal value of those greatest wonders. We all know that societal offence, means a perception of hurt has been felt by society as a whole. But by many minority beliefs banding together as a political force, each little thing that offends individuals or small groups is to them, a wound to be addressed. For many, the line of punishable offense, is insanely personal. A Godly society, follows the Scriptural ways of love, so as not to offend GOD. And in GOD setting his morals over us, there must be the threat of accountability for betrayals. GOD'S ideals and standards must be lived by all, before a shared eternal peace can exist. In eternity, there must be the threat of complete loss, to stop each and every act of imperfection. Imperfection always spreads, because weakness now spreads easier and faster than strength. Why? Because the *choice* for GOD and truth is weak in most souls, and their choice for self and confusion is strong. For a Godly choice you must fight, but for the other you simply surrender and go along. We must understand responsibility, accountability, the threat of punishment, and actual punishment. We know evil must be imprisoned to protect the good. The great repulsion mankind has for GOD'S Lake of Fire, comes from a sense of unbearable pain in being burnt; but this searing physical pain will not happen. Scriptures describe an angel's *ethereal* body's experience of pain or pleasure, as being on a negligible order. This raises that desire *fallen* Angels have towards the earth and flesh, to experience the very strong feelings of physical pleasure through the dense powerful information sensed from a human body. Angels strongly sense the spiritual realm, but their physical sense is negligible. So, this is how we must understand these things. What could this lake of separation for all destructive souls be like? What should be the price for rejecting in all careless disdain, the created eternal gifts offered by GOD. Scripture says this lake of fire, is a dry thirsty place. The word thirst, as most things in Scripture, refers more to the soul than the body. Though here, it means both, a strong conscious, *and* dry physical thirst. A Godly soul is said to thirst for truth, love, and omniscience to understand and enjoy eternal life. But in the belief of timely temporary life, it is the physical sensations most thirst for. In our wanting to sustain eternal life, all physical things are seen in a different way. Scripture refers to the physical waters on the earth, saying that if we drink from these, we will thirst again. Then we read of living waters, described as bringing the indwelled truth of GOD. Saying if we drink of these, they bring all purpose and reason for wanting and living life, and we will never again thirst, as they will be a spring welling up inside us. We know physical water stops the destruction of physical flames, being that greatest physical misery; but physical waters cannot end conscious misery. In Scripture we see this lake of fire is a dry thirsty place, void of both physical and living spiritual waters, as souls there exist in complete loss of understanding all things. And to each lost soul in this lake, by their *own* measures of chosen desired hell of confused beliefs, comes their differing deserved eternal sentence. They are void of truth, set in an individual prison of evil confused denials of forever wanting but never having. Eternally denied every physical expression and satisfying feeling for all the evil things they desire. And haunted and taunted by their memories and desires for the same. This lake is a place where all un-forgiven created perceptions of evil can be confined. A place filled with every image and desire of every evil conscious creation of perception ever conceived by all those there. Yet lacking all acts of fulfillment and pleasure. So, their memories of sin and pleasure forever tempt and torment each soul by their own individual confusion. So, to differing degrees, this is a place of never-ending misery, anguish, loss, and regret. Souls forever descending into a nightmare of not understanding where they are, or why they are there. Now through the eternal perspective, we ask why an interfacing body and aether, are pre-requisites to our shared existence. Well, in pure Dualism, we see a soul *and* body are the equation of FATHER, in having both his forceful soul of awareness, adjoined to his body of physical singularities allowing his created particle/wave-based perceptions. All awareness of reality has eternally existed this way. And by this eternal truth, all who now exist by FATHER'S gift of a soul, must exist in the same dualistic way. Seeing our created physical perceptions, and our soul's awareness of them, form the required and acquired sum of our own conscious existence. For GOD to provide a loving *eternal* reality for many souls, requires an *eternal* way for many souls to exist together within that *one ex*ternal shared reality. All laws of shared eternal peace, joy, and love, are born of GOD, and set in omniscient sacrifice and faithful responsibility to those lawful ways of eternal love; as nothing less can endure eternal. Seeing souls increase all chaos as they fall into *individual* insanity, proves this point. Minus these oracles of eternal love, leaves eternal conflict and disorder. These ordained laws of eternal reality, after the endless denials of GOD, remain intact, exact, and as initially ordered. In this world, these laws are denied, abandoned, and loathed, in insanity. In CHRIST'S

New Heaven, they are eternally treasured, knowing perfect love equals eternal spiritual elation and peace. Scriptures reference the Angels and our LORD walking among the stones and flames of fire, not being burned or feeling pain. Also, Daniel's three Hebrew friends he called Shadrach, Meshach, and Abednego were cast into the fiery furnace, yet the flames did not harm or kindle upon them. Why? GOD in that moment, lifted the veil of his ultimate control. Not for those three, but to humble King Nebuchadnezzar. GOD altered their physical properties to not allow the flames to affect a single cell of their body or clothing. Now we must ask, what is the *eternal* value of a lost reprobate soul? What is their eternal value to the whole of GOD'S creation and family? What is the good eternal purpose that remains in a soul, minus their natural fear and eternal love for GOD? Their only remaining purpose, is to spread the confusion of hell and destroy all creations. And to GOD, being all light love and life, each reprobate soul, becomes the value of a virus. Each soul's *eternal* value is qualified and quantified by our strength of desire towards our shared loving *eternal* purpose. Is this not our highest essence of being sentient and eternal? Is life's purpose for the many, now about taking, consuming, and feeding off the life's force of others, which in the end, kills all? Is a reprobate mind, a desirous predator against every eternal purpose, having relentless physical desires? Such souls, by Holy Decree, are destined to be forever trapped in all their own same unmet physical desires. Reprobates, while here, deny GOD in all ways and beliefs. At their end here, Scripture says their conscious has been seared, as if by a hot iron. Meaning their minds are scarred over, and numb to all truth. They are permanently mired in their confusion of hell, against ever being one in the truth of life they were created in hope to be. What percent of souls alive today, are reprobates. We can't know a number, but it is certainly a majority. We witness a fast-growing spiritual coldness in the earth, not just towards GOD, but love and life itself. Mankind is now the highest order of predatory animals. The further from GOD a soul goes, the more perverse and destructive their desires must become. Many at the end of physical life, would fearfully draw the last breath from the lungs of their own newborn child for another second of insanity. There being within them, no love for GOD or another, greater than their unrecognized love of self, and fear of death. All reprobates are vacant empty dry souls, seeking a physical way out. But there is no physical way out of this spiritual reality. How else do we explain that so many, not having any eternal love, trade away everything good that is to come, for what has always amounted to a current fading physical echo. At this point of anguished thoughts, we must define eternal *worth*. The only *eternal* worth a soul can have, is in our eternal responsible faithful love for GOD, each other, and the created eternal ways. There is our eternal spiritual worth, and our lesser required physical worth. Remember GOD *is* love, so, the highest form of love is eternal, holy, and righteous. But by fallen mankind, love is now defined as temporary, irresponsible, unfaithful, and physical, which GOD calls lust (meaning fallen fatal love). All lust comes by evil free-will desirous choice. Faithful responsible eternal love is the only good, and is also currently the most painful costly desire to find and keep for GOD and all who hold this eternal hope. So, if you see this fading world as something that really never was, you do well. For the destiny of this world's ways, and all its things temporary, is to be forgotten. Who being a believer, blinds themselves, and says we can never stop sinning, and sees ongoing sin as GOD'S standard. If it is a sin to dash a stone upon the head of a newborn, and you say you could never do that, and the very thought scares you. Ok, but why? Is it because <u>you</u> define it as murder, loss, destruction, and pain, felt in an innocent awareness that can't defend itself? Or because it is defined by GOD as sin? Now we see in this world, many are capable of this horrific act, being equal to abortion. How can anyone ever do such things, except they are blind and numb to the pain and destruction they bring to others and GOD. Is it our repulsion and aversion, found in our understanding the destruction and pain we each bring to love, truth, and life, that causes us to *not sin* (meaning to *not* destroy). Why not let your eyes and ears be opened by GOD to not just your great sin, but every tiny pain and destruction brought to love, truth, and life, by your every lesser sinful offense? I ask you again, if you say you can't stop sinning; why? Why would you hurt and destroy any good thing, even a tiny bit? Do you desire to sin because you are still as Eve and Adam, now blinded to the pain, hurt, confusion, and destruction that one tiny sin will bring to the creation of GOD? A simple impure belief in GOD will not yield the needed strength to seek, find, and obey the pure truth. This requires more, being an *eternal* loving fear of losing GOD'S family and creation. Remember the endless destruction that Adam's one tiny sin introduced into it all.

GOD’S law defines sin according to his omniscient love. Do you know better than GOD? Most believers in GOD, believe in them to a lesser extent, and minimized GOD’S truth concerning the way they live their lives. They see *a* god, who without effort or ongoing forceful work, just wishes all physical things to be eternal like a cartoon genie in a lamp. Such belief, denies FATHER’S eternal labors and long suffering to find and secure eternal love. All good work is now seen contrasted against mankind’s increasing destruction of it all. Why did GOD say he rests every Seventh Day from all his labors. It is hard for deniers and weak believers to see FATHER’S never-ending work to sustain reality. But *we* see FATHER is far more real and knowable in soul and body than most believe. So, please see that for GOD, like us, unnecessary repetitive work is unpleasant. Weak believers deny the full picture that says each sin brings damage, needing to be forcefully fixed. So, souls reject the perfect path back, which rests solely in our loving obedience to GOD. We know GOD needs to greatly *rest* from his labor, and just enjoy communion with “his family” in eternal love. If your child abusively increases your work to sustain your family and home, you feel abused and unloved by their rebellion. Yet you hope they open their eyes before the last chance for love between you both is gone. We know we are made in GOD'S ways and image, and in the sufficient measure of our soul to have an eternal loving intimate perfect relationship with them. This declares the great heights and measures we can know and attain *with* GOD. So, the size and properties of the body/brain interface we now sense this physical reality through, reveals nothing of our ability to grow or know. Can we consciously grow to learn and live at the lower end of a holy angelic level while still in this world? Yes. We are called to climb out of this world, to meet GOD where they are. What if we had new ethereal body-interfaces, capable of formatting inter-soul emotions at many times our current levels, like the Angels. After the Judgment, we all receive a new less physical body. And with GOD we will share one faith of understanding with all souls around us. Together we will watch GOD perceive and create their New Kingdom and our home, right before our eyes. This will elate, elevate, and validate our faith, in ways we can't imagine. Our force projections creating our perceptions, is the way and truth of our existing. And this fact does not belong in the Philosophical and Theological arenas. If *unprovable* theories like Multi-verse, Quantum gravity,

Abiogenesis, and others are a physical science, then forceful created particle-wave based perceptions, is a true physical theory of science. GOD'S conscious force model predicts every observed aspect, and GOD'S faster Laws Principles and truths are within a certain eventual proof, and are currently beyond disproving. Making it as valid a theory as theirs, under their current standards. Physical relativity is entangled to all creative conscious relativity, regardless of denials. We expand here again on the vast quantum ethereal existence, in that all that has been said to this point, might leave an impression it is mostly a realm of conscious things. Yet out of the aether, entangled to exist literally everywhere, there is a created ethereal yet substantive reality that dwarfs in size, this firmer reality we strongly sense. There is GOD'S vast ethereal Kingdom and its boundaries, filled with immense ethereal Thrones, orders of bodies, dwellings, and vast souls experiencing reality under more ethereal laws and physical consistency. Imagine *not* being increased in mass or weight, but having our exact current body and soul inflated in a relative and entangled way, to be the size of the Sun, and to exist under different laws and principles. We would experience reality in a far faster, larger, spiritual, ethereal way, and far less a physical way. And we will need a faster light particle to see. Scripture says the greatest in Heaven is a servant or provider of all. It also says the last shall be first, and the first shall be last; these are far more faithful responsible spiritual consideration than physical considerations. We see many in this fallen world, want to be closer to GOD, but do not want to change *all* their desired physical ways. They can't stop all sinning, so they turn to a doctrine that allows for their lesser sins, or they give up and fall away beyond reach. A few may fall *far* from GOD, yet turn to be led back, while others may fall less and fail to come all the way back. We see not one *physical* thing was created to satisfy our eternal love, as GOD is to be that greatest provision. If we love GOD above all creations, in their returned love, GOD cannot fail to fulfill our spiritual needs to over-flowing. The great extent of GOD'S sacrificial love will be revealed to all in that great day when eternity reigns over their children. No one can truly or fully define the physical reality, minus our conscious reality; nor the conscious, minus the physical. GOD is bicameral, existing consciously and physically. So, I hope you see the truth that reality, like a photon, cannot be explained in any lesser way than Dualism, and all our reality comes by this marriage. Our eternal joyful existence rests in loving and understanding the realities of GOD, as each step away from GOD'S truth, is a step into insanity, being detached from reality and truth. Most souls can't see GOD'S truth is the glue that sustains all things, including sanity. Most souls see sanity as a worldly subjective consensus of belief. Strange and ever-changing is *this* world's sanity, and definitions of all things. To live the truth of eternal life, we must examine and define a few words that describe our true Spirituality. The first is *faith*, two is *faithful*, three is *faithfulness*, four is *unfaithfulness*. In GOD'S Family, we are talking *only* about faithfulness in following and living the eternal loving ways of GOD. FATHER refers to the Tribes of Israel and Judah, as being a harlot whenever they turn from living his ways of life. A spiritual harlot is one *unfaithful* to GOD'S ways, so being unfaithful. GOD also says many souls do religiously honor and confess him with their mouths, but their hearts desire and ways are far from theirs. Let's use Scripture and eternity, minus any slow small worldly view, to spiritually define these four terms. Faithfulness means a determined way of living, an action of a rare love, revealed in a passionate way, not founded in spoken words and beliefs alone. The exhibited *faithfulness* of CHRIST towards FATHER, is in CHRIST standing beside him, unwavering, unfailing, through all works and all things throughout all eternity. To *never* ever once leave, to *never* once hurt, or turn from, or offend, or deny. To *forever* be with and remain as one with. To *never* turn to another as being one's first love for selfish cause, want, or pleasure. Word two, unfaithfulness, which is to *say* with our lips we love someone above all, yet look upon and seek another also. To cling to one, but turn to another also. To declare love above all to one, yet turn to love another in a similar or greater way. To know something hurts the one you love, but do it anyway. To have a love for one that is too weak to stay true. To love and desire another so much, one cannot remain true. To be split in one's love with no pure allegiance. Unfaithfulness must also have a selfish willingness to deceive the one you love *and* yourself into believing you are faithful, while not being faithful. The unfaithful must convince their self that they truly love the one above all, when they just want multiple things that cannot be had in the same way, seeing two cannot be the first. Word three is faithful; this is an attribute of a soul who lives in their greatest regard without fail, according to all of those just described attributes of faithfulness. Word four is faith; and here we see the great problem in vain religious doctrines. All worldly religions caught up in deception and covetous, work hard to deny all unbreakable bonds. They move the word *faith,* further and farther from the meaning of faithful and faithfulness, and tie it to unfaithfulness. We see a modern definition of *faith* is now, in one's lack of desire to live in GOD'S ways, and one's *in*ability to be faithful, yet with one's mouth, to always declare words of belief, regret, and love as one fails to be faithful. To declare love for GOD with one's mouth, while then hurting and betraying GOD. This is modern lukewarm religious faith, and the just defined evidence of being unfaithful, in having abundant faith on their lips, but not in their life. What is the full Scriptural meaning? What is the substance of this word *faith* we are called to have as it pertains to GOD? By HOLYSPIRIT we receive the true interpretation of *faith* to be this: to emulate CHRIST, who loves FATHER so much he exhibits all the attributes of faithfulness, and remains eternally faithful. A true strong faith is a powerful driving force of loving convictions. The strength or weakness of our love and faith are the definers of our beliefs that steer our ways of life, bringing victory against every evil desire, or our defeat thereby. Scripture says GOD calls the true and *faithful*. One cannot be true and faithful, yet also false and unfaithful in the same soul. Though many in error claim they can. Based on all these inarguable facts just laid out, let's interpret the following statement; I believe with all my heart, GOD knows I love them more than *anything* in this world, but they accept that I can't quit sinning against them, because of my desires for this world. Do you see the contradiction and deception towards GOD and one's self in that statement? We can do this exercise, showing the same results, using the words: believe, belief, believer, and unbeliever. Now according to GOD'S truth and omniscient love, who exactly is evil in a moment? It is any soul who projects contrary imperfection into GOD'S perfection, turning his great perfect work of creation into something fallen, corrupt, and temporary. Evil is a conscious unmooring from GOD, where all find an infinite abyss of hell's never-ending disseminations of endless individual truths and beliefs. All such destructive forces, must in the end be sealed away in confinement at the fulfillment of GOD'S reasoned purposed plan. GOD does not desire shared eternal misery; and if GOD tried to be *eternally* merciful to all, he would be eternally merciful to none. We would see no Heaven and no Lake of Fire, just eternal waring beliefs. GOD calls all souls who violate their created ways, sinners. But mankind sees all these violations define their prized individuality. So, GOD'S love now exists at levels and heights this world cannot comprehend. What *eternal* value do you see in any soul who says truth is moving, evolving, and changing, and finds us each in its own way. That truth settles upon us all in different colors, sounds, textures, and emotions. We can only explain such a confused stance by saying that in the mind and flesh of the fallen, there is never a finding of truth, or wanting the same. They desire a search for *a* truth, to forever be before them, and never behind them. They want and need to search in a subjective unencumbered ever-changing way, for something they never want to find; which allows them total conscious freedom. You see, finding the one truth of reality, reveals all subjective ways of life are false. So, in their search, they need the truth to forever ebb and flow with their ever-changing desire. But for the omniscient truth of GOD, how might its perfection ever change? It can't. It *can* grow, but never change. GOD is far out in front and above all, being perfect, and there can be no Godly compromise or change towards lesser souls. GOD is the pinnacle of perfection who can never diminish or move towards imperfection. Even if the annihilation of all external reality occurred, it could not affect GOD. We know and trust GOD as being immutable, eternal, and yet merciful to all who return to them. GOD cries out over all deception, yet who can hear them. All creation, containing the choice of love and hate, now has a bi-polar aspect. The vainly religious believe GOD *can* do anything, and *will* do everything for them; all as they do less and less for GOD. We can say without offence that GOD can*not* tell a lie or sin. Meaning GOD can*not* say or do anything destructive to the good force of eternal life. This is a huge restriction against doing everything soul's desire. Isn't it? And GOD can *never* make any fee-willed soul love them or anybody else; another big restriction on GOD. GOD can*not* create up without down, big without small, or love without hate; a big restriction. FATHER created the bounds of his life sharing conscious forces, then gifted many free-will souls, which introduced the Uncertain Probability Principle. And for a moment GOD allows all antithetical free-will-destructive forces of evil and hate. Evil is defined as anything outside GOD'S good created ways. Again, Isaiah 45:7, "*I FORM THE LIGHT, AND CREATE DARKNESS: I MAKE PEACE, AND*

*CREATE EVIL: I THE* ***LORD*** *DO ALL THESE THINGS.*" There are GOD'S internal *and* external created principles. GOD'S internal creat*ive* principle applies to their persons. And FATHER'S external principles controlled *his* perfect external creation until fallen free-will souls brought their waring choices to live by their own slow small contrary physical observational frames of reference. No external individual principles, can ever affect FATHER'S internal person. This may sound a bit circular, yet it is true. There is the one mind of GOD, who cannot be a different GOD for all different people. This too is a huge restriction. FATHER is real and defined, being who and what he is, period. And our eternal love for who and all he is, must be our life and light given to him of our own desire. No one can be forced to ever love anyone. Yes, a huge restriction. But where FATHER'S external restrictions stop, is that only he eternally defines love, and physically creates the external reality we need. So, any or all lesser *rebellious* forceful souls, can never rule over GOD. Yet by free-will and created choice, the lesser can temporarily believe and act as if they can. FATHER eternally set *his* line that divides good from evil at the point where on one side, all conscious projected forces of desire, grow and lift up love and life between GOD and his Family forever. This line, being the loving, suffering *surface* of the spherical soul of JESUS, where we may become one with the beliefs and ways of HOLYSPIRIT, JESUS, and FATHER. Yes, the intimate *surface*, but not one with the internal entirety of GOD where all the things of GOD beyond us and all creations dwell. These we can never be one with, being GOD'S self-awareness and *experiential* qualia of their existence. And minus an eternal connection to CHRIST'S surface, we see a soul is set adrift in evil desires, pulling them down in confused divisions of their individual hell and death. Now please see how every projected divisive force, is evidence of a universe created for choice under conscious determinism. I see free-will choice, and each soul's choosing. I see each soul's desires and new force generations, that had no prior existence. I see created perceptions are born of three aspects. All such creations of reality require a desired forceful intent to exist, and so create; being the only aspect in reality to come from nothing physical. It is born within our soul as a never existing before, newly desired generated force patterned projection. And in having its physical half requirement, this other-worldly everywhere in every soul phenomenon exists without beginning or end or restraint by anything outside our gifted size and generated desires within. Now how can all the afore mentioned restrictions on GOD, begin to diminish their greatness. To reiterate, if FATHER says he cannot do a certain this in physical reality, without it causing a certain that in reality. This does not offend or diminish GOD at all. This brings only a defining reality to GOD. GOD and reality are both bicameral, and have both always existed. To know that FATHER is bound by the realities of who he consciously and physically is, and that by his omniscience and desire, he has many other self-set strict boundaries for defining and finding love, which he will never cross. And so, we see all reality comes to exist by who and what FATHER knows himself to be, and his knowing what he seeks to find and accomplish. GOD is not defined by what others want or need them to be. All self-imposed ways and limitations upon FATHER, JESUS and HOLYSPIRIT by them, are what distinguish and lift them ever so high above us in perfection. The lack of one contrary imperfection, defines perfection. FATHER'S reasoned purposed loving self-restrictions, define his righteousness. And place him so far above every lesser perfection, and apart from all contrary imperfection. We are one with GOD, only by our emulation of their perfection. GOD'S love is eternally defined, and beyond redefining by any and all lesser souls. I have seen GOD'S great work in profound detail, yet never once thought of GOD as unknowable in size, speed, reach, force, or perfection. I see a spiritual substantive Trinity. Being the loving provision of all Creation, intimately knowable, lovable, honorable, and true. Scriptures say we can see and know, and be seen and known by GOD. And reality is then revealed to us by GOD. Romans 1:16-22, "*FOR I AM NOT ASHAMED OF THE GOSPEL OF* ***CHRIST****: FOR IT IS THE POWER OF* ***GOD*** *UNTO SALVATION TO EVEYONE THAT BELIEVETH; TO THE JEW FIRST, AND ALSO TO THE GREEK. FOR THEREIN IS THE RIGHTEOUSNESS OF* ***GOD*** *REVEALED FROM FAITH TO FAITH: AS IT IS WRITTEN, THE JUST SHALL LIVE BY FAITH. FOR THE WRATH OF* ***GOD*** *IS REVEALED FROM HEAVEN AGAINST ALL UNGODLINESS AND UNRIGHTEOUSNESS OF MEN, WHO HOLD THE TRUTH IN UNRIGHTEOUSNESS; BECAUSE THAT WHICH MAY BE KNOWN OF* ***GOD*** *IS MANIFEST IN THEM; FOR* ***GOD*** *HATH SHEWED IT UNTO THEM. FOR THE INVISIBLE THINGS OF HIM FROM THE*

*CREATION OF THE WORLD ARE CLEARLY SEEN, BEING UNDERSTOOD BY THE THINGS THAT ARE MADE, EVEN HIS ETERNAL POWER AND* ***GODHEAD****; SO THAT THEY ARE WITHOUT EXCUSE: BECAUSE THAT, WHEN THEY KNEW* ***GOD****, THEY GLORIFIED HIM NOT AS* ***GOD****, NEITHER WERE THANKFUL; BUT, BECAME VAIN IN THEIR IMAGINATIONS, AND THEIR FOOLISH HEART WAS DARKENED. PROFESSING THEMSELVES WISE, THEY BECAME FOOLS, AND CHANGED THE GLORY OF THE UNCORRUPTABLE* ***GOD*** *INTO AN IMAGE MADE LIKE TO CORRUPTIBLE MAN, AND TO BIRDS, AND FOURFOOTED BEAST, AND CREEPING THINGS.*" 1st Corinthians 2:6-16, "*HOWBEIT WE SPEAK WISDOM AMONG THEM THAT ARE PERFECT: YET NOT THE WISDOM OF THIS WORLD, NOR OF THE PRINCES OF THIS WORLD, THAT COME TO NOUGHT: BUT WE SPEAK THE WISDOM OF* ***GOD*** *IN A MYSTERY, EVEN THE HIDDEN WISDOM, WHICH* ***GOD*** *ORDAINED BEFORE THE WORLD UNTO GLORY: WHICH NONE OF THE PRINCES OF THIS WORLD KNEW: FOR HAD THEY KNOWN IT, THEY WOULD NOT HAVE CRUCIFIED THE* ***LORD*** *OF*

*GLORY. BUT AS IT IS WRITTEN, EYE HATH NOT SEEN, NOR EAR HEARD, NEITHER HAVE ENTERED INTO THE HEART OF MAN, THE THINGS WHICH* ***GOD*** *HATH PREPARED FOR THEM THAT LOVE HIM. BUT* ***GOD*** *HATH REVEALED THEM UNTO US BY HIS* ***SPIRIT****: FOR THE* ***SPIRIT*** *SEARCHETH ALL THINGS, YEA THE DEEP THINGS OF* ***GOD****. FOR WHAT MAN KNOWETH THE THINGS OF A MAN, SAVE THE SPIRIT OF MAN WHICH IS IN HIM? EVEN SO THE THINGS OF* ***GOD*** *KNOWETH NO MAN, BUT THE* ***SPIRIT*** *OF* ***GOD****. NOW WE HAVE RECEIVED NOT THE SPIRIT OF THE WORLD, BUT THE* ***SPIRIT*** *WHICH IS OF* ***GOD****; THAT WE MIGHT KNOW THE THINGS THAT ARE FREELY GIVEN TO US OF* ***GOD****. WHICH THINGS ALSO WE SPEAK, NOT IN THE WORDS WHICH MAN'S WISDOM TEACHETH, BUT WHICH THE* ***HOLY GHOST*** *TEACHETH; COMPARING SPIRITUAL THINGS WITH SPIRITUAL. BUT THE NATURAL MAN RECEIVETH NOT THE THINGS OF THE* ***SPIRIT*** *OF* ***GOD****: FOR THEY ARE FOOLISHNESS UNTO HIM: NEITHER CAN HE KNOW THEM, BECAUSE THEY ARE SPIRITUALLY DISCERNED. BUT HE THAT IS SPIRITUAL JUDGETH ALL THINGS, YET HE IS*

*JUDGED OF NO MAN. FOR WHO HATH KNOWN THE MIND OF THE **LORD**, THAT HE MAY INSTRUCT HIM? BUT WE HAVE THE MIND OF **CHRIST**.*" If CHRIST knows it, we are meant to learn and know it. Colossians, 1:14-17, "*IN WHOM WE HAVE REDEMPTION THROUGH HIS **BLOOD**, EVEN THE FORGIVENESS OF SINS: WHO IS THE IMAGE OF THE INVISIBLE **GOD**, THE FIRSTBORN OF EVERY CREATURE: FOR BY HIM WERE ALL THINGS CREATED, THAT ARE IN HEAVEN, AND THAT ARE IN THE EARTH, VISIBLE AND INVISIBLE, WETHER THEY BE THRONE'S, OR PRINCIPLEITIES, OR POWERS: ALL WERE CREATED BY HIM, AND FOR HIM: FOR HE IS BEFORE ALL THINGS, AND BY HIM ALL THINGS CONSIST.*" 1st Timothy 6:20-21, "*O TIMOTHY, KEEP THAT WHICH IS COMMITED TO THY TRUST, AVOIDING PROFANE AND VAIN BABBLINGS, AND OPPOSITIONS OF SCIENCE FALSELY SO CALLED: WHICH SOME PROFESSING HAVE ERRED CONCERNING THE FAITH. GRACE BE WITH THEE. AMEN.*" We are told by GOD thousands of years ago, to avoid *not* the scientific evidence of GOD, but this world's fallen slow small purely physical interpretations of that evidence. Hebrews

11:3, "*THROUGH FAITH WE UNDERSTAND THAT THE WORLDS WERE FRAMED BY THE* ***WORD*** *OF* ***GOD****, SO THAT THINGS WHICH ARE SEEN WERE NOT MADE OF THE THINGS WHICH DO APPEAR.*" 2nd Peter 3:10-13, "*BUT THE DAY OF THE* ***LORD*** *WILL COME AS A THEIF IN THE NIGHT; IN THE WHICH THE HEAVENS SHALL PASS AWAY WITH A GREAT NOISE, AND THE ELEMENTS SHALL MELT WITH FERVENT HEAT, THE EARTH ALSO AND THE WORKS THAT ARE THEREIN SHALL BE BURNED UP. SEEING THAT ALL THESE THINGS SHALL BE DISSOLVED, WHAT MANNER OF PERSONS OUGHT YE TO BE IN ALL HOLY CONVERSATION AND GODLINESS, LOOKING FOR AND HASTING UNTO THE COMING OF THE DAY OF* ***GOD****, WHERIN THE HEAVENS BEING ON FIRE SHALL BE DISSOLVED, AND THE ELEMENTS SHALL MELT WITH FERVENT HEAT?*" These Scriptures and many more, say we can learn and know all things through HOLYSPIRIT, to become one spiritually in the powerful knowing mind of CHRIST. All souls are servants to desire, as even GOD seeks to fulfill their perfect loving eternal desires. Many will serve the shared eternal abundant provision of GOD, yet far, far more, serve themselves alone, and will forever die in their individual loneliness. We know *every* word of Scripture holds our understanding of FATHER, his Creation, and the chronicled history of GOD'S efforts and pain among mankind's rejections. We know GOD *both* preserves their truth, and also allows that truth to be subtly, widely, and greatly subverted. GOD'S, and endless other versions of truth are in this world, so be sure to choose his preserved truth. For GOD'S truth, trust the Authorized King James, or Geneva Bibles, without added comments or foot notes. Strength is only a good thing, if it is used for good; so, we need only good things to be strong. This is not the case in this world. *Eternal* peace can only be found among all its strong spiritual elements. Let's ask a question which can shed great light on the very heart of this matter. How could there ever be, not a momentary, not a very long temporary, but true *eternal* peace, safety, and love, among the current billions of conscious free-will souls of mankind. A perfect peace among mankind's irreparable fractures of differing strong beliefs, can never truly start, let alone last. Neither can *eternal* peace be found between the perfect love and strength of GOD, and the rebellious weakness of mankind together. So, how can there be an eternal place called the

New Heaven. Knowing GOD'S first Heaven and Earth, were destroyed by different desired beliefs. We see great weakness and loss of continuity in reality arises when one soul lives one way, and another soul lives a contradicting way. One wanting something, and one not wanting it. One having something, and one not having it, all in swirling chaotic pools of pluralistic individual physical desires. The secular world strives to lay the blame for most strife on deity religions. But the true blame rests on every tiny to large imperfect forceful desire, projected into GOD'S creation. Humanism says we can all live together in peace, if we can all compromise on all things. Where is their evidence, or *their* contributing compromise? The world goes increasingly insane with their trying. And hate only grows greater with their efforts. If everyone compromises on all that they strongly desire, life becomes meaningless, undesirable, and miserable. Open your eyes and see it is humanism, liberal politics, and *all* worldly religions; not GOD'S truth that destroys peace and love. In a fallen blindness, most hate and blame the true definer of love and truth for all their problems, saying how can GOD *let* this happen. The truth of GOD is found at the farthest highest right side of reality, while across the middle are all degrees of compromise and blending of lies and truths, and at the far-left lowest side is pure evil, where all Satan's liberal insanity, self-righteousness, hatred, terrorism, perversion, abomination, and the spewing of all great false accusations dwell as the pure spirit of antichrist. Liberals live to repeat and add to any lie that denies GOD. I do not say democrat, republican, or independent, these have no meaning. We see true Godly conservatives, we see zombie liberals, and the unsure across the middle. "Godly conservative" means to correct ourself and hold to GOD'S familial foundations that preserve people and their faith. A Zombie liberal sees morality as a godless subjective desire, seeing no good or truth in the past. Radicals, believe good can come only in shaming and burying the *entire* past. Radicals can't see that people cannot govern themselves, or ever compromise and hold all things equally. All souls hold certain things loosely, and hold others ever so dearly. Without all souls surrendering to live by GOD'S one pure truth, there will always be the contrast and strife where one's most dear desires, differ from another's. All souls can't compromise their core beliefs for others who stands in opposition. In doing that, *all* are irreparably hurt. The term compromise is a tool the devil uses to creates a cesspool called politics.

I do not equally disparage politicians, but *any* compromise of GOD'S truth is a trap used to lure, weaken, and turn life into death. The monster called liberal political correctness is a psychological warfare used by Satan's authoritarians. It is an uncontrollable undefinable force, which must turn to eat its own arm and cut its own throat in its last insane act of hate. The eternal truth can never be compromised, and the one omniscient GOD cannot spend eternity trying to reason with billions of forever lost souls in their endless confused refusal to hear the truth; and all wanting GOD to compromise towards *their* own fluid beliefs and desires. GOD cannot forever try to teach them a truth they can never want. So here and now is our own fading moment to hear and accept GOD'S truth; it will never be easier than now. What do you see occurring from compromise? I see weakness, destruction, and hell's confusion. Humanistic ways are the height of confused denial, and their imperfect blind management of ever-growing crisis; which only adds an accrued increase to its coming final catastrophe. If the world refuses to unite under the one true omniscient GOD, offering eternal life, how can it ever unite under foolish bickering individuals, holding their own ways up as *the* true way of life. These all have flawed self-serving minds, offering nothing dear to all, and asking all to surrender what is dear to them. Might some insane liberal politicians be possessed by demons? Yes, and I suspect some I won't name. Visualize the one perfect strongest light of truth and love that can exist, which once filled the universe before choice and chance were gifted and created. Imagine each dark belief of false confused politics, religion, and every evil action, being a bit of darkness, projected into the light. By each tiny to huge sin from each soul, the light of reality fades. Religions around the world are led by compromised souls, being influenced or demon possessed as they bow to mankind's liberal societal desires and demands. Satan's confederates seek out positions of leadership; and all say no one can be perfect with GOD. 2nd Cor. 11:10-15, "*AS THE TRUTH OF CHRIST IS IN ME, NO MAN SHALL STOP ME OF THIS BOASTING IN THE REGIONS OF ACHAIA. WHEREFORE? BECAUSE I LOVE YOU NOT? GOD KNOWETH. BUT WHAT I DO, THAT I WILL DO, THAT I MAY CUT OFF OCCASION FROM THEM WHICH DESIRE OCCASION; THAT WHEREIN THEY GLORY, THEY MAY BE FOUND EVEN AS WE. FOR SUCH ARE FALSE APOSTLES, DECEITFUL WORKERS, TRANSFORMING THEMSELVES INTO*

*THE APOSTLES OF CHRIST. AND NO MARVEL; FOR SATAN HIMSELF IS TRANSFORMED INTO AN ANGEL OF LIGHT. THEREFORE IT IS NO GREAT THING IF HIS MINISTERS ALSO BE TRANSFORMED AS THE MINISTERS OF RIGHTEOUSNESS; WHOSE END SHALL BE ACCORDING TO THEIR WORKS.*" Satan subtly fills the halls of religious learning with all worldly compromise, calling it righteousness. Throughout history, have all souls grown stronger in GOD'S truth, or weaker? 1st John 3:4-10, "*WHOSOEVER COMMITTETH SIN TRANSGRESSETH ALSO THE LAW: FOR SIN IS THE TRANSGRESSION OF THE LAW. AND YE KNOW THAT HE WAS MANIFESTED TO TAKE AWAY OUR SINS; AND IN HIM IS NO SIN. WHOSOEVER ABIDETH IN HIM SINNETH NOT: WHOSOEVER SINNETH HATH NOT SEEN HIM, NEITHER KNOWN HIM. LITTLE CHILDREN, LET NO MAN DECEIVE YOU: HE THAT DOETH RIGHTEOUSNESS IS RIGHTEOUS, EVEN AS HE IS RIGHTEOUS. HE THAT COMMITTETH SIN IS OF THE DEVIL; FOR THE DEVIL SINNETH FROM THE BEGINNING. FOR THIS PURPOSE THE SON OF GOD WAS MANIFESTED, THAT HE MIGHT DESTROY THE WORKS OF THE DEVIL. WHOSOEVER IS BORN*

*OF GOD DOTH NOT COMMIT SIN; FOR HIS SEED REMAINETH IN HIM: AND HE CANNOT SIN, BECAUSE HE IS BORN OF GOD. IN THIS THE CHILDREN OF GOD ARE MANIFEST, AND THE CHILDREN OF THE DEVIL: WHOSOEVER DOETH NOT RIGHTEOUSNESS IS NOT OF GOD, NEITHER HE THAT LOVETH NOT HIS BROTHER*." When born into this world, not desiring to choose the eternal perspective and ways of GOD. One is born of a wild narrowed slowed physical mind. And from such perspectives, souls desire to propagate and perpetuate their individual political and religious desires. This is exactly the same lying contagion Satan infected Eve with in Eden's Garden, as she quickly turned to infect Adam. Religious individuals can believe they are trying to make the world better and more just, but by being wrong about reality itself, they only make the world divisive and insane. Partially good is not GOD'S standard. The eternal perspective is a simple question and choice; do we *want* to love and follow a loving GOD, or something less; and to what extent. Is it not a true mystery that within one reality, many souls strongly believe and live in so many contrary ways; which at their beginning, or their middle, or at their end, *can't* all be true or right at the same time. Only one true way can align with an eternal reality. Most all are content to exist in a state of consensus confusion. So, I will not define again, the fallen temporary perspective; but let *it* be seen in contrast to defining again, the eternal perspective. If we are filled by HOLYSPIRIT, which only the few are, we see all the things and ways of this world as the divisive evil they are seen to be by GOD. The pure truth is clearly seen in each moment, by seeing the way things were once created and intended to be, and the way they will be again in the New Heaven and Earth, compared to all we see now. In Scripture, souls filled with HOLYSPIRIT ask how long till FATHER ends this ever-growing conscious evil. Souls born of HOLYSPIRIT see divisions coming from all the choices made by each lost soul. We see all beauty wonder and fascinations of GOD'S way of life grow evermore real, as those ways in this world fade with each day. Imagine if you can, every soul having learned and grown to live in pure eternal love and truth, forever feeling elation in caring and looking to the spiritual needs of all. The vainly religious, deny GOD'S command to be a separate peculiar people. This world hates GOD'S true Children because we do not accept their evil. And if we *do not* reject their evil, we too are evil. Living among evil, you *can believe* CHRIST accepts your ongoing sin; yet CHRIST has never accepted sin. CHRIST condemns all sin, and we do the same. In Heaven, the need of love in every Child of GOD is filled by the caring of GOD and his children before we ever feel that need, it being a timeless existence in all ways. The strongest basic aspect of *eternal* love, is faithful responsible caring. So, we know we are responsible for each child's protection and provision, and our great joy resides in faithfully fulfilling every need. This is the only form of love between all, that can deliver eternal peaceful loving life among many. If we see GOD or their child, and do not hold a faithful responsibility to preserve their Spiritual love, how can we say we eternally love them. The flesh has a few needs to continue as our operating platform, such as air, water, raiment, shelter, and modest simple food. Yet our soul has its far greater needs to be met. If our soul dies away from GOD, as the body is strong and adored, we exist in our false physical delusions. Now are we to love a reprobate in a hope for them to be found and saved when they never can. Matt. 7:6, "*GIVE NOT THAT WHICH IS HOLY UNTO THE DOGS, NEITHER CAST YE YOUR PEARLS BEFORE SWINE, LEST THEY TRAMPLE THEM UNDER THEIR FEET,*

*AND TURN AGAIN AND REND YOU.*" We are to have compassion for the dire physical needs of all, but we are not to put at risk the hurt or loss of any Child of GOD in doing so. The Spiritual realm, is in all ways contrary to all self-centered flesh and this world. Not one aspect of this fallen world is reflective of FATHER, or Heaven. This world is built on lust and covetous of individuals, lacking all natural-fear of GOD. We can clearly see how all selfish physical desires are related to the fear of physical death, and a pride of physical life. Can you not see the spiritual difference between Adam and Eve before, and after they fell by their individual desires. Even reprobates, if they admit it or not, when nearing physical death, they hope for something better to be there beyond this life. Those who practice vain religions, picture in their heart, that coming into Heaven is like the morning of Christmas. Wondering more about family, friends, and the physical things they will receive, see, feel, and eat. Wanting a place that fulfills all physical dreams and aspirations; yet they don't see this as lustful or worldly. They want and expect an impossible unrighteous heaven. They expect a heaven that is not holy or spiritual. Most speak of their formed images of heaven, as being a physical place made perfect for each of them, and not a spiritual place that we each are commonly made perfect for. For most, giving up the temporary physical pleasures they adore, to receive instead a greater Spiritual joy and reality they do not know, is not the heaven they dream of. They see a worldly heaven and a compromising GOD who accepts their physical desires, imperfections, and irresponsible unfaithfulness. The real *eternal* peace and joy GOD promises his Children is beyond our current knowing, our having no experiential reference. We have GOD'S promise we will live eternal in a spiritual state of awe and splendor. Most don't consider that it is never the physical act itself that pleases them, but the experiential conscious awe and joy that follows. But awe and joy can be had in GOD'S stronger spiritual ways. GOD'S good is received in patient honoring and worship. But what GOD gives us in a body, or environment, or sustenance, can never be what leads us to GOD. It must be by our desired passioned love for *who* GOD is, being thankful in our love for their created eternal ways of life, and our desire to eliminate all pain, hell, and death, by never sinning. If we truly are one in GOD, it brings joy to live this way now in eternal faithfulness, not something we periodically feel, or want, or expect to receive in Heaven. GOD'S Love is truly *first* received and felt in our soul's center. But if it is not returned in *fullness*, it is returned in an insufficient loving purpose; so that love is diminished. We cannot love GOD less and from afar, and we can never go to GOD physically or sinfully. It is a spiritual journey, not a physical journey, and we must draw ever so close. Consider what it means to *be* one in the Family of GOD, or an individual, who by rejecting the pure omniscient truth says, but I believe this. We all will be revealed as one of these two examples. Only FATHER'S plan, CHRIST'S Blood, and HOLYSPIRIT'S teaching, make us one with GOD, and souls not adjoined to HOLYSPIRIT, are just one unto themselves with their own internal god (personal belief). Remember that to say one does not believe in GOD, is just a strong way to say one does not desire GOD and nothing more. In our considering that moment long ago in the life of Adam and Eve; what if GOD instead of calling and asking where *they* were, after *they* sinned? What if GOD called your name right now? Asking where are you, asking what is this *you* have done? Asking, do *you* understand the destruction brought by *your* sins? Asking why would *you* do such a thing, in that GOD'S loving desire is to provide for his and your every *eternal* need? This was the question posed to

Adam and Eve after their individual sinful choices. Most, when they read of that first sin in Scripture, say all they did was eat a piece of fruit, what's the big deal? The piece of fruit had nothing to do with it. It was not poison or anything special as a piece of fruit. But it represented a core choice of desire. It was their choice to commit, know, and be evil. They chose to venture out of GOD'S perfect ordered Creation and change the work, ways, and creation of GOD. Looking around at the divisive state of the world as we have a perspective Adam and Eve did not have. We see how their first sin grew and continually spread. It all started with that one small rebellious choice, unleashing trillions of small to greatly destructive divisive choices. They knocked over that first domino, starting the ever-increasing destruction of GOD'S creation. All GOD'S vast perfect work is now destroyed. Sin has a butterfly and chaos affect. There can be *no* accepted rebellious disorder in perfect eternal peace, just as there can be no acceptable amount of any other deadly highly infectious contagion. If we see sin as the most-deadly infectious contagion there could ever be, which it is. And we see it spread by speech, action, perversion, and even provocative appearances of the body, that draw eyes from GOD'S Holy purpose. Sin has a mortality rate far beyond all fear, and only one cure. The cure is in our great desire through forgiveness of our past, to here and now learn and live in GOD'S created ways of eternal life. We all see two choices: one for GOD, and one against. Choice one is defined by GOD, in being who he is. The other choice is defined by each individual soul. Can GOD'S pure way of life bring eternal peace under his care? Yes. Can *all* contrary ways of life bring eternal peace for *everyone*, under GOD'S care? No. That is what we currently see around us. Can one way of life be found and forced under the rule of mankind to bring a true peace? No. That being socialism, humanism, communism and a forced compromise and compliance. There can never be a compromise between imperfection and perfection. So, minus GOD, imperfection must remain; and all that is imperfect must become even more imperfect. We cannot compromise between living GOD'S ways, and not living GOD'S ways; such confusion is an illusion. At each tiny compromise of GOD'S ways, evil wins. Mankind's fallen godless humanistic efforts at world peace are all futile. Compromise requires *all* parties give up the things they need and love the most that divides them. Secularism says their liberal immoral beliefs, being contrary to ours, must be lived openly to be real. They then say following GOD in his ways, is something to stay within the mind, and not to be expressed outwardly, as it offends them. This world's weak believers, continually compromise on GOD'S ways. And when GOD'S children reject their compromises, we are labeled fanatical and prejudicial. Throughout my life, I have watched the liberals demand compromise, and weak believers compromise each time. But never in my entire life, have I seen liberals give or allow one compromise towards allowing the lively ways of GOD into our public life; not once. Compromise is a liberal weapon of deceit. Compromise is impossible in situations of strong contrary core beliefs, as one must cut their soul apart to do such a thing. In GOD'S eternal separation, liberal humanist will be judged as sinister enemies of GOD, who destroy souls who might have chosen GOD. We see their hate against allowing true believers to live a peaceful life in fullness under GOD; just like evil did in Sodom and Gomorrah. We are called hateful for a biblical life, though the right to do so, is first in the U.S. Constitution. Our Constitutional Rights are set in an order of showing intent of preeminent protection above all lesser implied or inferred rights, souls might argue for. We see religious speech and liberty are placed high above others, written explicitly that the lively free public exercise of our ancient religion is illuminated and protected high above, and before all others. After our enumerated original and amended federal rights, we see all lesser rights and laws are left to state and local jurisdiction. National Rights come by constitutional amendment; but no new right can be given status above our free religious exercise. Liberals cannot say I have free religious exercise, yet my sacred belief is hateful to their individual belief, and must be kept to myself. Any restriction of my outward public expression is the very opposite of free-exercise. A liberal interpretation of the Constitution, as with Scripture, and all words, continually changes and alters the original definition and intent to suite the *noisome* lies from their mouths, born of inner rotting desires. We have expressed, what is the essence of our love for GOD, but what is the essence of our fear of GOD? And how great can our fear be. If you have thus far rejected a belief in, or the rightful rule of GOD over your life, please indulge me. Imagine it is that soon day of Judgment and final separation, where all souls must literally stand before GOD, consumed as never before by an emotion taking your every breath. Even GOD'S beloved children will be overcome by a fear and awe we never dreamed we could feel or know. None will remain standing before GOD on this day, as none will have the strength to stand. Scripture says, all will be on their knees, all will tremble and cry, eyes as wide as possible, hearts pounding so hard it shakes the whole body and blurs the vision. None will run to hug JESUS as many say, and none will stand and bravely curse GOD as others say. Those resurrected to promise will feel great sadness for not doing more, seeing all the self-deceived who hold a false weak insufficient belief in GOD. What a realization of horror for the lost. Such a fearful thing to find out you are lost. That by your disbelief in GOD, or believing yourself saved by false doctrine and the words of your *own* mouth, and not a true love for GOD'S ways of life. Matthew 7:21-23, "*NOT EVERY ONE THAT <u>SAITH</u> UNTO ME, **LORD**, **LORD**, SHALL ENTER INTO THE KINGDOM OF HEAVEN, BUT HE THAT DOETH THE WILL OF MY **FATHER** WHICH IS IN HEAVEN. <u>MANY</u> WILL SAY TO ME IN THAT DAY, **LORD**, **LORD** HAVE WE NOT PROPHESIED IN THY NAME? AND IN THY NAME CAST OUT DEVILS? AND IN THY NAME DONE MANY WONDERFUL WORKS? AND I WILL PROFESS UNTO THEM, <u>I NEVER KNEW YOU</u>; DEPART FROM ME, YE*

*WHO WORK INIQUITY.*" Iniquity is sin. And again, this speaks of those, who while living in *this* world, professed belief in CHRIST, and *believed* they were forgiven by living according to charitable *physical* works. But they did not live by the first great commandment. Rev 2:4-5, "*NEVERTHELESS I HAVE SOMEWHAT AGAINST THEE, BECAUSE THOU HAST LEFT THY FIRST LOVE. REMEMBER THEREFORE FROM WHENCE THOU ART FALLEN, AND REPENT, AND DO THE FIRST WORKS; OR ELSE I WILL COME UNTO THEE QUICKLY, AND WILL REMOVE THY CANDLESTICK OUT OF HIS PLACE, EXCEPT THOU REPENT.*" We see *physical* charity and compassion coming from the secular community the same as from the believing community. GOD is first pleased in our *knowing* and loving who they are intimately; not in acknowledging their distant obscure existence. If a soul knows and praises the Holy name of the SON of GOD, but does not *know* and love the SON in all his true intimate attributes, they do not *know* who they praise. If they believe in his name, but then by their own belief and desires, assign to the SON the attributes they think and hope he has. These do not praise the SON. They instead, praise their own projected desires and wants. This is vain religion, shining a dark false light only the flesh will see, and speaking a false empty hope only the flesh will hear. Again, mankind gives up a *personal* relationship with GOD, for their relationship with a *personalized* god. And again, what should strike us all like a bolt of lightning, is that JESUS tells these in that last verse from Mathew, that he *never* knew them. While these say they professed their love for JESUS in their lives, and asked forgiveness. But if JESUS *never* knew them, then JESUS *never* forgave them at their request. But why? Are we forgiven because we *say* we love GOD, and pray to be forgiven? No. Our first commandment and work is to love GOD above all and everything, and repent and turn from *all* evil, which is sin. By that last Scripture, we see one could profess CHRIST with their mouth every day, attend church, and do physical charitable works for eternity. All while holding a *false* personal image of a god who allows their sinful ways, pleasures, and comforts. Many believe this in their self-deception, seeing a wider gate to eternal life, instead of searching, finding, and going through the narrow sinless gate to eternal life. Many believe GOD always comes to them, instead of their going to GOD. What will be their paralyzing fear when it sets in? What horror are such souls capable of experiencing in a moment, what terror is a soul capable of comprehending? We can surely visualize the many Scriptures saying there will be wailing and gnashing of teeth at their realizations. I imagine many set before GOD in un-describable fear, will faint repeatedly. No one has ever felt it, or can try to imagine it. This fear, is on a level we can only know before the final judgment throne of GOD. Most are lost, living within their own definitions of good and evil, so they cannot see why GOD rejects their sinful ways. And more sadly, GOD cannot explain this to them, as their hearts are hardened and minds scarred over in being reprobates (beyond reach). They can only plead for compromise, as that is all they want of GOD. It is wise, and we do well to *now* each day hold as great a fear of falling as we can. We hear the cries saying, but we can't help what we love. Yes, we all choose what to love in our knowledge or ignorance. But in a new knowledge, we *can* alter what we once loved. Through fervent effort, we *can* learn to see the destruction to eternal life, that comes by each sin, and find our reasoned purpose and strength to change our desires. We must see and stop *every* hurt we cause to GOD'S eternal work. Once we see our desired eternal path, our every next perfect or imperfect choice, strengthens or weakens our faith of belief. We all interpret everything in our learned entangled desired way. Do we each have an experience, and search one shared unbiased catalog of information to form our conclusions? No. Each soul's current belief, is a complex web of all their chosen desires, filling their soul's sphere. A soul's interpretation of routine mundane experiences, are quickly and blindly formed in their current biased moment. But for uncommon more serious experiences, we ponder these for a stronger conformation. Each false biased belief, is part of a mind web we must dismantle and reform in GOD'S eternal way. But weak believers will blame and cry, saying how could a truly loving GOD let them desire and be tempted by evil choices. Again, to find *eternal* free-will love, how could GOD not allow exactly this. We see temporary love has its temporary costs, but to reject or lose eternal love has its eternal cost. The lost are those never wanting to be there below or for GOD, and not wanting GOD to be there above and for them. They want GOD to create and sustain their desired physical Heaven and be there for, and beside them; but not above them. They have little to no desire to rise up to be one with GOD. Their great desire is for GOD to come down with them, and fulfill their desires. The definition of

GOD, is not set by a lost soul's imperfect standard. GOD is far above us all as an immutable fact. And until we can accept this fact in love without our slightest reservation, there can never be truth and love between our soul and GOD. We know GOD is there, and according to Scripture, they want us to know them intimately. GOD is *not* known to most souls in any way, because most do not want to know and emulate GOD in any true way. By their desires for this world, and self-delusion, they see that the less they truly know of GOD, the more physical freedoms they retain. Many souls struggle with Scripture because they want to believe two false things: one is the error that GOD loves everyone equally; and two, that GOD is unrestrained in any way, believing GOD wants to give them everything they physically desire. But according to GOD'S words, that is not true at all. You and I will *not* do everything for everyone, or love everyone in this world equally. And this does not preclude us from being a loving soul. So being loving according to his own Holy Righteousness, cannot preclude FATHER from eternally loving his children; yet hating others. Prov. 8:17, "*I LOVE THEM THAT LOVE ME; AND THOSE THAT SEEK ME EARLY SHALL FIND ME.*: John 14:20-21, "*AT THAT DAY YE SHALL KNOW*

*THAT I AM IN MY FATHER, AND YE IN ME, AND I IN YOU. HE THAT HATH MY COMMANDMENTS, AND* *KEEPETH THEM, HE IT IS THAT LOVETH ME: AND HE THAT LOVETH ME SHALL BE LOVED OF MY FATHER, AND I WILL LOVE HIM, AND WILL MANIFEST MYSELF TO HIM.*" We must accept that *eternal* love as GOD defines it, is an increasing rarity among mankind. Most live by pride envy and desires for the things they perceive around and below them, and not that which is far above them. If FATHER could make all souls love him in a true eternal way, he would. But love is born, not created, so, he can't. Many say that by my saying these great needs sufferings and restrictions exist within GOD, that I humanize GOD. This is more rebellion, for I do the reverse, by saying we are created in FATHER'S image, and created to be Godly. We are created to be a lower spiritual, bodily, and emotional image of GOD. We see the destruction individuals bring to this world, and they now crave this drama, danger, suspense, and gamble, associated with temporary life. It is strange that from a slow small temporary perspective, individuals see life as a game, having physical winners and losers. But from the eternal, faster, larger perspective, there are no physical winners, and it is not a game. There is no physical reward for the lost, there is only the physical loss of everyone and everything one had, and would have ever had for eternity. We are all created to spiritually be one together with GOD; or remain individuals apart and alone if one desires and chooses. The need for love, and all the feelings of pain, sorrow, and loneliness, were there within FATHER, long before we were gifted and created in his conscious-physical-emotional image. The Creation's purpose, is for all free-will gifted souls to spiritually fill that loving part of GOD'S and our existence. But most within GOD'S creation, have followed Lucifer. And the only real attempt at humanizing GOD, is by those expecting GOD to be like fallen mankind, instead of mankind becoming like GOD. We preach to strengthen and lift mankind up to be one with GOD. False apostles preach to bring their false-ideas of a god down to be one with an imperfect mankind. If these ever open their eyes, they will see their current father is Satan. There are multitudes who write about who GOD is, and these many to certain degrees contradict each other. We ask, who among all those preaching many differing beliefs, preach their own desires? And who is truly called and instructed to write or speak? I have no earthly desire or formal education to write about these things. It has been both fearful and joyous in writing words that are not of my design. But in being shown how all spiritual and physical realities unfold, my emotions have ranged from tears of joy, to harsh nauseous inner turmoil, fear, anguish, and loneliness, seeing what this all means for me, and all I love. Yes, I must wonder where I, and my loved one's will be spiritually found in that last day. Yet, I must strive, and not weaken one word of GOD'S truth within myself for my temporary comfort. Not one of these words is new or born out of my understanding or desires. I saw around forty years ago, that *all* my prior desires and understanding of GOD, myself, and this world, were born of a confused temporary perspective, and by listening to false preachers, teachers, politicians, and friends, offering their worldly understanding. HOLYSPIRIT showed me why this is pure foolishness. So, in 1974, I set aside former understanding. Every understanding except basic reading, spelling, and math. So, knowing nothing, sitting in a field in the middle of the night with only GOD and Scripture to find the truth, I began my journey. From that moment, I waited and listened and found learning from GOD is just as with all learning. We must start at its foundation to then build

FATHER'S understanding in us. Over many years, HOLYSPIRIT *showed* me many immense things, but I never heard any words to express them. So, I strive in my weakness to relay what I am shown. I have my greatest assurance and trust in GOD who cannot fail. I have not that same assurance in myself or others, who can fail. YESHUA, please pray for and protect my house, and my work of love. Amen. GOD goes to every possible good extent to save each soul to GOD'S ways and love. Now if you are with us to this point, I pray in this moment of hope, your soul sees the one true way clearly. Free-will is an absolute necessity for love to exist, but it is a dangerous two-edged sword. It can sneak up inside us and bring us down if not fearfully watching. And if any soul can ever find a formula to bring such a wonder as *eternal* love, into one shared physical reality, with many free-will souls at a reduced cost. Our loving FATHER would surely accept it. But there is no other way, it was thoroughly searched for in FATHER'S omniscience, omnipotence, omnipresence, and no other way exists. We are all called by *our* gifted-created existence, to examine and understand the lengths to which GOD has gone, and the pain and effort GOD has endured in all of this. The pain and effort endured by

FATHER, JESUS, or HOLYSPIRIT alone, out-way by magnitudes, every pain mankind combined has felt. Now we see this concept of FATHER being eternal in every direction. But we see newly gifted differing distinctions of eternal awareness, bring a new concept of being eternal from a first gifted and created moment. We see JESUS and HOLYSPIRIT had their very first awareness of their now eternal spherical moment, and we had our first awareness of our eternal spherical moment. But FATHER is the omniscient soul who had no first awareness of his eternal moment. We see FATHER'S eternal moment is without a single consideration or aspect of time, and is not seen in any linear sense of eternally backwards and forwards, requiring a center or beginning. FATHER exists in a spherical dimensional sense of eternally folding and rolling both inwardly and outwardly. And we see in FATHER'S every contemplation, deliberation, reflection, and each consideration going beyond the boundaries of rumination, that his every good pure effort has been to be loving. He waits in his hope for a real FATHER-Child relationship with each of us. GOD can forgive past sins, in true repentance and growth, but not in endless sinful destruction. The high crown of knowledge and hope, and all the good that is offered to us, was there in CHRIST'S loving Sacrifice. There are the few souls who look deeply at what GOD has truly done, born, and suffered. And how after evil entered reality, our love *for* GOD as with all love, must be born of pain, suffering, sacrifice, and loss. Yet what could life be without love but a hollow, meaningless, torturous, never-ending hell. What communion can light have with darkness, and what can the children of GOD ever have with individuals in eternal love and life. The falsely religious will say, if this is *your* GOD, he is cruel. These say *their* GOD would never create life, *foreknowing* he might have to separate out an unknown probability of souls to be imprisoned in their own chosen conscious anguish forever. Many say *their* GOD can make love without required pain, death, and hell. But their words have zero support. And I warn all to be very weary of souls who *often* imply a *possessiveness* in saying, "*their god*" or "*your god*" as these speak of allusions. GOD is not defined or held by any. But there is all abundant proof for what *we* write, coming from Scripture and from the breath of every life. In that we do see greater and greater death pain and hell do exist in ever-growing abundance. So, if their personal created idolized god, could create love without pain suffering and death. then it is their god and idol of their imaginings who is cruel. In that if death and pain were not a necessity for finding love, yet their god still brought it. But why? If not for choice and finding love, for what other reason and purpose were pain hell and death allowed. It can only be for love to exist, or only for cruelty. Isaiah 5:4, "*WHAT COULD HAVE BEEN DONE MORE TO MY VINEYARD, THAT I HAVE NOT DONE IN IT? WHEREFORE, WHEN I LOOKED THAT IT SHOULD BRING FORTH GRAPES, BROUGHT IT FORTH WILD GRAPES*?" To help us climb up, GOD reaches down; but only as far as we reach up. Yet we should not expect GOD will, or should reach down to us more than they already have. As GOD'S "grace" meaning their enduring merciful effort towards us, is already full, complete, and sufficient. So, whether we recognize and admit it or not, GOD has already given us all they are and have, being all we need to find life, when we deserve only death. We just need to see it and receive it in fervently desiring GOD. We must rise up in FATHER'S beautiful truth, and run to his eternal love. We must strive and fight together as one body with prayer and fasting to overcome and endure this world until its end. And again, heed in fear, the warning*s* of JESUS.

Matthew 22:14, "*FOR MANY ARE CALLED, BUT FEW ARE CHOSEN.*" 1st Peter 4:18, "*AND IF THE RIGHTEOUS SCARCELY BE SAVED, WHERE SHALL THE UNGODLY AND THE SINNER APPEAR.*" Matthew 7:13-16, "*ENTER YE IN AT THE STRAIT GATE: FOR WIDE IS THE GATE, AND BROAD IS THE WAY, THAT LEADETH TO DESTRUCTION, AND MANY THERE BE WHICH GO IN THEREAT: BECAUSE STRAIT IS THE GATE, AND NARROW IS THE WAY, WHICH LEADETH UNTO LIFE, AND FEW THERE BE THAT FIND IT.*" Luke 13:23-24, "*THEN SAID ONE UNTO HIM,* ***LORD****, ARE THERE FEW THAT BE SAVED? AND HE SAID UNTO THEM, STRIVE TO ENTER IN AT THE STRAIT GATE: FOR MANY, I SAY UNTO YOU, WILL SEEK TO ENTER IN, AND SHALL BE UNABLE.*" These verses speak not of believers and non-believers. Those trying to enter at the wide gate are the multitudes of vain professed believers who try to enter without the eternal truth or any loving fearful obedience to GOD. That last verse instructs us to *strive*, meaning to give our greatest efforts, as many will see and know there is the straight and narrow gate, and strive to go in, yet many are unable to fit because of their strong desires for the flesh and the world; this being the Judas equation. Their spiritual *and physical* desires together, do not fit through the narrow gate. We all must strive, and see ourselves as *possibly* being represented in this last verse. I see the good and narrow gate, and run to it. But until I pass through that gate into Heaven, I strive in my as yet unfulfilled hope and faith. Can a man go left and right in the same moment? If so, where will his path be changed. Scripture does not say hold onto and live your life professing JESUS. Scripture says profess JESUS, by giving up your life here to emulate his by allowing his righteousness, truth, love, and example, to live in you. Do not say you are perfect, by professing JESUS. Profess JESUS by emulating his obedient loving perfection. Do not say we hope the world can see JESUS in us; but hope they see us in JESUS. Do not live to speak *for* JESUS; let JESUS live in and speak through you. Worldly Christians seem to see these things inversely. Who should we be in our love for such a GOD. Is not JESUS the KING over all lesser Kings, and the LORD over all lesser Lords. Then he is also GOD over all lesser Gods. Let us see our LORD as the Spiritual example he gave to us. Let us emulate the Holy Angels, Prophets, and Saints, who faithfully follow him. We must ask *and* receive every spiritual force granted by GOD to overcome this world and its illusions. We must strive in desire to be seen *now* as a Spirit Angel of the LORD. If we love FATHER, his WORD, his SPIRIT, and his ways and truth above all things, there stands nothing between us and oneness with GOD. But if we partake of the covetous perversions and pleasures that fill this world. Then evil and sin stand between us and GOD. Free-will choice was gifted *and* created for us to find love in our soul. The broad current spectrum of love is born out of all GOD has gifted, created, loved, hated, and temporarily allowed. Choice is *given* to all, and our choice *must* be and is made by all. Our unavoidable choices are made, and our just reward for our choices must also come. There is no other free-willed way to have eternal love and life. And we have been clearly told there is no other example, or name, under which we *might* be saved. Only a loving emulation of JESUS can save any of us. And our judgment of accountability is thereby so set forth and will be executed. And on each side of the soon eternal divide, everyone will have chosen and affirmed the ways of life they love. Surely all must see and know this is occurring, and that it must be. This is a hard, yet great fascinating wonder of the great design. Do you see GOD'S plan and grasp both the simplicity and the complexity of it all. It truly is a most fascinating force of labor, design, and purpose, even for the greatness of FATHER. What measure of love and understanding have you currently received from GOD? Not what great wonderful things have you read or heard in your ear; but what great intimate revelations has GOD revealed to you in your soul? Do you talk *at* GOD, or truly communicate back and forth in your strong intimate ways *with* GOD? What do we each in all fullness, know of GOD'S personage? Can we say this is not our place, it is for a Priest or Pastor? This is surely not so. If we are called to express our true sum knowledge of GOD in oath and vow of truth. To declare all things tried and proven to us of GOD'S ways, love, needs, and creation. Holding aside, all things we have read that GOD will give us. Expressing only our intimate knowledge of GOD'S Persons, love, emotions, wants, hopes, and what GOD needs and expects from us. What is our created purpose? What does GOD declare love is? What is the one true belief for all created life? If we had to stand on our solemn knowledge received *from* GOD alone, what could we each say? Does our full summation, reflect not just our total knowledge of GOD; but the sum of our *desires and efforts* to search GOD out and intimately know them? Is GOD our highest concern? How much can we say about our parents, spouse, siblings, children or friends, about who they all are in such rich detail? How about our job, and possessions? How much could we say about the things and ways of *this* world we desire that bring us pleasure, comfort, and joy? Would our words about our intimate knowledge of GOD, be far less, and far more-vague? Would our words about this world be many, and filled with descriptive nuanced beauty and intimacy? Are the things of this world our highest concerns and greater understanding? These illuminating scary questions are hard to ask. But should we ask these of ourselves now, or wait for GOD to ask us these questions in *his* day. Must most say, I do not know GOD at all? And the rest say, I do not know GOD nearly enough? I have barely scratched their surface, and I seek and can never know GOD enough to satisfy my soul. Our existence is a continuous exercise of *our* conscious desire, and in this all *our* hopes and dreams are seen and revealed to GOD. We each know what we want to know, and are ready to know. We each don't know anything we don't want, or are not ready to know. We can never truly know something we desperately don't want to know. One can never say to

GOD, I did not know; without saying I was busy trying to know something else I love more. We have our lifetime of learning associations, and where is or was our desire to intimately know GOD in it all? We hear in doctrine about our so-called "seasons of doubt". These seasons are never about doubting GOD, but doubting our love for GOD. Such doubts reveal we are moving off the true path, and into danger. We read in Scripture, after the cleansing flood of Noah's day, how individualism began to spread again. So, GOD looked to the faithful in the earth, finding the few who desired, knew, and loved GOD above the world. GOD saw a need and purpose, and chose to establish their ways in the world and establish their earth-bound family under one banner. This to give us *their* strength in *our* adoption through FATHER'S holy standards and a familial gathering point. By certain names of old, GOD established and nourished as our anchoring standard, his written and ordained commands to sustain, spread, shine, and call out to the world in his truth and love. GOD wrote in the stones of the Earth his elemental law, to declare his family of oneness and order will henceforth be both born, or adopted through CHRIST into the Tribes of Israel, to faithfully live in his written Holy ways. GOD'S covenant passed through Moses, Noah, Abraham, Isaac and Jacob; whom GOD renamed Israel. Moses received and brought down the ten firm Oracles ordained by GOD. But mankind tried to weaken his words, to define a wider gate. So, GOD gave Moses a more precise Levitical elaboration. Now to serve GOD, we must never think of ourselves as adults, which gives to us a false sense of wisdom. We are only and forever children before GOD. Look at a five-year-old child who thinks they are smart and grown up; this is *us* under GOD. In examining CHRIST'S Ministry, we see he came to turn the lost and all who had strayed to a wide gate, and whose doctrine devolved over generations, back to their narrow gate. CHRIST said they thought they were wise, and replaced GOD'S commands with the religious traditions of mankind. He warned we can never find GOD in that way, it being a lie and not the path of truth. Mark 7:9, "*AND HE SAID UNTO THEM, FULL WELL YE REJECT THE COMMANDMENT OF GOD, THAT YE MAY KEEP YOUR OWN TRADITION.*" JESUS told them *and* us to return to the belief and worship FATHER had ordained. CHRIST did not once in his earthly ministry, imply he brought a new religion or truth. What confusion would that be in that FATHER said he never changes.

CHRIST said FATHER has given *his* commands and promises to his Children in *perpetuity*. From the fall of mankind, to the final trumpet and vial, Scripture calls all who breathe, and not reprobate, to become what we were meant to be. The first coming of CHRIST was in that manner and purpose of bringing the truth and love of FATHER'S unchanging eternal strength, guidance, and love, in fullness. CHRIST came to lead us back up to the oneness mankind fell from and lost. Not to stay among lost souls in their hell of individual division. Matt. 5:18-20, "*FOR VERILY I SAY UNTO YOU, TILL HEAVEN AND EARTH PASS, ONE JOT OR ONE TITTLE SHALL IN NO WISE PASS FROM THE LAW, TILL ALL BE FULFILLED. WHOSOEVER THEREFORE SHALL BREAK ONE OF THESE LEAST COMMANDMENTS, AND SHALL TEACH MEN SO, HE SHALL BE CALLED THE LEAST IN THE KINGDOM OF HEAVEN: BUT WHOSOEVER SHALL DO AND TEACH THEM, THE SAME SHALL BE CALLED GREAT IN THE KINGDOM OF HEAVEN. FOR I SAY UNTO YOU, THAT EXCEPT YOUR RIGHTEOUSNESS SHALL EXCEED THE RIGHTEOUSNESS OF THE SCRIBES AND PHARISEES, YE SHALL IN NO CASE ENTER INTO THE KINGDOM*

*OF HEAVEN*." Build the righteous loving obedient mind of CHRIST within you. We cannot forever cast ongoing sins upon CHRIST. Most who believe in JESUS, visualize a coming Heavenly King, ruling over many sinful worldly subjects. This is a false belief that blinds and comforts those who desire it, and all such belief brings destruction. GOD is looking for those who see how this world's ways take them away from all that is good. GOD searches for every hand reaching up to them in love and humility, to be pulled up and out. But we add, the longer and deeper one is caught in a false belief, the more blind that soul will become, and closer to the point of being reprobate and permanently scarred over. 1st Tim 4:1-2, "*NOW THE SPIRIT SPEAKETH EXPRESSLY, THAT IN THE LATTER TIMES SOME SHALL DEPART FROM THE FAITH, GIVING HEED TO SEDUCING SPIRITS, AND DOCTRINES OF DEVILS; SPEAKING LIES IN HYPOCRISY; HAVING THEIR CONSCIENCE SEARED WITH A HOT IRON*;" Titus 1:16, "*THEY PROFESS THAT THEY KNOW GOD; BUT IN WORKS THEY DENY HIM, BEING ABOMINABLE, AND DISOBEDIENT, AND UNTO EVERY GOOD WORK REPROBATE*." CHRIST is our example of righteousness and love, who came to fulfill *all* the Commandments, Laws, and Prophets of GOD. Now if you are currently faithfully saved; exactly what price did MESSIAH pay for you personally as an accountable soul. It is written MESSIAH'S suffering was beyond our ability to know. Scripture says JESUS was "*MARRED*" (cut, torn, bruised) "*MORE THAN ANY MAN*" ever. This was written so we can understand that no other human body ever felt, suffered, and endured, this amount of external and internal damage and pain before succumbing to physical death. We also see that before his great final suffering, CHRIST lived around 33 years of suffering, and grief. Our LORD found *no* joy in this sinful fallen world. He searched for eternal love, but this world rendered little of it. As a baby and a child, the mind of CHRIST was not a child's mind. We do not see JESUS running and playing as a child. At twelve years old, he was boldly about FATHER'S work in the Holy Temple. CHRIST'S temporary body was born into this world a babe, adjoined to the mature soul of CHRIST. Scripture reveals CHRIST put off his *eternal* body of GOD in Heaven, coming here to live in a physical body to feel and know the worst of what we feel on earth. CHRIST walked in sorrow and grief, witnessing a world drowning in sin against his FATHER. Our LORD in each day as his mortal body grew, saw great constant rebellion, apostasy, deception, and hells confusion. He left Heaven behind to walk through a world of evil desires and a cauldron of hell and death. Every moment, CHRIST rejected the physical desires of that mortal body, FATHER prepared for his sacrifice. Drawing nearer to his greatest of many reasons he came, being to pay a great ransom. He was in that moment, both the body of a man, and a Holy Soul of GOD. At CHRIST'S experience of physical death, he spanned and bridged that great gulf between GOD and fallen mankind. Can you understand what it must have been like for MESSIAH to actually die and for three full days be completely separated, as FATHER had to look away. For MESSIAH to never commit, but accept and have many sins against FATHER, put upon his soul. This *did not* in any way mean CHRIST sinned, or became a sinner; but a Perfect Sacrificial Bearer of Sin. What great horror to endure, knowing *he* must be accountable and horrendously suffer and die for our sins of unfaithfulness to FATHER; whom *he* had faithfully loved and obeyed above life itself. We can also know the exact moment this massive weight of sin was literally placed upon our KING. It was at the moment MESSIAH said it is finished, and he gave up the ghost. We know this, because Scripture say's the price of sin is death. So, our sinless KING could never die until sin was placed upon him. His marred mortal body could not be allowed by FATHER to die at any point in his sinless life on earth, no matter its physical damage. Also, that weight of so many sins could not be placed upon CHRIST, until that full Holy price was paid for those sins, or he would have the deadly effects of sin placed upon him prematurely. CHRIST could only bear those sins, after the price was paid to absolve them. This is *why* CHRIST endured more physical damage and pain than humanly possible. Again, the price of sin is death, so no matter how long CHRIST was beat, cut, torn, and bruised, FATHER could not sever his body from his soul without sin. By FATHER'S force of life sustaining that body, JESUS continued to be punished and paid for and felt and suffered the pain and death of many. The men who beat JESUS relentlessly for those many hours, surely wondered by what power does this man live and endure. They were surely bewildered and wondering how does he hide from deaths hand, when will death find and take this man. They were set beyond their belief, formed by their entire past experience. I imagine, having heard of his reputation and fame of miracles of life, that one by one some became afraid at what they were doing, and to whom were they doing it. Mel Gibson's movie "The Passion" is said to be controversial because of the physical brutality it showed JESUS endure. I tell you, that went as far as could be done in a movie, but that was not the true extent of it all. In Scripture, a centurion who was there to witness most if not all of this suffering from the first strike in the Garden at Gethsemane, to CHRIST giving up the ghost. A centurion who had witnessed many past crucifixions and lashings, had no explanation for a man to have such force of life, far beyond his belief. So, it was he, who when JESUS said it is finished, and the ground quaked as he gave up his ghost, who said in Matt 27:54, said; "*TRULY THIS WAS THE **SON** OF **GOD***." We try to comprehend what it was like for CHRIST, to for his first moment ever, feel a complete falling separation from FATHER'S sight. Pulled down by the weight of the many betrayals of others he accepted upon himself. For JESUS to be in a place and need of having to cry out in fear of eternal death to FATHER for mercy. Having done no wrong, crying out to FATHER, so the world would hear him ask "*WHY HAST THOU FORSAKEN ME*" having done no wrong. He was forsaken for us. JESUS then experienced the darkness, bringing his light down into the pit for three day and nights, comforting the dead who died loving GOD. We must see the SON did not die painlessly unaware in his sleep for our sin. *Why*? Why did MESSIAH suffer beyond all imagination? *Why* was such a great price of pain and suffering required to remove a multitude of sins? CHRIST bore the pain and punishment of sins *truly* repented of and forgiven. Not every sin committed, but each sin to be *forgiven,* was paid for by his suffering. If you ask MESSIAH to endure your painful punishment and forgive *your* sins, and he truly forgives you; he was beaten for each of your forgiven sins. How can you then continue to sin, and ask for your LORD to suffer and bleed for more sins. Isaiah 53:5, "*BUT HE WAS WOUNDED FOR OUR TRANSGRESSIONS, HE WAS BRUISED FOR OUR INIQUITRIES: THE CHASTISEMENT OF OUR PEACE WAS UPON HIM; AND WITH HIS STRIPES WE ARE HEALED.*" How can we continue to sin again and again, and repent again and again, knowing JESUS had to suffer for each sin. Isn't it that personal to claim our LORD'S blood and pain for our *past* sins. Romans 3:25, "*WHOM **GOD** HATH SET FORTH TO BE A PROPITIATION THROUGH FAITH IN*

*HIS BLOOD, TO DECLARE HIS RIGHTEOUSNESS FOR THE REMISSION OF SINS THAT ARE PAST, THROUGH THE FORBEARANCE OF **GOD**;*" Please note and never forget, the KJV includes this word "*past*". Which for evil intent has been removed, or reinterpreted in some modern weak versions of Scripture. GOD did not use the word "past" (meaning former) by mistake, but for a great purpose and warning. If we sin again, after our first great forgiveness, we must repent again. Every *forgiven* sin on our Day of Judgment, was paid for by an amount of CHRIST'S pain and blood. Was CHRIST'S Sacrifice, meant to be a license to continue to sin, a license for us to keep beating and hurting JESUS? GOD forbid. If we continue to sin and repent, we hurt and spit in the face of our LORD again. Scripture says it pleased, (meaning satisfied) FATHER, to see JESUS'S great loving sacrifice to cleanse and save many. In Isaiah 52 and 53, we see the most perfect picture of beauty and love. Yet, Scripture says CHRIST grew up as uncomely, having no *physical* beauty or strength that any soul would desire him for his appearance; for this would be the lusts of this world. Why do movies always portray a strong handsome JESUS? Well, because most believers are still lustful worshippers of worldly beauty, who would be put off by an emaciated uncomely JESUS. If we cannot hold the full understanding of what MESSIAH was and did for all who come to him, then how can *we* go to him? And how can CHRIST present us before FATHER? So, we must see our every forgiven sin was part of the endured suffering of JESUS. You must see the sum total weight of your every forgiven sin on one side of the scale, and the sum of our LORD'S suffering for you, on the other side of the scale. Do not believe or say JESUS did this for all the sins of all mankind as a whole, that he died for every sin of every person in the world. For if so, all debts are paid in full, and every soul in the Earth is saved. But we are told *many* times in Scripture, that all are not saved, but only the *fewer* who truly love GOD. JESUS died for their sins, and no others. If you, or I, are saved in that last day, our every sin up to our judgment was on that scale. The sins of the lost were not on that scale. For each time we truly repent and *are* forgiven, our pain and shame were endured by CHRIST. Just for your strength, picture JESUS still on his cross (he is not) to suffer for your next sin. So, will you still sin again. It is good to admonish each other in this way. The truth is, FATHER commands JESUS, HOLYSPIRIT, the Holy

Angels, and *us* to live and worship in a constant intimate spiritual way, not just corporately from afar. We must each see *our* part in CHRIST'S painful death for our prior, and any coming forgiven sins. False doctrines, misinterpret the Bible to allow sin and evil into Heaven; yet JESUS says he does not. Which do you believe? If we are a true Child of GOD we must come before FATHER on our knees, knowing as he looks on us, he is looking at our reverence and eternal care and love for JESUS, and hating us if he sees we have abused his SON'S Love. By understanding and accepting such a gift, we gain love and receive strength and fear to overcome our sins of the flesh. When JESUS took those many sins upon the Cross, saying "it is finished" the unimaginable price of all who will be saved, was paid. But GOD still suffers all the additional ongoing pain of watching all the rebellious souls curse, blaspheme, and mock them. And FATHER'S pain of many claiming his SON'S Blood without turning from their sin. Do you think a loving FATHER'S heart does not ache, when after all they has done, GOD'S Holy names are used vainly as slanderous interjections of exclamation and mockery. By such acts, souls do mock, and deny any true desire for GOD. We were all created in FATHER'S emotional physical image. So, from our own perspective, we can relate to the pain GOD has endured in being denied by most. I dare not think what FATHER may feel towards each soul that does not give all holy righteous reverence. But instead, lifts up their self to dishonor, disrespect, abuse, and disregard the love and suffering of GOD. When a soul hears of the things of GOD and is drawn, they are wise to understand that at being called, they should know, consider, and count the eternal cost before asking CHRIST to suffer in their place. For within their flesh abides the knowledge and desires for the things of this world that they must overcome. Again, John, 2:15-17, "*LOVE NOT THE WORLD, NEITHER THE THINGS THAT ARE IN THE WORLD. IF ANY MAN LOVE THE WORLD, THE LOVE OF THE FATHER IS NOT IN HIM. FOR ALL THAT IS IN THE WORLD, THE LUST OF THE FLESH, AND THE LUST OF THE EYES, AND THE PRIDE OF LIFE, IS NOT OF THE FATHER, BUT IS OF THE WORLD. AND THE WORLD PASSETH AWAY, AND THE LUST THEREOF*:" If one tries to go to GOD, knowingly holding onto a certain lust, believing their lust are allowed and separate from an eternal communion with GOD. This soul is not growing towards GOD, and at best remains a babe standing still spiritually in peril of falling away. Before CHRIST'S supreme sacrifice, mankind was under the Commandments *and* the Animal Sacrifice alone. And were told to give ten percent of all they have to GOD. But after all GOD'S love was given us in CHRIST, we are now to give our all to him. Luke 12:48, "*FOR UNTO WHOMSOEVER MUCH IS GIVEN, OF HIM SHALL BE MUCH REQUIRED: AND TO WHOM MEN HAVE COMMITTED MUCH, OF HIM THEY WILL ASK THE MORE.*" GOD alone will judge us. And the last place that awaits every soul is at GOD'S final separation; being things hard to imagine. To all saved by GOD, there will be things too wonderful to describe or comprehend, and all souls there will forever be a lesser reflection of the more. But for the lost, the spiritual darkness of loss and anguish that await, already grows inside them as the firm body that fulfills their desires nears its end. Our created internal existence of awareness comes by *our* interpretations of all we once knew, what we now know, and what we will know in our growing, *or* fading desires. There can never be anything good or eternal for us beyond our soul, that GOD has not *first* known and perceived as being good for us. GOD and every soul must and will bear the pains of both finding and losing love. So valuable is this quale called love, that we accept great pain and sorrow burrowing through our lives and generations to find it. Nothing but love could be worth such a great cost. And we all know it is worth the cost, for it alone gives our life, shared meaning. Those who can't see the higher meaning of *eternal* love, can never know the higher meaning of *eternal* life. There is no life or existence in the past or the future; not even for GOD. All living existing awareness, is an aspect of the present moment alone. Any imagined quality called a linear structure of time, that holds all recorded past, present, and future realities and experience in a segmented continuity is impossible. The information of all physical reality, exists in the experiential present moment alone. Yes, the created aether did, does, and eternally exists, filled with patterned particle structures representing every former experience. Yes, the earth itself bears scars, remnants, and evidence of prior moments to be interpreted in ways of faith to support true or false beliefs about those former moments. And yes, libraries are filled with written accounts of history. But no quality or construct of a non-spherical linear time-space, exists to store in perpetuity, even one tiny aspect of physical reality.

Though what once was, did once exist, and what will be, will exist, there is no copying or saving every experiential *and* firm aspect of existence in any linear form of record. Now FATHER alone, could re-perceive the duration of a certain past moment of physical life and earth, using all souls stored experiential information. But this would still be a recreation of a certain past moment, in the present moment, and not traveling back to visit any actual former moment that no longer exists. There is no paradox called time, it is a ghost believed to exist by scared children. Can time and eternity both exist in the same moment? No. They are abstractly in opposition in all understanding, qualities, and quantities. They would interact like abstract matter and anti-matter. Each consciously deterministic moment of experience that covers each soul's spherical surface within the aether, consist of subjective thoughts, beliefs, and emotions. And each created moment of ours, is a unique sum and size of subjective existence, within the objective created reality of FATHER. We inherently need HOLYSPIRIT in his fullness of love and truth, and not in part. Truth and love in part, is GOD in part. Please ponder the pure eternal love from YESHUA for FATHER, in that for his entire existence, his every choice and desire have been to please FATHER without one departure. Can you imagine such loving responsibility and faithful depths of loving desire. Who among us wants and waits to see such a miracle and wonder? Who waits to see a Majestic Eternal Throne of Glory? One Throne, having the semblance of none before; and of which there shall never be another to compare. How many want and wait to see our LORD, for whom this rarest of Thrones awaits? Who desires to see CHRIST, who in the earth, overcame and traversed the breadth and gaps of a fallen creation, and the evil divide of time and eternity to bring and join a lost end with a new saved beginning? Who longs to see the one who suffered all things in a faithfulness not seen before, since, or forever more? JESUS who has never once taken, and only given in every measure and thought without fault. The SON, whose love for FATHER and us is of the purest form, radiant, fervent, and a consuming eternal fire. Our LORD whose desire has never turned his eye from his Holy desire for FATHER'S truth and love. JESUS whose fiery gaze is steeled, immovable, impenetrable; and whose shoulders press forward from his footing, firm and victorious against every strong evil wind. Who waits to see the longed-for moment when the sum entirety of the greatest beauty of love and truth for FATHER above, and all below, forever sits and reigns? That ONE who is the purest joy, the very core and want of FATHER'S heart, and all who are Holy. The SON who is the brightest Righteousness and Holiness. FATHER, HOLYSPIRIT, and the Holy Angels cry out who in the earth, where are the desired of GOD? Who waits to see *this* beyond and high above all things that are good and to be desired? How can we speak his name, and call him our KING when we have not yet heard our name from his lips? Only by love, faith, hope, and obedience alone. You ask, how is it possible to see reality in such a pure selfless way? It is only possible if you know you are of GOD. JESUS knew his eternal soul came forth from, so he is a true eternal part of FATHER. JESUS also *said*, all lesser souls who are saved, see, know, and accept their eternal souls came forth from, and so they too are a true eternal part of FATHER, all being lesser Gods. All having a faithful Godly responsibility to sustain a loving perfect eternal reality. I hope you have the same tears of joy in reading these words, I had writing them. All that you have read in this book, comes from the trusted Original King James Bible, and its true interpretations and revelations opened by HOLYSPIRIT. Now seeing you have come this far with us, we hope you clearly see it is both the spiritual loving reason and purpose of FATHER, and his true Science of Creation, that together reveal and confirm his truth to us. We do hold hope. Yet it is never our writing, but your reading, that moves you. It is never GOD'S loving, but your sense of being loved by GOD. It is never our Godly speaking, but your heard interpretation which matters. It is never our Godly teaching, it is your learning. Minus your correct half of any communication, GOD'S and our first half, are *as* dead to you. If our soul coveys something of love and beauty, but another hears to the contrary, it is fruitless. The fruit of all spiritual labor, rests not in planting seeds, but in the final harvest of souls who hear, see, and live the omniscient truth. We hope all do not regret reading this, but are hopeful *and* fearful. We hope strength and purpose have come, yet what eternal end are we each going to find. We must know that change within us will surely be by our own choice, born of our own free-will desire. It is our gifted created purpose in this world of choice, to as best we can, share the truth of eternal life with each soul who searches. What understanding is required to describe the sum of all reality? And is not such a great understanding found and defined only in knowing GOD'S created boundaries and parameters. The defining of reality must account for all things known, and reveal those not yet known, aligning in pure consonance. Are all things now known to mankind? No. Just 6% of FATHER'S physical reality at most is being searched, and from mankind's slow small rebellious perspective. Mankind's knowledge of the electromagnetic spectrum, holds at most, the weakest third of the full electromagnetic spectrum of reality. This is all the fallen can ponder or try to comprehend. This is all the fallen can believe in from such a limited sense of reality. Please know beyond all doubt, that the true sum of reality is far beyond mankind's comprehension, *and* 6% of one half of reality, can never be properly interpreted. How can souls see the wonder of life and reality in all its glory, yet believe these all come physically by chance, and see it as a settled issue. Modern science cannot *begin* to explain our everyday experience occurring everywhere every-moment and in everyone. There are our created point particle perceptions, our created awareness of bright qualia, and our acquired beliefs of desired understanding. From our eternal perspective, we see each Materialist is a science denier and isolationist; being souls descending into great and greater illogical insanity. Such individuals, can never see or understand GOD'S purposed mechanisms that rule over all our lawful *un*awareness in death; not until they understand GOD'S purposed mechanisms that rule over all our lawful awareness of eternal life. We know we can never do *anything* to or for ourselves alone. Everything we do or think, are our forceful acts or thoughts for or against our oneness with GOD. This is how all conscious forceful entangled reality works under GOD. All that exists, is in all its ways consciously entangled with and to all other things. Mankind accepts as natural law, whatever is seen to consistently physically occur through their slow small perspective. So, their theories of reality do vaguely overlap the one true reality. They are of course, just a far lesser part of the full layered truth. Visualize the many rising layers of a pyramid, having one final unifying pyramid shaped headstone on top. And for this visual, that capstone is GOD and their created defining understanding of all below. But fallen mankind, denies GOD, the Angels, and Heaven, so, they refuse to see the top third of this pyramid exists at all, and they can only sense the firmest 6% of each stone and layer below. Now if you can visualize all of this, please see the 6% mankind senses, is the firm outer aspects of all that comes from where the conscious half of reality creates the physical half. And from a narrow slow small perspective, mankind actually denies, or can't see the top third, and 94% of everything below it. So, sensing only 6% of the half they do study, they theorize an ambiguous false top and bottom to what they see. They try to build theories to fit the 6% of the half of reality they believe exists. Whilc under GOD we see and accept all 100% of that pyramids gifted souls and created physical layers. On the physical side alone, there is that which has been created, and that which has been destroyed. Then there is all which *shall be* created and endure for eternity. We must see how all fallen mankind's anomalies, paradoxes, hidden variables, and magical strong emergences, fall outside of any true physical cause and effect. Please, this plus that, can never equal more than that plus this in any realm of laws presented under cause and effect. And created matter and its perfectly aligned forces together, can be forever conserved. But contrary waring forces tear at all created realities. A paradox that materialist can't even define, is that physical determinism cannot allow or co-exist with true chance, love, or free-will choice. Do you believe you have no free-will, and that love and chance do not exist. Created chance, and gifted free-will choice are constantly seen throughout the probabilities and uses of quantum mechanics where all conscious cause and effect originate. On our conscious soul's side, we have our desired creative awareness, understanding, and belief of it all. And on the physical side, there is seen at *every* biological step of life, far more than weak or strong-emergence. All souls are composers of creation, and nothing is a miracle. Until both sides of reality are seen and known together, understanding can be no more than vague hope and desire. The universe, earth, love, life, our souls of belief and awareness, and GOD; how can all these things be, yet they all exist. Now we have found *both* Scriptural and scientific reasons for why the Garden of GOD in Eden, was set in the lowest place on the surface of the Earth. The surface then being valleys, hills, seas, and mountains. And all plants were watered by morning dew alone. The Great Flood, left the abandoned Garden of eternal life, as the lowest most salty place on earth, completely dead and submerged. Search for aerial views of the Dead Sea from space. See the huge area of what was the Garden of Eden. Now a dead place, no plants, fish, or animals, live there. The Earth's crown of life, is now its crown of death. There were once only perfect conscious *and* physical aspects, but now we see all antithetical forceful aspects that form false answers to all the false question raised in mankind's antithetical views of reality. There are still the conscious and physical halves of reality, as all true good forced physical order is attacked by all chosen forced *disorder*. Write down every godless antithetical, and see what causes the physical half of this reality to dissolve. Now how and why do order and disorder exist, and which existed first. Is *dis*order found in one soul, and order in another. See a dynamic complex guided physical environment filled with souls, minus all disorder. Order and disorder, can these be seen as an agent of cause? No. So, they are secondary aspects of affect, and not a primary aspect of created reality. All great wonders, like free-will mobile awareness, and expressions of life and love can exist eternally within a perfect physical order. While no free-will mobile expressions of life and love can exist forever in physical *dis*order. What understanding do we require to see the sum of reality. Such understanding can only be held in seeing the *sum* of reality's points and conscious forces. But we see reality's smallest and largest boundaries are not detected by fallen mankind, they see but a small slow sliver of it. Now a definition consists of objective informational words, terms, and phrases, which describe something in a way and to a point, that one can understand what it is without seeing or interacting with the actual thing. But we don't see this in theory. So, theories are individually interpreted. Reality itself, is a term used to describe the full summation of all that exists as real and true. So, do we all consciously exist? Yes. We are all consciously real. So that *must* be allowed into the equation defining reality. A true model of reality must be built using all that exists in perfect continuity and consonance. Now Einstein said "Science without Religion is lame; and Religion without Science is blind". He also said "Reality is merely an illusion, albeit a very persistent one". We add that all patterned force equals information, and patterned information equals guidance; and lastly, each soul's desire is a generated force. So, desired projected 3D structural patterns of forceful information, are the cause for every consciously aligned creation. We see Einstein's Special and General Relativity appear to work from a slow small physical perspective, yet they are quite incomplete; which science is finding out. Triple mankind's sensed physical qualities and quantities, and add the variable cause and effect of all conscious size frames of forceful reference, and you can begin to study reality. We see the three short ranged stronger of the four classical forces, *almost* survive a definition within mankind's small slow physical model of a Big Bang. But gravity refuses to cooperate at all. A Materialist model contains a lot of ambiguous speculation and paradoxes. It has exclusions and accepts actual particle annihilation, unknown anomalies, variables, and perturbations. It has unresolved particle entanglement, and unexplainable physical miracles; and it excludes GOD and so much more. Such a problematic model can never be right. Not one of these problems exist in GOD'S model. Now after science tested for, and did not find the aether wind they predicted to fit their standard model, they just denied an aether medium underlying all reality even exists. They then introduced the unqualifiable enigmatic aspect of nothing, as being the background for their equation of reality. Some do ascribe to what they call the Great Higgs Field, that raises and fails to answer even more questions. The idea that complex reality exists by this Higgs physical variable, is impossible. They just keep creating more new intractable problems. Their malleable fourth dimension called time, sends their beliefs in many wrong directions. The subjective aspects of time raise many real objective problems, for each *subjective* solution it creates. They never see the problems they build that await them farther down the line. And in their abandonment of all reason, they throw in an *infinite* inflationary creation of endless matter, born of a mid-20$^{th}$ century math error. Again, an infinite mathematical answer is never a valid answer. Yet for this instance they grant exception. They disregard the impossible odds, and other *dis*qualifiers of purely physical Abiogenesis and Evolution. And their twenty+ ambiguous slow small interpretations of quantum mechanics, allow endless false theories. Now imagine a complex *patterned* program of forceful guiding information that shrinks within each reiteration, recurring until this forceful pattern is too small to disseminate. Imagine forceful patterned orchestrations of life coming down from FATHER, being designed to order all complex work within cells. Also imagine every pattern of force, projected from GOD to orchestrate physical life, as being like DNA made of force quanta, ordering and entangling every complex cell system; a quantum process existing beyond the bounds of chaos theory. In discovering and building mathematics, fallen mankind has built their understanding upon, and out from the whole number one. But once we add 0, and 2-9, like the Observer's Frame of Refence Paradox, the number one becomes a paradox of scale, as we divide and multiply the concept of one into infinitely smaller and larger reiterations. So, "one" is now just a point of reference, but not any true beginning. To which we then see no beginning or end. Our point of this is; to overcome having to perceive every minute forceful guidance of life one at a time, GOD conceived repeating patterned programs to force all cell's many lively functions. *All* patterns within reality and physical life are designed by GOD to do repetitious tasks. And such patterns cause us to stop and stare. Examples like the Mandelbrot set, the Golden ratio, Prime numbers patterns, and others. These created tools carry patterns of forceful guidance down into minute levels, to orchestrate the physical functions of life. But mankind sees these strange entangled rising patterns build out from the number one, not seeing their reversed creative purpose. The eternal perspective sees all these repeating downward patterns as funnels carrying forceful patterned guidance downward towards *one* programmed purpose. The body's immense complexity, requires many patterns of forceful guidance, which could never occur in chaos by chance. *All* rising complexity is sustained by small designed sets, working within larger and larger designed sets. Look at a Mandelbrot set as lines of code, beyond binary, being a patterned forceful command. Now, Science knows the physical senses of fallen mankind's are quite limited. And that *no* two bodies sense anything in the exact same way, bringing to all a minute relative uncertain disagreement. Reality looks like so many things, from so many slow small incomplete individual perspectives. But it can only look one way to all, if seen from the largest fastest omniscient perspective. But why is that? Quantum mechanics is a set of qualifications and quantification, that fallen mankind devised to try to define what they *believe* they barely sense and observe through their narrow small slow perspective. And their 20+ interpretations of what they believe they see, try in different ways to describe the forceful process that creates every current moment of physical reality. Mankind needs a unified theory to bring consonance between their defined classical laws and quantum laws. But they see all laws from a wrong perspective, and refuse to see a lively quantum realm that remains intact, while the firmer realm is dissolving. Their classic view of reality is their more blurred illusion over the quantum. They don't understand gravity at all, as they hopelessly struggle to reconcile it into their beliefs as it refuses. Now in a consciously deterministic reality, there are no random fluctuations, or unseen variables and perturbations. Even created chance is not purely random. There is no physical uncertainty, until we see waring conscious desires introduced. So, as science says things must be this way or that way, they can be neither. Now the classical/quantum divide we see in reality, appears where the projected forceful desires of FATHER'S absolute rule stop; and where all projected free-will desires of GOD and lesser souls interact to hurt or help what GOD desires. Thus, creating uncertainty and entropy in the firm physical reality. Science sees their physical sensing limitations, but they can't see that their enhanced technical sensing is still limited by the slow small perspective of their created experience. The fallen body and soul's perspective are the weakest link. Science knows we never experience a first-hand physical awareness of reality. They know we don't physically see, hear, taste, smell, touch, or measure anything. They know we sense only our biased second-hand created perceptions of doing so. If any scientist says otherwise, I stand in the presence of GOD and all science to affirm they are insane or lying. A profound physical and conscious inability is set in *fallen* mankind, tethered to their small slow temporary-perspective, and their limited physical sensing. Again, Eccl. 3:10-11, "*I HAVE SEEN THE TRAVAIL, WHICH GOD HATH GIVEN TO THE SONS OF MEN TO BE EXERCISED IN IT. HE HATH MADE EVERY THING BEAUTIFUL IN HIS TIME: ALSO HE HATH SET THE WORLD IN THEIR HEART, SO THAT NO MAN CAN FIND OUT THE WORK THAT GOD MAKETH FROM THE BEGINNING TO THE END.*" Consciously and Spiritually, FATHER'S children, are capable of knowing all things. But from their physical perspective alone, the fallen can never sense or know reality on their own. Such confused souls turn inward in self-inflicted blindness and deafness, denying their great limitations exist. Declaring they can know reality without one sane rational cause or reason to believe so. Now our objective truth of reality can drive a few into a rebellious madness, but denying reality's truth is pure madness. Free-will choice is about desire and control. So, what do you want, and who do you want to be in control. Most want this world, and want to control their own way of life. These cannot live among the ways of GOD that they will forever not want. Madness has no eternal spiritual bonding aspect, and madness is synonymous with set adrift, loneliness, hate, sorrow, and anguish. As most souls become more individualized, having no eternal anchor of truth, they are more confused, hateful, fearful, and lonely. The truth and ways of GOD are the only eternal anchor to free-will love, peace, and sanity. All the great truths of reality are right in front of us, yet only the few see or desire them. Once we know the truth that eternal love and life require perfection, we see that before we learned this, we held no desire to be perfect. 1st Corinthians, 2:16, states, *"FOR WHO HATH KNOWN THE MIND OF THE LORD, THAT HE MAY INSTRUCT HIM? BUT WE HAVE THE MIND OF CHRIST"*. What is the mind of CHRIST we are called to have? How can we be one in and of this mind? We are gifted, yet we are born in ignorance of CHRIST, and from conception to all mitosis and meiosis, we experience a process of our experiencing life, learning, desiring, and making our choices, having a broad spectrum of outcomes. But if we truly desire eternal loving life, we follow JESUS who leads us all to FATHER. And in this, rests *our* eternal purpose and loving mind of CHRIST and HOLYSPIRIT. Again, every prior belief mankind once held as empirical fact, such as a Flat Earth, the Sun orbiting the Earth, and so many more, were all superseded by their next new and limited empirical understanding. And this failure process continues. From what perspective were mankind's laws and constants of nature formed? They are all just current temporary beliefs, subjectively formed. And accordingly, their laws and constants are problematic and unresolved in many ways. Their laws all come by their fallen subjective *observations* of the small slow classical and quantum systems that fall within their purview. In this way they seek to know the sum substance and cause of reality. Not seeing reality runs far beyond their purview, and is ruled by forceful processes beyond their belief. They do not now know how to seek and understand reality, so their efforts can never be truly rewarded. The only way to know reality is to see every aspect of it all at once. Short of this, each soul is partially and subjectively blind, and left to share their subjective blindness with others. For millennia, fallen godless science has tried to see, know, and define reality without its creator. This is their true purpose, even as they eternally fail at every effort. Please consider that a perfect definition of anything, requires a complete understanding of all relative entangled aligned objective terms, equation, qualities and quantities of information that describe and reveal *that* thing in a way and to the point, one can see, know, and understand it in every way required, without needing to see or interact with the actual thing itself. So, please see that in all ways and considerations, the one true defining essence of force itself has never been found, measured, or defined, in any way beyond its seen physical affects. So, like our self-awareness, desire, and our beliefs, force is a purely non-physical phenomenon, having all the required qualities, effects, and associations, to be seen as originating from the conscious side of reality; hence the real true observer affect. So, all force quanta patterns are creat*ive* information, and all creat*ed* physical information is aligned by those force quanta patterns. And again, the truth is that all empirical law and logic state, our created perceptions that we *physically*, consciously, and spiritual sense, must be *substantive* for us to sense them. How can we physically sense an illusion of conscious awareness that is not there. In reality, even a conscious illusion, being itself a created perception, must exist in this same way to be sensed. Please show me any alternate empirical creative process that can enable our awareness. Show me anything real, beyond fallen mankind's illusory godless miraculous undefinable inexplicable impossible untestable, too slow, too small, abstract strong emergence. Now in all their abstract essence, we see time and eternity share in the real aspects of forced motion, speed, and sensed experience. Even fallen mankind's slow small limitations see that if things move or oscillate fast enough, they disappear from their small slow comprehension and move up to a higher eternal perspective. They also know, if anything stops or goes slow enough, it also disappears to them. So, we see the narrow sliver of their abilities are set in that they can't sense aspects of reality that are too slow or small, or too fast, or too ethereal. So, what does disappear mean to materialist? All it means is something they once sensed, they can no longer sense in a perceivable way. But they can't say it no longer exist, or was annihilated. Even with all their future technical means, mankind can never sense, measure, or know, the sum of reality. Has mankind *even* fully or accurately sensed and measured the tiny minute six percent sliver they know best? No. And if they ever accurately sense and measure the fullness of that sliver, how can they correctly interpret it? From their fallen too small and slow perspective, all fallen mankind can do is measure, and compare the apparent statistical contrast they see within the part they sense. They can only know that among the aspects they sense, some are bigger or smaller, some go faster, some slower; some are firmer, some are not, some are more energetic and some are less so. But they can never know the vast spectrum of what exists beyond their limitation. They have no true foundation to form any first immutable knowledge. Which shows their every physical question and physical answer are all wholly inadequate. So, fallen mankind will never be able to truly measure or understand the history of this creation, yet they theorize and debate about it among their many too small, too slow individual interpretations of it all. I say focus upon all the physical sciences do not, and can never truly know, and all they cannot know and refuse to admit about reality. As this is where you see their godless scientific religious beliefs. Science finds that most souls who go insane, don't realize it is happening. And *we* see two main causes for souls not realizing this. One, is if it occurs gradually; but two is it being a result of strong false desires. Many desire and chase false abstract beliefs about reality, and such souls actually go insane by choice. Our perspective is everything. Every moment we exist, we are seeing, hearing, perceiving, desiring, choosing, believing, and creating. And by all of this we build our perspective and belief. But remember, though GOD'S creation provides us strong sensed information through the bodies to which our souls are adjoined, we do not, and can never-ever exist in any real physical way. Our first, last, and only reality is our soul and awareness. Now if using technology, I mix equal parts of pure blue and green paint together, forming a perfect bi-color. Because of differences in all eyes, the world will split into 3 groups of seeing it as more green than blue, more blue than green, or equally blue and green. But on spirituality and all our abstract realities, the world splits into endless individual perspectives. We know the Trinity of GOD'S perspective of belief and reality is one, pure, and never-changing. So, throughout the ranks, layers, and ordered souls of GOD, this must also be true. How much love and need for GOD is required for a soul to dedicate all, and change their beliefs from their own individual construct, to GOD'S pure belief? And how important is this for eternal life with GOD? I have been at this very task since 1974, and I have had my share of this world's troubles and set-backs along the way. I had to fight my way back from an early backslide after a family death which was hard and took a lot of prayerful work to recover from. But for the last 30 or so years, I have been on the path, having my work to finish. Are you on *this* path? It matters not where you are on the path, all that matters is are you firmly on the right path. And if not, please wake up and turn. Why does the world only grow colder towards GOD? Do they not see an increasing population, followed only by increasing failure and death? Is it Satan's plan and goal for souls to see life as futile and godless, and give up and surrender to despair, hell, and death? Now eternal life of Godliness only exists in emulating CHRIST. And evil only exists in rejecting GOD'S ways. Live strong in the ways of GOD, and never under-estimate the dual natures of life or deny GOD'S gift. Science fiction movies and books and even science run wild with ideas about what god-like breakthrough things a physical brain might be able to physically do or create in the future, and many believe this. Yet to believe we are gifted eternal souls, having an inherent ability to create particle-wave-based perceptions, comes at too high a cost. GOD'S grace and our loving faithful responsible emulation, are both crucial. If we could all exist in GOD'S one love, belief, and understanding, all would receive, give, act, speak, hear, and write, in that one belief. Now this book does not call all, but some scientists to be preachers. But this truth of creation calls *all* preachers to become educated in GOD'S science. Please read "The Eternal Perspective", "All Conscious Formulations", and "Live to know the GOD of Prophecy" in order. Which gradually speak less about the science, and more on the spiritual reason, purpose, and true eternal ways, belief, and love, of GOD, ending with a narrative of the dreadful Prophesied things to come. If any hear their certain calling to know GOD'S love, then know each sin can kill the flesh, and more so the spirit of many. How are souls separated from GOD, and how is GOD'S truth destroyed in them? Only by lies and omissions. The Jews long ago, awaited their Prophesied Messiah; do they still wait? Many do, yet Isaiha 53 and other Prophecies of Messiah, are avoided in Synagogues, like many profound New-Testament verses are avoided in Churches. Let's set things straight. JESUS was a Jew, his Apostles, all Jews. JESUS lived and taught as the Jews Prophesied Messiah. Messianic Jew and Judeo Christian both include all blood born Jews, *and adopted* gentiles who follow CHRIST as their Jewish Savior. If you see a line of separation in your doctrine between Messianic Jews and Christians, you are not a real Disciple of CHRIST. Gentile Christians, *according to Apostle Paul*, are adopted Jews. These two names when set in truth, have no distinction. But change the underlying truth, and you raise evil distinctions. It is only after you see and understand what Heaven truly is and means, that you can then see and understand what is required from us to live there eternal. If while reading this you felt pulled back and forth between GOD'S understanding and this world's misunderstandings, between Godly belief, and a worldly pull of disbelief. Please pray, fast, and read this, and the KJV Bible again and again. GOD prepares a perfect place for all his perpetuating forces and his highest properties of light. *And* another place for all the divisive forces and properties of darkness. One place will be eternally separated from GOD'S love by his wrath. One, eternally separated from his wrath by his love. We will soon be in one of these two places. If you hear this truth, let GOD have mercy upon us; as he will end this soon. Amen. If you desire to share GOD'S love, or any *constructive helpful positive* critique, or ask any meaningful true questions. Or if you can't afford the cost, and *need* a reduced-price copy of our books for you or another; we hope to respond. In all such loving needful cases, please email Michael, @ livetoknowtheGODofprophecy@gmail.com

Made in the USA
Monee, IL
11 November 2026

68342119R00400